Volume V
Solvents 3

Handbook of Environmental
FATE
and
EXPOSURE
DATA
For Organic Chemicals

Editor
Philip H. Howard

Associate Editors

Edward M. Michalenko   Gloria W. Sage

Dipak K. Basu   Amanda Hill

Dallas Aronson

LEWIS PUBLISHERS

Boca Raton   New York

Publisher:	Joel Stein
Project Editor:	Les Kaplan
Marketing Manger:	Greg Daurelle
Direct Marketing Manager:	Arline Massey
Cover Design:	Denise Craig
Manufacturing:	Sheri Schwartz

Library of Congress Cataloging-in-Publication Data

Howard, Philip H. (Philip Hall), 1943–
  Handbook of environmental fate and exposure data for organic chemicals.

  Includes bibliographical references and index.
  Contents: v. 1. Large production and priority pollutants -- v. 2. Solvents --
v. 3. Pesticides. -- v. 4. Solvents and Chemical Intermediates.
  1. Pollutants--Handbooks, manuals, etc. 2. Environmental chemistry--Handbooks, manuals, etc.
  I. Title
  TD176.4.H69 1989 363.7′38 89-2436
  ISBN 0-87371-151-3(v.1)
  ISBN 0-87371-204-8(v.2)
  ISBN 0-87371-328-1(v.3)
  ISBN 0-87371-413-X(v.4)
  ISBN 0-87371-976-X(v.5)

No claim to original U.S. Government works
International Standard Book Number 0-87371-976-X
Library of Congress Card Number 89-2436
Printed in the United States of America 1 2 3 4 5 6 7 8 9 0
Printed on acid-free paper

Associate Editors for Volume V

The following individuals from the Syracuse Research Corporation's Environmental Sciences Center either were authors of the individual chemical records prepared for the Hazardous Substances Data Bank or edited the expanded and updated chemical chapters in this volume. The order of names, which will vary in each volume, is by the number of chemicals for which the individual was responsible.

Gloria W. Sage, Ph.D.

Edward M. Michalenko, Ph.D.

Dipak K. Basu, Ph.D.

Amanda Hill

Dallas Aronson

Philip H. Howard joined Syracuse Research Corporation in 1970 and has served as project director for numerous environmental fate and effects projects for federal agencies and industry. Dr. Howard's current research projects include development of structure/biodegradability correlations, development of estimation techniques for environmental fate physical properties and rate constants, and databases of information to support these efforts. He received a B.S. degree in chemistry from Norwich University in 1965 and a Ph.D. in organic chemistry from Syracuse University in 1970.

Gloria W. Sage joined Syracuse Research Corporation in 1980, at which time she was the codeveloper of the Environmental Fates Data Bases, DATALOG and CHEMFATE. Later on, she developed the environmental fate section of the Hazardous Substances Data Bank and contributed profiles to this section. Dr. Sage has been involved in many projects at Syracuse Research Corporation concerned with assessing the environmental fate of chemicals, making exposure assessments, and developing information tracking systems. She is also interested in using mosses to measure atmospheric deposition of metals and organics. Dr. Sage holds an A.B. with distinction and honors in chemistry from Cornell University, an A.M. in chemistry from Radcliffe College, and a Ph.D. in physical chemistry from Harvard University.

Edward M. Michalenko has been an associate of Syracuse Research Corporation since 1988 where he has written books and numerous documents on the environmental fate and transport of chemicals for U.S.EPA, National Library of Medicine and Agency of Toxic Substances and Disease Registry. Dr. Michalenko currently serves as head of the Ecological Resource Group at O'Brien & Gere Engineers, Inc. His current research involves assessing the fate and effects to non-target birds and insects of aerially applied pesticides used for mosquito control in wetlands, and evaluating base wide risks to ecological receptors associated with chemical release from Cape Canaveral Air Station to the Banana River. He received a B.S. in Biology/Chemistry education from SUNY College at Cortland in 1979, a M.S. in environmental communication/education and a Ph.D. in soil ecology from SUNY College of Environmental Science and Forestry in 1984 and 1991, respectively.

Dipak K. Basu joined Syracuse Research Corporation in 1975. As the principal investigator of projects funded by the U.S. EPA, Dr. Basu developed analytical methods for the determination of micropollutants in drinking water. He has a number of publications in analytical chemistry and toxicology. Dr. Basu has co-authored several documents for the Agency of Toxic Substances and Disease Registry, U.S. EPA and National Institute for Occupational Safety and Health. He received his M.S. degree from Calcutta University, India in 1959 and Ph.D. with specialization in physical photochemistry from Case Western Reserve University in 1970. He is presently with the USAF at PAFB, FL.

Amanda Hill joined Syracuse Research Corporation in 1995. Since joining Syracuse Research Corporation, Ms. Hill has participated in the evaluation of data for environmental and exposure properties including an assessment of chemical/physical properties, commercial uses, production, environmental fate, monitoring, and occupational exposure. This work was performed for entry into the National Library of Medicine's Hazardous Substances Data Bank. She has also used SRC's proprietary Chembase database for examination of chemical structures and various chemical properties such as water solubilities, vapor-pressures, and octanol-water partition coefficients for chemical compounds. She received her B.S. in chemistry and her M.S. degree in environmental chemistry from SUNY College of Environmental Science and Forestry in Syracuse, NY in 1991 and 1994, respectively.

Dallas Aronson joined Syracuse Research Corporation in 1995. Since this time, she has researched and written chemical profiles evaluating the fate of chemicals in the environment for the Hazardous Substances Data Bank and has provided technical support for projects for the U.S. EPA and other sponsors. Ms. Aronson received a B.A. in biology from Cornell University and a M.S. in environmental science from SUNY-College of Environmental Science and Forestry in Syracuse, NY.

Preface

Many articles and books have been written on how to review the environmental fate and exposure of organic chemicals (e.g., [15] and [30] - citations at end of Explanation of Data). Although these articles and books often give examples of the fate and exposure of several chemicals, rarely do they attempt to review large numbers of chemicals. These "how to" guides provide considerable insight into ways of estimating and using physical/chemical properties as well as mechanisms of environmental transport and transformation. However, when it comes to reviewing the fate and exposure of individual chemicals, there are discretionary factors that significantly affect the overall fate assessment. For example, is it reasonable to use regression equations for estimating soil or sediment adsorption for aromatic amine compounds? Is chemical oxidation likely to be important for phenols in surface waters? These discretionary factors are dependent upon the available data on the individual chemical or, when data are lacking, on chemicals of related structures.

This series of books outlines in detail how individual chemicals are released, transported, and degraded in the environment and how they are exposed to humans and environmental organisms. It is devoted to the review and evaluation of the available data on physical/chemical properties, commercial use and possible sources of environmental contamination, environmental fate, and monitoring data of individual chemicals. Each review of a chemical provides most of the data necessary for either a qualitative or quantitative exposure assessment.

Chemicals were selected from a large number of chemicals prepared by Syracuse Research Corporation (SRC) for inclusion in the National Library of Medicine's (NLM) Hazardous Substances Data Bank (HSDB). Chemicals selected for the first two volumes were picked from lists of high volume commercial chemicals, priority pollutants, and solvents. The chemicals in the first two volumes include most of the nonpesticidal priority pollutants and many of the chemicals on priority lists for a variety of environmental regulations (e.g., RCRA and CERCLA Reportable Quantities, Superfund, SARA). Pesticides selected for Volume III were identified from a list of significantly produced pesticides (Resources for the Future, National Pesticide Usage Data Base, 1986) and the priority pollutants. Chemicals selected for Volume IV were mostly solvents and chemical intermediates. Chemicals in Volume V are mostly solvents and cover many of the natural product and HCFCs and HFCs that are being considered as replacements for chlorinated solvents and CFC uses. More solvents, chemical intermediates, plasticizers, polycyclic aromatic hydrocarbons, pesticides, and other groups of chemicals will be included in later volumes.

The chemicals are listed in strict alphabetical order by the name considered to be the most easily recognized. Prefixes commonly used in organic chemistry which are not normally considered part of the name, such as ortho-, meta-, para-, α-, β-, γ-, n-, sec-, tert-, cis-, trans-, N-, as well as all numbers, have not been considered for alphabetical order. Other prefixes which normally are considered part of the name, such as iso-, di-, tri-, tetra-, and cyclo-, are used for alphabetical positioning. For example, 2,4-Dinitrotoluene is under D and tert-Butyl alcohol is under B. In addition, cumulative indices are provided at the end of each volume to allow the reader to find a given chemical by chemical name synonym, Chemical Abstracts Services (CAS) number, and chemical formula.

Acknowledgments

The following authors of the initial chemical records for the Hazardous Substances Data Bank were or are staff scientists with the Environmental Sciences Center of the Syracuse Research Corporation: Dipak Basu, Julie A. Beauman, Erin K. Crosbie, Amy E. Hueber, William F. Jarvis, Jeffery Jackson, Jeffrey Robinson, Edward M. Michalenko, William M. Meylan, Gloria W. Sage, and Jay Tunkel. We wish to thank several individuals at the National Library of Medicine (NLM) for their encouragement and support during the project. Special thanks go to our project officer, Vera Hudson, and to Bruno Vasta and Dalton Tidwell.

Explanation of Data

In the following outline, each field covered for the individual chemicals is reviewed with such information as the importance of the data, the type of data included in each field, how data are usually handled, and data sources. For each chemical, the physical properties as well as the environmental fate and monitoring data were identified by conducting searches of the Environmental Fate Data Bases of Syracuse Research Corporation (SRC) [17].

SUBSTANCE IDENTIFICATION

Synonyms: Only synonym names used fairly frequently were included. A more extensive list of names and synonyms can be found in the *Dictionary of Chemical Names and Synonyms* [18].

Structure: Chemical structure.

CAS Registry Number: This number is assigned by the American Chemical Society's Chemical Abstracts Services as a unique identifier.

Molecular Formula: The formula is in Hill notation, which is given as the number of carbons followed by the number of hydrogens followed by any other elements in alphabetic order.

SMILES Notation: In previous volumes of this Handbook, we have included the Wiswesser Line Notation. This notation was developed back when the maximum number of characters that a computer could accept was 80, and it is very difficult to learn and interpret. Thus, starting with this volume, we will be substituting the SMILES notation, a more modern and easy to learn chemical structure notation [42]. For example, n-butanol is CCCCO. A good source of SMILES notation for the most commonly used chemicals is the *Dictionary of Chemical Names and Synonyms* [18]. This input is often used by SRC investigators to run estimation programs for melting point, boiling point, and vapor pressure [37], octanol/water partition coefficient [23], water solubility [26], Henry's Law constant [25], soil adsorption coefficients [27], atmospheric oxidation rates [24], hydrolysis rates [38], and biodegradation probability [6].

CHEMICAL AND PHYSICAL PROPERTIES

The Hazardous Substances Data Bank (HSDB) of the National Library of Medicine was used as a source of boiling points, melting points, and molecular weights. The dissociation constant, octanol/water partition coefficient, water solubility, vapor pressure, and Henry's Law constant were judiciously selected from the many values that were identified in SRC's DATALOG file. All values selected were referenced to the primary literature source when possible.

Boiling Point: The boiling point or boiling point range is given along with the pressure. When the pressure is not given it should be assumed that the value is at 760 mm Hg.

Melting Point: The melting point or melting point range is given.

Molecular Weight: The molecular weight to two decimal points is given.

Dissociation Constants: The acid dissociation constant as the negative log (pKa) is given for chemicals that are likely to dissociate at environmental pHs (between 5 and 9). Chemical classes where dissociation is important include, for example, phenols, carboxylic acids, and aliphatic and aromatic amines. Once the pKa is known, the percent in the dissociated and undissociated form can be determined. For example, for an acid with a pKa of 4.75, the following is true at different pHs:

> 1% dissociated at pH 2.75
> 10% dissociated at pH 3.75
> 50% dissociated at pH 4.75
> 90% dissociated at pH 5.75
> 99% dissociated at pH 6.75

The degree of dissociation affects such processes as photolysis (absorption spectra of chemicals that dissociate can be considerably affected by the pH), evaporation from water (ions do not evaporate), soil or sediment adsorption, and bioconcentration. Values from evaluated sources such as Perrin [32] and Serjeant and Dempsey [35] were used when available.

Log Octanol/Water Partition Coefficient: The octanol/water partition coefficient is the ratio of the chemical concentration in octanol divided by the concentration in water. The most reliable source of values is from the Medchem project at Pomona College [12,13] and the database of Sangster [34]. When experimental values are unavailable, estimated values have been provided using a fragment constant estimation method, CLOGP3, from Medchem or the SRC LOGKOW Program [23]. Chemical octanol/water partition coefficients can always be calculated for organic chemicals using the LOGKOW Program because the fragments that are used are so small that missing fragment constants are not a problem. The octanol/water partition coefficient has been shown to correlate well with bioconcentration factors in aquatic organisms [21,41] and adsorption to soil or sediment [21,27], and recommended regression equations have been reviewed [21].

Water Solubility: The water solubility of a chemical provides considerable insight into the fate and transport of a chemical in the environment. High water soluble chemicals, which have a tendency to remain dissolved in the water column and not partition to soil or sediment or bioconcentrate in aquatic organisms, are less likely to volatilize from water (depending upon the vapor pressure - see Henry's Law constant) and are generally more likely to biodegrade. Low water soluble chemicals are just the opposite; they partition to soil or sediments and bioconcentrate in aquatic organisms, volatilize more readily from water, and are less likely to be biodegradable. Other fate processes that are, or can be, affected by water solubility include photolysis, hydrolysis, oxidation, and washout from the atmosphere by rain or fog. Water solubility values were taken from either the ARIZONA dATABASE [43] or from articles identified in SRC's DATALOG or CHEMFATE files [17]. The values were reported in ppm at a temperature at or as close as possible to 25 °C. Occasionally, when no values were available, the value was estimated from the octanol/water partition coefficient using recommended regression equations [21] or the SRC WS/KOW Program [26].

Vapor Pressure: The vapor pressure of a chemical provides considerable insight into the transport of a chemical in the environment. The volatility of the pure chemical is dependent upon the vapor pressure, and volatilization from water is dependent upon the vapor pressure and water solubility (see Henry's Law constant). The form in which a chemical will be found in the atmosphere is dependent upon the vapor pressure; chemicals with a vapor pressure less than 10^{-6} mm Hg will be mostly found associated with particulate matter [5,11]. When available, sources such as Boublik et al [7], Riddick et al [33], and Daubert and Danner [9] were used, since the data in these sources were evaluated and some of them provided recommended values. Vapor pressure was reported in mm Hg at or as close as possible to 25 °C. In many cases, the vapor pressure was calculated from a vapor pressure/temperature equation. Occasionally, the vapor pressure was calculated from structure using the SRC MPBPVP Program [37].

Henry's Law Constant: The Henry's Law constant is really the air/water partition coefficient, and therefore a nondimensional Henry's Law constant relates the chemical concentration in the gas phase to its concentration in the water phase. The dimensional Henry's Law constant can be determined by dividing the vapor pressure in atm by the water solubility in mole/m^3 to give the Henry's Law constant in atm-m^3/mol. The Henry's Law constant provides an indication of the partition between air and water at equilibrium and also is used to calculate the rate of evaporation from water (see discussion under Volatilization from Water/Soil). Henry's Law constants can be directly measured, calculated from the water solubility and vapor pressure, or estimated from structure by the method of Hine and Mookerjee [14] using the SRC HENRY Program [25], and this same order was used in selecting values. Some critical review data on Henry's Law constants are available (e.g., [22]).

ENVIRONMENTAL FATE/EXPOSURE POTENTIAL

Data for the following sections were identified with SRC's Environmental Fate Data Bases. Biodegradation data were selected from the DATALOG, BIOLOG, and BIODEG files. Abiotic degradation data were identified in the Hydrolysis, Photolysis, and Oxidation fields in DATALOG and CHEMFATE. Transport processes such as Bioconcentration, Soil Adsorption/Mobility, and Volatilization, as well as the monitoring data, were also identified in the DATALOG and CHEMFATE files.

Summary: This section is an abbreviated summary of all the data presented in the following sections and is not referenced; to find the citations the reader should refer to appropriate sections that follow. In general, this summary discusses how a chemical is used and released to the environment, how the chemical will behave in soil, water, and air, and how exposure to humans and environmental organisms is likely to occur.

Natural Sources: This section reviews any evidence that the chemical may have any natural sources of pollution, such as forest fires and volcanos, or may be a natural product that would lead to its detection in various media (e.g., methyl iodide is found in marine algae and is the major source of contamination in the ocean).

Artificial Sources: This section is a general review of any evidence that the chemical has anthropogenic sources of pollution. Quantitative data are reviewed in detail in Effluent Concentrations; this section provides a qualitative review of various sources based upon how the chemical is manufactured and used as well as the physical/chemical properties. For example, it is reasonable to assume that a highly volatile chemical which is used mostly as a solvent will be released to the atmosphere as well as the air of occupational settings, even if no monitoring data are available. Information on production volume and uses was obtained from a variety of chemical marketing sources including the *Kirk-Othmer Encyclopedia of Chemical Technology*, *Ullmann's Encyclopedia of Industrial Chemistry*, and the Chemical Profiles of the *Chemical Marketing Reporter*.

Terrestrial Fate: This section reviews how a chemical will behave if released to soil or ground water. Field studies or terrestrial model ecosystems studies are used here when they provide insight into the overall behavior in soil. Studies which determine an individual process (e.g., biodegradation, hydrolysis, soil adsorption) in soil are reviewed in the appropriate sections that follow. Quite often, except with pesticides, field or terrestrial ecosystem studies either are not available or do not give enough data to make conclusions on the terrestrial fate of a chemical. In these cases, data from the sections on Biodegradation, Abiotic Degradation, Soil Adsorption/Mobility, Volatilization from Water/Soil, and any appropriate monitoring data will be used to synthesize how a chemical is likely to behave if released to soil.

Aquatic Fate: This section reviews how a chemical will behave if released to fresh, marine, or estuarine surface waters. Field studies or aquatic model ecosystems are used here when they provide insight into the overall behavior in water. Studies which determine an individual process (e.g., biodegradation, hydrolysis, photolysis, sediment adsorption, and bioconcentration in aquatic organisms) in water are reviewed in the appropriate sections that follow. When field or aquatic ecosystems studies are not available or do not give enough data to make conclusions on the aquatic fate of the chemical, data from the appropriate degradation, transport, or monitoring sections will be used to synthesize how a chemical is likely to behave if released to water.

Atmospheric Fate: This section reviews how a chemical will behave if released to the atmosphere. The vapor pressure will be used to determine if the chemical is likely to be in the vapor phase or adsorbed to particulate matter [5,11]. The water solubility will be used to assess the likelihood of washout with rain. Smog chamber studies or other studies where the mechanism of degradation is not determined will be reviewed in this section; studies of the rate of reaction with hydroxyl radical or ozone or direct photolysis will be reviewed in Abiotic Degradation and integrated into this section.

Biodegradation: The principles outlined by Howard and Banerjee [15] are used in this section to review the relevant biodegradation data pertinent to biodegradation in soil, water, or wastewater treatment. In general, the studies have been separated into screening studies (inoculum in defined nutrient media), biological treatment simulations, and grab samples (soil or water sample with chemical added and loss of concentration followed). Pure culture studies are used only to indicate potential metabolites, since the artificial nutrient conditions under which the pure cultures are isolated provide little assurance that these same organisms will be present in any quantity or that their enzymes will be functioning in various soil or water environments. Anaerobic biodegradation studies, which are pertinent to whether a chemical will biodegrade in biological treatment digestors, sediment, and some ground waters, are discussed separately.

Abiotic Degradation: Nonbiological degradation processes in air, water, or soil are reviewed in this section. For most chemicals in the vapor phase in the atmosphere, reaction with photochemically generated hydroxyl radicals is the most important degradation process. Occasionally, reaction in the atmosphere with ozone (for olefins), nitrate radicals at night, and direct photolysis (direct sunlight absorption resulting in photochemical alteration) are significant for some chemicals [3,4]. For many chemicals, experimental reaction rate constants for hydroxyl radical are available (e.g., [1,4]) and are used to calculate an estimated half-life by assuming an average hydroxyl radical concentration of 5 x 10^{+5} molecules/cm^3 in nonsmog conditions (e.g., [3]). Some other average hydroxyl radical concentrations have been suggested [4], but we have stayed with 5 x 10^{+5} molecules/cm^3 to allow for a common comparison. If experimental rate constants are not available, they have been estimated by applying the fragment constant method of Atkinson [1] using SRC's Atmospheric Oxidation Program [24] and then a half-life estimated using the assumed hydroxyl radical concentration. The reaction rate for ozone reaction with olefins may be experimentally available or can be estimated using the Atmospheric Oxidation Program [24] from SRC which is based upon the work of Atkinson and Carter [2]. Using either the experimental or estimated rate constant and an assumed concentration of $7.2 \times 10^{+11}$ molecules/m^3 [3], an estimated half-life for reaction with ozone can be calculated. Nitrate radicals are significant only with certain classes of chemicals such as higher alkenes, dimethyl sulfide and lower thiols, furan and pyrrole, and hydroxy-substituted aromatics [3].

The possibility of direct photolysis in air or water can be partially assessed by examining the ultraviolet spectrum of the chemical. If the chemical does not absorb light at wavelengths provided by sunlight (>290 nm), the chemical cannot directly photolyze. If it does absorb sunlight, it may or may not photodegrade depending upon the efficiency (quantum yield) of the photochemical process, and unfortunately, such data are rarely available. Indirect photolysis processes may be important for some chemicals in water [28]. For example, some chemicals can undergo sensitized photolysis by absorbing triplet state energy from the excited triplet state of chemicals commonly found in water, such as humic acids. Transient oxidants found in water, such as peroxy radicals, singlet oxygen, and hydroxyl radicals, may also contribute to abiotic degradation in water for some chemicals. For example, phenols and aromatic amines have half-lives of less than a day for reaction with peroxy radicals; substituted and unsubstituted olefins have half-lives of 7 to 8 days with singlet oxygen; and dialkyl sulfides have half-lives of 27 hours with singlet oxygen [28].

Chemical hydrolysis at pHs that are normally found in the environment (pHs 5 to 9) can be important for a variety of chemicals that have functional groups that are potentially hydrolyzable, such as alkyl halides, amides, carbamates, carboxylic acid esters, epoxides and lactones, phosphate esters, and sulfonic acid esters [29]. Half-lives at various pHs are usually reported in order to provide an indication of the influence of pH.

Bioconcentration: Certain chemicals, due to their hydrophobic nature, have a tendency to partition from the water column and bioconcentrate in aquatic organisms. This concentration of chemicals in aquatic organisms is of concern because it can lead to toxic concentrations being reached when the organism is consumed by higher organisms such as wildlife and humans.

Such bioconcentrations are usually reported as the bioconcentration factor (BCF), which is the concentration of the chemical in the organism at equilibrium divided by the concentration of the chemical in water. This unitless BCF value can be determined experimentally by dosing water containing the organism and dividing the concentration in the organism by the concentration in the water once equilibrium is reached, or if equilibration is slow, the rate of uptake can be used to calculate the BCF at equilibrium. The BCF value can also be estimated by using recommended regression equations that have been shown to correlate well with physical properties such as the octanol/water partition coefficient and water solubility [21]; however, these estimation equations assume that little metabolism of the chemical occurs in the aquatic organism, which is not always correct. Therefore, when available, experimental values are preferred.

Soil Adsorption/Mobility: For many chemicals (especially pesticides), experimental soil or sediment partition coefficients are available. These values are measured by determining the concentration in both the solution (water) and solid (soil or sediment) phases after shaking for about 24 to 48 hours and using different initial concentrations. The data are then fit to a Freundlich equation to determine the adsorption coefficient, Kd. These Kd values for individual soils or sediments are normalized to the organic carbon content of the soil or sediment by dividing by the organic content to give the Koc, since of the numerous soil properties that affect sorption (organic carbon content, particle size, clay mineral composition, pH, cation-exchange capacity) [20], organic carbon is the most important for undissociated organic chemicals. Occasionally, the experimental adsorption coefficients are reported on a soil-organic matter basis (Kom) and these are converted to Koc by multiplying by 1.724 [21]. When experimental values are unavailable, estimated Koc values are calculated using either the water solubility or octanol/water partition coefficient and some recommended regression equations [21]. The measured or estimated adsorption values are used to determine the likelihood of leaching through soil or adsorbing to sediments using the criteria of Swann et al [39]. Occasionally, experimental soil thin-layer chromatography studies are also available and can be used to assess the potential for leaching.

The above discussion applies generally to undissociated chemicals, but there are some exceptions. For example, aromatic amines have been shown to covalently bond to humic material [31] and this slow but nonreversible process can lead to aromatic amines being tightly bound to the humic material in soils. Methods to estimate the soil or sediment adsorption coefficient for dissociated chemicals which form anions are not yet available, so it is particularly important to know the pKa value for chemicals that can dissociate so that a determination of the relative amounts of the dissociated and undissociated forms can be determined at various pH conditions. Chemicals that form cations at ambient pH conditions are generally thought to sorb strongly to clay material, similar to what occurs with paraquat and diquat (pyridine cations).

Volatilization from Water/Soil: For many chemicals, volatilization can be an extremely important removal process, with half-lives as low as several hours. The Henry's Law constant can give qualitative indications of the importance of volatilization; for chemicals with values less than 10^{-7} atm-m^3/mol, the chemical is less volatile than water and as water evaporates the concentration will increase; for chemicals around 1×10^{-3} atm-m^3/mol, volatilization will be rapid. The volatilization process is dependent upon physical properties of the chemical (Henry's Law constant, diffusivity coefficient), the presence of modifying materials (adsorbents, organic films, electrolytes, emulsions), and the physical and chemical properties of the environment (water depth, flow rate, the presence of waves, sediment content, soil moisture, and organic content) [21]. Since the overall volatilization rate cannot be estimated for all the various environments to which a chemical may be released, common models have been used in order to give an indication of the relative importance of volatilization. For most chemicals that have a Henry's Law constant greater than 10^{-7} atm-m^3/mol, the simple volatilization model outlined in Lyman et al [21] was used; this model assumes a 20 °C river 1 meter deep flowing at 1 m/sec with a wind velocity of 3 m/sec and requires only the Henry's Law constant and the molecular weight of the chemical for input. This model gives relatively rapid volatilization rates for this model river and values for ponds, lakes, or deeper rivers will be considerably slower. Occasionally, a chemical's measured reaeration coefficient ratio relative to oxygen is available, and this can be used with typical oxygen reaeration rates in ponds, rivers, and streams to give volatilization rates for these types of bodies of water. For chemicals that have extremely high Koc values, the EXAMS-II model has been used to estimate volatilization both with and without sediment adsorption (extreme differences are noted for these high Koc chemicals). Soil volatilization models are less validated and only qualitative statements are given of the importance of volatilization from moist (about 2% or greater water content) or dry soil, based upon the Henry's Law constant or vapor pressure, respectively. This assumes that once the soil is saturated with a molecular layer of water, the volatilization rate will be mostly determined by the value of the Henry's Law constant, except for chemicals with high Koc values.

Water Concentrations: Ambient water concentrations of the chemical are reviewed in this section, with subcategories for surface water, drinking water, and ground water when data are available. In general, the number of samples, the percent positive, the range of concentrations, and the average concentration are reported when the data are available.

Effluent Concentrations: Air emissions and wastewater effluents are reviewed in this section. In general, the number of samples, the percent positive, the range of concentrations, and the average concentration are reported when the data are available.

Sediment/Soil Concentrations: Sediment and soil concentrations are reviewed in this section. In general, the number of samples, the percent positive, the range of concentrations, and the average concentration are reported when the data are available.

Atmospheric Concentrations: Ambient atmospheric concentrations are reviewed in this section, with subcategories for rural/remote and urban/suburban when data are available in such sources as Brodzinsky and Singh [8] and Singh and Zimmerman [36]. In general, the number of samples, the percent positive, the range of concentrations, and the average concentration are reported when the data are available.

Food Survey Values: Market basket survey data such as found in Duggan et al [10] and individual studies of analysis of the chemical in processed food are reported in this section. In general, the number of samples, the percent positive, the range of concentrations, and the average concentration are reported when the data are available.

Plant Concentrations: Concentrations of the chemical in plants are reviewed in this section. If the plant has been processed for food, it is reported in Food Survey Values.

Fish/Seafood Concentrations: Concentrations in fish, seafood, shellfish, etc. are reviewed in this section. If the fish or seafood have been processed for food, the data are reported in Food Survey Values.

Animal Concentrations: Concentrations in animals are reviewed in this section. If the animals have been processed for food, the data are reported in Food Survey Values.

Milk Concentrations: Since dairy milk constitutes a high percentage of the human diet, concentrations of the chemical found in dairy milk are reviewed in this section and not in Food Survey Values.

Other Environmental Concentrations: Concentrations of the chemical found in other environmental media that may contribute to an understanding of how a chemical may be released to the environment or exposed to humans (e.g., detection in gasoline or cigarette smoke) are reviewed in this section.

Probable Routes of Human Exposure: The monitoring data and physical properties are used to provide conclusions on the routes (oral, dermal, inhalation) of exposure.

Average Daily Intake: The average daily intake is a calculated value of the amount of the chemical that is typically taken in daily by human adults. The value is determined by multiplying typical concentrations in drinking water, air, and food by average intake factors such as 2 liters of water, 20 m^3 of air, and 1600 grams of food [40].

Occupational Exposures: Monitoring data, usually air samples, from occupational sites are reviewed in this section. In addition, estimates of the number of workers exposed to the chemical from the two National Institute for Occupational Safety and Health (NIOSH) surveys are reviewed in this section. The National Occupational Hazard Survey (NOHS) conducted from 1972 to 1974 and the National Occupational Exposure Survey (NOES) conducted from 1981 to 1983 provided statistical estimates of worker exposures based upon limited walk-through industrial hygiene surveys.

Body Burdens: Any concentrations of the chemical found in human tissues or fluids is reviewed in this section. Included are blood, adipose tissue, urine, and human milk.

REFERENCES

1. Atkinson RA; Internat J Chem Kinet 19: 799-828 (1987)
2. Atkinson RA, Carter WP; Chem Rev 84: 437-70 (1984)
3. Atkinson RA; Chem Rev 85: 60-201 (1985)
4. Atkinson R; J Chem Phys Ref Data Monograph 1 (1989)
5. Bidleman TF; Environ Sci Technol 22:361-7 (1988)
6. Boethling RS et al; Environ Sci Technol 28: 459-65 (1994)
7. Boublik T et al; The Vapor Pressures of Pure Substances. Amsterdam: Elsevier (1984)
8. Brodzinsky R, Singh HB; Volatile organic chemicals in the atmosphere: an assessment of available data. SRI Inter EPA contract 68-02-3452 Menlo Park, CA (1982)
9. Daubert TE, Danner RP; Data Compilation Tables of Properties of Pure Compounds. Amer Inst Chem Engr pp 450 (1985)
10. Duggan RE et al; Pesticide Residue Levels in Foods in the U.S. from July 1, 1969 to June 30, 1976. Washington, DC: Food Drug Administ. 240 pp (1983)
11. Eisenreich SJ et al; Environ Sci Technol 15: 30-8 (1981)
12. Hansch C, Leo AJ; Medchem Project Issue No 26. Claremont, CA: Pomona College (1985)
13. Hansch C et al; Explori ng SAR Hydrophobic, Electronic, and Steric Constants ACS Professional Reference Book. Heller SR, consult ed, Washington, DC: Amer Chem Soc pg. 3 (1995)
14. Hine J, Mookerjee PK; J Org Chem 40: 292-8 (1975)
15. Howard PH, Banerjee S; Environ Toxicol Chem 3: 551-562 (1984)
16. Howard PH et al; Environ Sci Technol 12: 398-407 (1978)
17. Howard PH et al; Environ Toxicol Chem 5: 977-88 (1986)
18. Howard PH, Neal M; Dictionary of Chemical Names and Synonyms. Lewis Publishers/CRC Press, Boca Raton, FL (1992)
19. Karickhoff SW; Chemosphere 10: 833-46 (1981)
20. Karickhoff SW; In Environmental Exposure from Chemicals Vol I. ed Neely WB, Blau GE, Boca Raton, FL: CRC Press p 49-64 (1985)
21. Lyman WJ et al; Handbook of Chemical Property Estimation Methods. McGraw-Hill, NY (1982)
22. Mackay D, Shiu WY; J Phys Chem Ref Data 10: 1175-99 (1981)
23. Meylan WM, Howard PH; J Pharm Sci 84: 83-92 (1995)
24. Meylan WM, Howard PH; Chemosphere 26: 2293-99 (1993)
25. Meylan WM, Howard PH; Environ Toxicol Chem 10:1283-93 (1991)
26. Meylan WM et al; Environ Toxicol Chem 15: 100-106 (1996)
27. Meylan WM et al; Environ Sci Technol 26: 1560-67 (1992)
28. Mill T, Mabey W; In Environmental Exposure from Chemicals Vol I. ed Neely WB, Blau GE, Boca Raton, FL: CRC Press p 175-216 (1985)
29. Neely WB; In Environmental Exposure from Chemicals Vol I. ed Neely WB, Blau GE, Boca Raton, FL: CRC Press p 157-73 (1985)
30. Neely WR, Blau GE; Environmental Exposure from Chemicals Vol I. Boca Raton, FL: CRC Press (1985)
31. Parris GE; Environ Sci Technol 14: 1099-1105 (1980)

32. Perrin DD; Dissociation Constants of Organic Bases in Aqueous Solution. IUPAC Chemical Data Series, London: Butterworth (1965)

33. Riddick JA et al; Organic Solvents: Physical Properties and Methods of Purification, 4th Edit. New York: J Wiley & Sons (1986)

34. Sangster J LOGKOW Database, Sangster Research Lab., Quebec, Canada (1993)

35. Serjeant EP, Dempsey B; Ionisation Constants of Organic Acids in Aqueous Solution. IUPAC Chemical Data Series No 23, New York: Pergamon Press (1979)

36. Singh HB, Zimmerman PB; Adv Environ Sci Technol 24:177-235 (1992)

37. SRC; MPBPVP Program from Syracuse Research Corp. (1995)

38. SRC; HYDRO Program from Syracuse Research Corp. (1995)

39. Swann RL et al; Residue Reviews 85: 17-28 (1983)

40. U.S. EPA; Reference Values for Risk Assessment. Environ Criteria Assess Office, Off Health Environ Assess, Off Research Devel, ECAO-CIN-477, Cincinnati, OH: U.S. Environ Prot Agency (1986)

41. Veith GD et al; J Fish Res Board Can 36: 1-40-8 (1979)

42. Weininger D; J Chem Inf Sci 28: 31-36 (1988)

43. Yalkowsky SH et al; ARIZONA dATABASE of Aqueous Solubility, U. Arizona, Tucson, AZ (1987)

Contents

2,2'-Bipyridine

SUBSTANCE IDENTIFICATION

Synonyms:

Structure:

CAS Registry Number: 366-18-7

Molecular Formula: $C_{10}H_8N_2$

SMILES Notation: n(c(ccc1)c(nccc2)c2)c1

CHEMICAL AND PHYSICAL PROPERTIES

Boiling Point: 272-273 °C

Melting Point: 69.7 °C

Molecular Weight: 156.18

Dissociation Constants: pK_a = 4.33 [15]

Log Octanol/Water Partition Coefficient: 1.38 (estimate) [13]

Water Solubility: 5000 ppm [3]

Vapor Pressure: 1.30 x 10^{-5} mm Hg at 25 °C (calculated from water solubility and Henry's Law constant)

Henry's Law Constant: 5.35 x 10^{-10} atm-m^3/mol at 25 °C (estimated using a group contribution method) [9]

ENVIRONMENTAL FATE/EXPOSURE POTENTIAL

Summary: 2,2'-Bipyridine is released to the environment at points of application of the quaternary salt of 2,2'-bipyridine, which is used as an herbicide (diquat) or via effluents at sites where it is produced or used as a chemical intermediate. Information pertaining to the biodegradation of 2,2'-bipyridine in soil and water was not located in the available literature. With a pKa of 4.33, 2,2'-bipyridine and its conjugate acid should exist in environmental media in varying proportions that are pH dependent. Ions generally do not volatilize. A Henry's Law constant of 5.35×10^{-10} atm-m^3/mol at 25 °C indicates that volatilization of 2,2'-bipyridine from environmental waters and moist soil should not be an important fate process. In aquatic systems, 2,2'-bipyridine is not expected to bioconcentrate. 2,2'-Bipyridine was shown to undergo slow oxidation with photochemically generated hydroxyl radicals in aqueous solution (half-life of about 129 days). A low Koc indicates 2,2'-bipyridine should not partition from the water column to organic matter contained in sediments and suspended solids; and it should be highly mobile in soil and it may leach to ground water. However, the bipyridylium cation of diquat has been shown to strongly adsorb humic acid and clays. In the atmosphere, 2,2'-bipyridine is expected to exist in both the vapor and particulate phases, and vapor phases reactions with photochemically produced hydroxyl radicals should be important (estimated half-life of 9.4 days). In addition, 2,2'-bipyridine has the potential to be physically removed from air by wet deposition. The most probable human exposure would be occupational exposure, which may occur through dermal contact or inhalation at workplaces where 2,2'-bipyridine is produced or used as a chemical intermediate or an herbicide.

Natural Sources:

Artificial Sources: 2,2'-Bipyridine is an important chemical intermediate in the production of the herbicide, diquat [6]. The quaternary salt of 2,2'-bipyridine is used as a contact herbicide [6]. Consequently, 2,2'-bipyridine may be released to the environment at points of its application as an herbicide or via effluents at sites where it is produced or used as a chemical intermediate.

Terrestrial Fate: Information pertaining to the biodegradation of 2,2'-bipyridine in soil was not located in the available literature. With a pKa of

4.33 [15], 2,2'-bipyridine and its conjugate acid should exist in soils in varying proportions that are pH dependent. Ions generally do not volatilize. Based upon the Henry's Law constant [9], volatilization of 2,2'-bipyridine from moist soils is also not expected to be an important fate process [11]. However, the bipyridylium cation of diquat has been shown to strongly adsorb to humic acid [2] and clays [7,8,10,17]. An estimated Koc of 40 [11] indicates 2,2'-bipyridine should be highly mobile in soil [16].

Aquatic Fate: Information pertaining to the biodegradation of 2,2'-bipyridine in aquatic systems was not located in the available literature. 2,2'-Bipyridine was shown to undergo slow oxidation with photochemically generated hydroxyl radicals in aqueous solution (half-life of about 129 days) [4]. With a pKa of 4.33 [15], 2,2'-bipyridine and its conjugate acid should exist among environmental waters in varying proportions that are pH dependent. The ratio of 2,2'-bipyridine to its conjugate acid should increase with increasing pH [11]. Ions are not expected to volatilize from water. Based upon the Henry's Law constant [9], volatilization of 2,2'-bipyridine from natural bodies of water is not expected to be an important fate process [11]. An estimated Koc of 40 [11] indicates 2,2'-bipyridine should not partition from the water column to organic matter [15] contained in sediments and suspended solids. However, the bipyridylium cation of diquat has been shown to strongly adsorb to humic acid [2] and clay [7,8,10,17]. An estimated bioconcentration factor (log BCF) of 0.70 indicates 2,2'-bipyridine should not bioconcentrate among aquatic organisms [11].

Atmospheric Fate: Based on the estimated vapor pressure, 2,2'-bipyridine is expected to exist in both the vapor and particulate phases in ambient air [5]. In the atmosphere, vapor phase reactions with photochemically produced hydroxyl radicals may be important. The rate constant for 2,2'-bipyridine was estimated to be 1.71×10^{-12} cm^3/molecule-sec at 25 °C, which corresponds to an atmospheric half-life of about 9.4 days at an atmospheric concentration of $5 \times 10^{+5}$ hydroxyl radicals per cm^3 [1]. The water solubility for 2,2'-bipyridine [3] indicates that physical removal from air by precipitation and dissolution in clouds may occur.

Biodegradation:

Abiotic Degradation: The rate constant for the vapor-phase reaction of 2,2'-bipyridine with photochemically produced hydroxyl radicals has been estimated to be 1.71×10^{-12} cm^3/molecule-sec at 25 °C, which corresponds to an atmospheric half-life of about 9.4 days at an atmospheric concentration of $5 \times 10^{+5}$ hydroxyl radicals per cm^3 [1]. The rate constant for the reaction of 2,2'-bipyridine with photochemically produced hydroxyl radicals in water has been measured to be $6.2 \times 10^{+9}$ M^{-1} sec^{-1} at 21 °C and a pH of 9.3 [4], which corresponds to an aqueous half-life of about 129 days at a hydroxyl radical concentration of 1×10^{-17} M/L [12].

Bioconcentration: Based upon the water solubility [3], a bioconcentration factor (log BCF) of 0.70 for 2,2'-bipyridine has been calculated using a recommended regression-derived equation [11]. This BCF value indicates 2,2'-bipyridine should not bioconcentrate among aquatic organisms.

Soil Adsorption/Mobility: Based on the water solubility [3], a Koc of 40 for 2,2'-bipyridine has been calculated using a recommended regression-derived equation [11]. This Koc value indicates 2,2'-bipyridine will be highly mobile in soil [16], and it should not partition from the water column to organic matter contained in sediments and suspended solids. With a pKa of 4.33 [15], 2,2'-bipyridine and its conjugate acid should exist among environmental media in varying proportions that are pH dependent. The bipyridylium cation of diquat has been shown to strongly adsorb to humic acid [2] and clays [7,8,10,17]. Diquat is adsorbed to the interlayer spaces of montmorillonite making it unavailable to micro-organisms [17]. Diquat is also highly immobile in soil [7,8].

Volatilization from Water/Soil: With a pKa of 4.33 [15], 2,2'-bipyridine and its conjugate acid will exist among environmental media in varying proportions that are pH dependent. Ions are not expected to volatilize from water. Based upon the Henry's Law constant of 5.35×10^{-10} atm-m^3/mol at 25 °C, which has been estimated using a group contribution method [9], volatilization of 2,2'-bipyridine from natural bodies of water and moist soils is not expected to be an important fate process [11].

Water Concentrations:

Effluent Concentrations:

Sediment/Soil Concentrations:

Atmospheric Concentrations:

Food Survey Values:

Plant Concentrations:

Fish/Seafood Concentrations:

Animal Concentrations:

Milk Concentrations:

Other Environmental Concentrations:

Probable Routes of Human Exposure:

Average Daily Intake:

Occupational Exposure: The most probable human exposure to 2,2'-bipyridine would be occupational exposure, which may occur through dermal contact or inhalation at places where it is produced or used as a chemical intermediate or an herbicide. NIOSH (NOES 1981-83) has estimated that 906 workers are potentially exposed to 2,2'-bipyridine in the USA [14].

Body Burdens:

REFERENCES

1. Atkinson R; Intern J Chem Kin 19: 799-828 (1987)
2. Choudry GG; Toxicol Environ Chem 6: 127-71 (1983)
3. Dean JA; Lange's Handbook of Chemistry (13th ed) NY: McGraw-Hill Book Co (1985)
4. Dorfman LM, Adams GE; Reactivity of the Hydroxyl Radical in Aqueous Solution. NSRD-NBS-46. (NTIS COM-73-50623) Washington, DC National Bureau of Standards p. 51 (1973)
5. Eisenreich SJ et al; Environ Sci Technol 15: 30-8 (1981)
6. Goe GL; Kirk-Othmer Encycl Chem Tech 3rd NY: Wiley Interscience 19: 454-83 (1982)
7. Helling CS; Soil Sci Soc Amer Proc 35: 743-8 (1971)

8.   Helling CS, Turner BC; Science 162; 562-3 (1968)
9.   Hine J, Mookerjee PK; J Org Chem 40: 292-8 (1975)
10. Kunze GW; pp. 49-70 in Pestic Eff Soil Water Symp Columbus, OH (1965)
11. Lyman WJ et al; Handbook of Chemical Property Estimation Methods NY: McGraw-Hill p. 4-9, 5-4, 6-3, 15-16 (1982)
12. Mill T; Science 207: 886-7 (1980)
13. Meylan W, Howard PH; J Pharm Sci 84: 83-92 (1995)
14. NIOSH; National Occupational Exposure Survey (NOES) (1989)
15. Perrin DD; Aust J Chem 17: 484-8 (1964)
16. Swann RL et al; Res Rev 85: 16-28 (1983)
17. Weber JB; Adv Chem Ser 111: 55-120 (1972)

Bis(2-chloroisopropyl) Ether

SUBSTANCE IDENTIFICATION

Synonyms: Dichloroisopropyl ether

Structure:

CAS Registry Number: 39638-32-9

Molecular Formula: $C_6H_{12}Cl_2O$

SMILES Notation: O(C(Cl)(C)C)C(Cl)(C)C

CHEMICAL AND PHYSICAL PROPERTIES

Boiling Point: 187.1 °C

Melting Point: <-20 °C

Molecular Weight: 171.07

Dissociation Constants:

Log Octanol/Water Partition Coefficient: 3.73 [2] (estimated)

Water Solubility: The compound hydrolyzes very fast in water [4] making solubility determination difficult.

Vapor Pressure:

Henry's Law Constant: 3.32×10^{-4} atm-m³/mol [5] (estimated)

ENVIRONMENTAL FATE/EXPOSURE POTENTIAL

Summary: If released to water or moist soil, bis(2-chloroisopropyl) ether will hydrolyze rapidly based on an estimated hydrolysis half-life of <38.4

sec in water. Therefore, biodegradation, bioconcentration in aquatic organisms and adsorption to soil and sediment are not expected to be significant fate processes for bis(2-chloroisopropyl) ether. If released to the atmosphere, vapor-phase bis(2-chloroisopropyl) ether is degraded by reaction with photochemically produced hydroxyl radicals with an estimated half-life of 25 days. However, the hydrolysis half-life of bis(2-chloroisopropyl) ether in moist air would be <25 hr. Therefore, hydrolysis may be the most important removal process from air.

Natural Sources: Bis(2-chloroisopropyl) ether is not known to occur naturally.

Artificial Sources: There is no evidence in the literature that bis(2-chloroisopropyl) ether is commercially produced in the U.S.

Terrestrial Fate: Based on a measured hydrolysis rate of 65/hr for bis(chloromethyl) ether [4], bis(2-chloroisopropyl) ether is estimated to have a hydrolysis half-life of <38.4 sec in water [4]. Therefore, hydrolysis is expected to be the predominant fate process in moist soil and it is unlikely that biodegradation or other chemical processes could compete with hydrolysis of bis(2-chloroisopropyl) ether in soil.

Aquatic Fate: Based on a measured hydrolysis rate of 65/hr for bis(2-chloromethyl) ether, it is estimated that bis(2-chloroisopropyl) ether has a hydrolysis half-life of <38.4 sec in water [4]. Therefore, hydrolysis will be the predominant fate process in aquatic systems and biodegradation and other chemical process may not be competitive with hydrolysis.

Atmospheric Fate: Based on an estimation method [1], bis(2-chloroisopropyl) ether is expected to undergo a gas-phase reaction with photochemically produced hydroxyl radicals at an estimated half-life of 24.8 days [1]. The gas phase hydrolysis of bis(chloromethyl) ether in humid atmosphere is estimated to be 0.00047/min, corresponding to a half-life of 25 hr [6]. If it is assumed that the hydrolysis of bis(2-chloroisopropyl) ether is faster than bis(chloromethyl) ether [4], the half-life for bis(2-chloroisopropyl) ether in the atmosphere would be <25 hr.

Biodegradation:

Bis(2-chloroisopropyl) Ether

Abiotic Degradation: The rate constant for the vapor-phase reaction of bis(2-chloroisopropyl) ether with photochemically produced hydroxyl radicals has been estimated to be 4.32 x 10^{-13} cm^3/molecule-sec at 25 °C which corresponds to an atmospheric half-life of 24.8 days at a 12-hr daylight atmospheric concn of 1.5 x 10^{+6} hydroxyl radicals per cm^3 [1]. The gas phase hydrolysis rate constant for bis(chloromethyl) ether was found to vary between 0.0016-0.00047/min depending on the moisture content in air and the type of reactor surface used in the experiments [6]. The upper limit for the hydrolysis of bis(chloromethyl) ether in humid atmosphere can be assumed to be 0.00047/min, corresponding to a half-life of 25 hr [6]. If it is assumed that the hydrolysis of bis(2-chloroisopropyl) ether is faster than bis(chloromethyl) ether [4], the half-life for bis(2-chloroisopropyl) ether in the atmosphere would be <25 hr. Based on a hydrolysis rate constant of 65/hr at pH 7 for bis(chloromethyl) ether [4], bis(2-chloroisopropyl) ether is estimated to have a hydrolysis half-life of < 38.4 sec in water [4].

Bioconcentration: Based on a measured hydrolysis for bis(2-chloromethyl) ether [4], bis(2-chloroisopropyl) ether has been estimated to have a hydrolysis half-life of <38.4 sec in water [4]. Since bis(2-chloroisopropyl) ether will hydrolyze very fast in water, aquatic organisms will not have a chance to bioconcentrate it.

Soil Adsorption/Mobility: Because of the fast hydrolysis of bis(2-chloroisopropyl) ether in moist soil [4], adsorption will not be important.

Volatilization from Water/Soil: Based on an estimated Henry's Law constant of 3.32 x 10^{-4} atm-m^3/mol [5] and relationship between Henry's Law constant and volatility [3], the half-life for the volatilization of bis(2-chloroisopropyl) ether from a model river of depth 1 m, flowing at 1 m/sec with an overhead wind speed of 5 m/sec has been estimated to be 4 hr. However, due to fast hydrolysis of the compound in water, volatilization will not be able to compete with hydrolysis.

Water Concentrations:

Effluent Concentrations:

Sediment/Soil Concentrations:

Bis(2-chloroisopropyl) Ether

Atmospheric Concentrations:

Food Survey Values:

Plant Concentrations:

Fish/Seafood Concentrations:

Animal Concentrations:

Milk Concentrations:

Other Environmental Concentrations:

Probable Routes of Human Exposure:

Average Daily Intake:

Occupational Exposure:

Body Burdens:

REFERENCES

1. Atkinson R; Int J Chem Kinet 19: 799-828 (1987)
2. GEMS; Graphical Exposure Modeling System, CLOGP3, Office of Toxic Substances, USEPA, Washington, DC (1986)
3. Lyman WJ et al; Handbook of Chemical Property Estimation Methods, Washington, DC: American Chemical Society pp. 15-21 (1990)
4. Mabey WR; Aquatic Fate Process Data for Organic Priority Pollutants. USEPA-440/4-81-014. USEPA. Washington, DC (1981)
5. Meylan WM, Howard PH; Environ Toxicol Chem 10:1283-1293 (1991)
6. Tou JC, Kallos GJ; Anal Chem 46: 1866-9 (1974)

Bis(2-chloro-1-methylethyl) Ether

SUBSTANCE IDENTIFICATION

Synonyms: 2,2'-Dichloroisopropyl ether

Structure:

CAS Registry Number: 108-60-1

Molecular Formula: $C_6H_{12}Cl_2O$

SMILES Notation: O(C(CCl)C)C(CCl)C

CHEMICAL AND PHYSICAL PROPERTIES

Boiling Point: 187.3 °C at 760 mm Hg

Melting Point: -96.8 to -101.8 °C

Molecular Weight: 171.07

Dissociation Constants:

Log Octanol/Water Partition Coefficient: log Kow = 2.49 (estimated) [9]

Water Solubility: 1,700 mg/L [19]

Vapor Pressure: 0.71 to 0.85 mm Hg at 20 °C [6]

Henry's Law Constant: 1.13×10^{-4} atm-m^3/mol (estimated from ratio of vapor pressure of 0.85 mm Hg and water solubility)

ENVIRONMENTAL FATE/EXPOSURE POTENTIAL

Summary: Bis(2-chloro-1-methylethyl) ether may be released to the environment in waste streams from propylene glycol production. If released to the atmosphere, vapor-phase bis(2-chloro-1-methylethyl) ether is expected to degrade by reaction with photochemically produced hydroxyl radicals with an estimated half-life of 3.26 days. Due to its significant water solubility, removal of bis(2-chloro-1-methylethyl) ether from the atmosphere may also occur by precipitation via rain and snow. If released to soil, bis(2-chloro-1-methylethyl) ether is expected to leach significantly (estimated Koc of 73) into ground water where it may persist for a long period of time. Based on biodegradation studies in water, biodegradation of bis(2-chloro-1-methylethyl) ether in soil may be slow. If released to water, volatilization and biodegradation are expected to be the principal removal processes; although these processes may be slow. Volatilization half-lives of 8.1 hr and 6.6 days have been estimated for a model river (one meter deep) and a model pond, respectively. Slow biodegradation may be a significant removal process in water where volatilization is not likely to be important. Hydrolysis, adsorption to sediment and bioconcentration in aquatic organisms are not expected to be environmentally significant removal processes in aquatic systems. Exposure to the general population is expected to occur through consumption of contaminated drinking water. In occupational settings, exposure to bis(2-chloro-1-methylethyl) ether may occur through inhalation of vapors and through eye and skin contact.

Natural Sources:

Artificial Sources: Bis(2-chloro-1-methylethyl) ether may be released to the environment in waste streams from propylene glycol production [3,7].

Terrestrial Fate: The concn of bis(2-chloro-1-methylethyl) ether in a river water showed little change after bank infiltration for 1-12 months [20]. One biodegradation study observed no biodegradation of bis(2-chloro-1-methylethyl) ether in a river water on incubation for 5 days at 20 °C [7]. Although these studies are not specific to biodegradation in soil, they suggest that biodegradation in soil may be slow. An estimated Koc value of 73 indicates high mobility for bis(2-chloro-1-methylethyl) ether in soil and as a result significant leaching from soil to ground water may occur [9,17]. Based on a vapor pressure of 0.85 mm Hg at 20 °C [6], bis(2-chloro-1-

methylethyl) ether may evaporate from dry soil surfaces; however, volatilization from moist soils may not be rapid, but still may be important.

Aquatic Fate: Based on an estimated Henry's Law constant derived from the ratio of vapor pressure over water solubility and an estimation method for determining volatility from water, the volatilization half-lives of 8.1 hr and 6.6 days have been estimated for bis(2-chloro-1-methylethyl) ether from a model river (one meter deep) and a model pond, respectively [9,18]. By analogy to bis(2-chloroethyl)ether (estimated hydrolysis half-life of about 20 yrs at 25 °C [10]), hydrolysis of bis(2-chloro-1-methylethyl) ether should not be important and independent of pH. An estimated Koc of 73 and BCF of 9.3 suggest that adsorption to sediment and bioconcentration in aquatic organisms may not be significant [9]; however, bis(2-chloro-1-methylethyl) ether has been identified in sediments [16]. Based on one river die-away test [7] and bank infiltration studies [13,20], the biodegradation of bis(2-chloro-1-methylethyl) ether in water may be slow. However, slow biodegradation may become a significant removal process in water where volatilization is not likely to be important. The estimated overall half-lives (due to a combination of loss process) of bis(2-chloro-1-methylethyl) ether in river and lake water have been estimated to be 3.1 and 59 days, respectively [20].

Atmospheric Fate: Based on a reported vapor pressure in the range 0.71-0.85 mm Hg at 20 °C [6], bis(2-chloro-1-methylethyl) ether is expected to exist almost entirely in the vapor-phase in the ambient atmosphere [2]. Vapor phase bis(2-chloro-1-methylethyl) ether is degraded in the ambient atmosphere by reaction with photochemically formed hydroxyl radicals; the half-life for this reaction in air can be estimated to be about 1.63 days [1]. Due to its significant water solubility [19], removal of bis(2-chloro-1-methylethyl) ether from the atmosphere may also occur by precipitation via rain and snow.

Biodegradation: Bis(2-chloro-1-methylethyl) ether showed no biodegradation after 5 days at 20 °C when incubated with Ohio River water at an initial concn of 33 mg/L [7]. Three bank infiltration studies from the Rhine River in The Hague showed a concn of approximately 0.5 ug/L, which was not reduced at all after infiltration [13]. In the Netherlands, 33% removal in <1 yr and 90% removal in <3 months was observed after bank and dune infiltration of Rhine water, respectively [20]. Based upon field monitoring data, degradation half-lives have been estimated to be 59 days and 3.1 days for bis(2-chloro-1-methylethyl) ether in a lake in the Rhine basin and in the Rhine River, respectively [20].

Abiotic Degradation: The rate constant for the vapor-phase reaction of bis(2-chloro-1-methylethyl) ether with photochemically produced hydroxyl radicals can be estimated to be 3.28×10^{-12} cm^3/molecule-sec at 25 °C, which corresponds to an atmospheric half-life of about 1.63 days at a 12-hr daylight hydroxyl concn of 1.5×10^6 radicals per cm^3 [1]. By analogy to bis(2-chloroethyl)ether (estimated hydrolysis half-life about 20 yrs at 25 °C [10]), hydrolysis of bis(2-chloro-1-methylethyl) ether should be slow and independent of pH. Photolysis is not expected to be significant since bis(2-chloro-1-methylethyl) ether contains no chromophores that absorb UV radiation > 290 nm.

Bioconcentration: Based on a measured water solubility of 1700 ppm at 20 °C [19] and a regression equation [9], the BCF for bis(2-chloro-1-methylethyl) ether can be estimated to be 9.3. This BCF value suggests that bis(2-chloro-1-methylethyl) ether would not bioconcentrate significantly in aquatic organisms.

Soil Adsorption/Mobility: Based on a measured water solubility of 1700 ppm at 20 °C [19], the Koc for bis(2-chloro-1-methylethyl) ether can be estimated to be 73 using a regression equation [9]. This Koc value suggests that bis(2-chloro-1-methylethyl) ether may be highly mobile in soil and has the potential to leach easily through soil [17].

Volatilization from Water/Soil: Based on a measured water solubility of 1700 ppm at 20 °C [19] and a vapor pressure of 0.85 mm Hg at 20 °C [6], the Henry's Law constant (H) for bis(2-chloro-1-methylethyl) ether can be estimated to be 1.13×10^{-4} atm-m^3/mol at 20 °C. This value of Henry's Law constant indicates that volatilization from water may not be rapid but possibly significant [9]. Based on the value of H, the volatilization half-life of bis(2-chloro-1-methylethyl) ether from a model river 1 m deep flowing 1 m/sec with a wind velocity of 5 m/sec has been estimated to be approximately 8.1 hr [9]. The volatilization half-life from a model environmental pond has been estimated to be 6.6 days [18].

Water Concentrations: DRINKING WATER: Bis(2-chloro-1-methylethyl) ether was identified in tap water from bank filtered Rhine water in the Netherlands at a maximum concn of 3,000 ng/L [14]. Bis(2-chloro-1-methylethyl) ether was detected in 8 of 113 samples at an average concn of 0.17 ug/L in drinking water from 113 community water supplies in the US from May-June 1976 [11]. Bis(2-chloro-1-methylethyl) ether was detected in 7 of 110 samples at an average concn of 0.11 ug/L in drinking water from 110 US community water supplies from November 1976-January 1977 [11]. It was identified in municipal drinking water at a concn of 0.8 ug/L in Evansville, IN on August 25, 1971 [7]. Bis(2-chloro-1-methylethyl) ether was detected in the water supply of Cleveland, OH [15]. Bis(2-chloro-1-methylethyl) ether was detected at a concn of 0.18 ug/L in finished water from the Carrollton water plant in the New Orleans, LA area [5]. SURFACE WATER: Bis(2-chloro-1-methylethyl) ether was detected in water at mean concns of 0.10 and 19 ug/L from the New Orleans/Baton Rouge, LA and Houston, TX areas, respectively [12]. Bis(2-chloro-1-methylethyl) ether was detected in water from the Rhine river in the Netherlands at an average concn of 4 ug/L with a maxima of 15 ug/L during 1978-1979; samples were taken every two weeks [13]. According to the STORET database, bis(2-chloro-1-methylethyl) ether was detected in ambient water from the US at a median concn of <10.000 ug/L [16]. GROUND WATER: At the Zwolle water utility in the Netherlands, bis(2-chloro-1-methylethyl) ether was detected in 7 observations made in 1978 at a maximum concn of 3 ug/L following bank infiltration of Rhine River water [13]. Three influent water tests from the Rhine River in The Hague showed a concn of approximately 0.5 ug/L, which was not reduced at all after infiltration [13].

Effluent Concentrations: Bis(2-chloro-1-methylethyl) ether was detected in U.S. rivers as a result of industrial outfall from propylene glycol production [3,7] at concentrations ranging from 0.2-5 ug/L (Ohio River) from August-September 1971 [7]. Bis(2-chloro-1-methylethyl) ether was detected in water samples from a specially constructed leachate treatment plant located at Love Canal (Niagara Falls, NY) [4]. According to the STORET database, bis(2-chloro-1-methylethyl) ether was detected in effluents from the US with a median concn of <10.0 ug/L [16].

Sediment/Soil Concentrations: According to the STORET database, bis(2-chloro-1-methylethyl) ether was detected in sediment from the US at a median concn of <500.0 ug/kg (dry) [16].

Bis(2-chloro-1-methylethyl) Ether

Atmospheric Concentrations:

Food Survey Values:

Plant Concentrations:

Fish/Seafood Concentrations:

Animal Concentrations:

Milk Concentrations:

Other Environmental Concentrations: According to the STORET database, Bis(2-chloro-1-methylethyl) ether has been detected in biota from the US at a median concn of <2.2 mg/kg [16].

Probable Routes of Human Exposure: Bis(2-chloro-1-methylethyl) ether has been detected in drinking water [5,8,11,14,15]; therefore, exposure to the general population is expected to occur through consumption of contaminated drinking water. In occupational settings, exposure of workers to bis(2-chloro-1-methylethyl) ether may occur through inhalation of vapors and through eye and skin contact.

Average Daily Intake:

Occupational Exposure: Workers in areas where bis(2-chloro-1-methylethyl)ether is produced as a byproduct of the chlorohydrin process and workers involved in propylene glycol production have the potential for exposure to bis(2-chloro-1-methylethyl) ether.

Body Burdens:

REFERENCES

1. Atkinson R; Int J Chem Kinet 19: 799-828 (1987)
2. Eisenreich SJ et al; Environ Sci Technol 15: 30-8 (1981)
3. Fishbein L; Sci Total Environ 11: 223-57 (1979)
4. Hauser TR, Bromberg SM; Env Monit Assess 2: 249-72 (1982)

5. Keith LH et al; pp. 329-73 in Ident Anal Organic Pollut Water, Keith, LH (ed), Ann Arbor, MI: Ann Arbor Press (1976)

6. Kirwin CJ, Sandmeyer EE; Ethers. In: Patty's Industrial Hygiene and Toxicology, 3rd ed., Vol IIA, Clayton GD, Clayton FE (eds), New York, NY: John Wiley & Sons p 2496-7 (1981)

7. Kleopfer RD, Fairless BJ; Environ Sci Technol 6: 1036-7 (1972)

8. Kool HJ et al; Crit Rev Env Control 12: 307-57 (1982)

9. Lyman WJ et al; Handbook of Chemical Estimation Methods, Washington, DC: American Chemical Society, p. 4-9, 5-10, 7-4, 15-15 to 15-32 (1990)

10. Mabey WR et al; Aquatic Fate Process Data for Organic Priority Pollutants USEPA-440/4-81-014 Washington, DC: USEPA (1981)

11. Mello W; Investigations of Selected Environ Pollutants USEPA-560/2-78-006, Washington, DC: USEPA (1978)

12. Pellizzari ED et al; Formulation of Preliminary Assessment of Halogenated Organic Compounds in Man and Environmental Media USEPA/560/13-79-006 Research Triangle Park, NC: USEPA (1979)

13. Piet GJ, Zoeteman BCJ; J Amer Water Works Assoc 72: 400-4 (1980)

14. Piet GJ, Morra CF; Behavior of organic micropollutants during underground passage. In: Artificial Groundwater Recharge (Water Resources Engineering Series) Huisman L, Olsthorn TN (eds), Marshfield, MA: Pitman Publ p. 31-42 (1983)

15. Sanjivamurthy VA; Water Res 12: 31-3 (1978)

16. Staples CA et al; Environ Toxicol Chem 4: 131-42 (1985)

17. Swann RL et al; Res Rev 85: 17-28 (1983)

18. USEPA; EXAMS II computer simulation, Athens, GA: USEPA (1987)

19. Yalkowsky SH et al; Arizona Database of Aqueous Solubility, Tucson, AZ: Univ of Arizona (1987)

20. Zoeteman BCJ et al; Chemosphere 9: 231-49 (1980)

2-Bromo-2-chloro-1,1,1-trifluoroethane

SUBSTANCE IDENTIFICATION

Synonyms:

Structure:

CAS Registry Number: 151-67-7

Molecular Formula: $C_2HBrClF_3$

SMILES Notation: FC(F)(F)C(Cl)Br

CHEMICAL AND PHYSICAL PROPERTIES

Boiling Point: 50.2 °C

Melting Point: -118 °C

Molecular Weight: 197.39

Dissociation Constants:

Log Octanol/Water Partition Coefficient: 2.30 [10].

Water Solubility: 3900 mg/L at 25 °C [22] 3450 mg/L [5]

Vapor Pressure: 302 mm Hg at 25 °C [7].

Henry's Law Constant: 0.0313 atm-m³/mol at 25 °C (estimated) [14].

ENVIRONMENTAL FATE/EXPOSURE POTENTIAL

Summary: 2-Bromo-2-chloro-1,1,1-trifluoroethane is an anthropogenic compound which is used as an inhalation anesthetic. It may be released to the environment as a fugitive emission during its production or use. If released to soil, 2-bromo-2-chloro-1,1,1-trifluoroethane will rapidly volatilize from both moist and dry soil to the atmosphere and will display moderate to high mobility. If released to water, 2-bromo-2-chloro-1,1,1-trifluoroethane will rapidly volatilize to the atmosphere. The estimated half-life for volatilization from a model river is 4.1 hr. 2-Bromo-2-chloro-1,1,1-trifluoroethane will not bioconcentrate in fish and aquatic organisms nor will it adsorb to sediment or suspended organic matter. Insufficient data are available to determine the importance or rate of biodegradation of 2-bromo-2-chloro-1,1,1-trifluoroethane in soil or water. If released to the atmosphere, 2-bromo-2-chloro-1,1,1-trifluoroethane will undergo a gas-phase reaction with photochemically produced hydroxyl radicals with an estimated half-life of 267 days. It may also undergo atmospheric removal by wet deposition processes; however, any removed is expected to rapidly re-volatilize to the atmosphere. Occupational exposure to 2-bromo-2-chloro-1,1,1-trifluoroethane may occur by inhalation or dermal contact during its production or use. The general population may be exposed to 2-bromo-2-chloro-1,1,1-trifluoroethane during surgical procedures where it is used as an anesthetic.

Natural Sources: 2-Bromo-2-chloro-1,1,1-trifluoroethane is of anthropogenic origin, and it is not known to be produced by natural sources.

Artificial Sources: 2-Bromo-2-chloro-1,1,1-trifluoroethane is an anthropogenic compound which is used as an inhalation anesthetic for humans and animals [2,9,17,18,21,24]; therefore, it may be released to the environment as a fugitive emission during its production or use. It also may be released to the atmosphere in waste anesthetic gases [9,18].

Terrestrial Fate: If released to soil, the experimental vapor pressure for 2-bromo-2-chloro-1,1,1-trifluoroethane, $3.02 \times 10^{+2}$ mm Hg at 25 °C [7], indicates that it will rapidly volatilize from dry soil to the atmosphere. Estimated soil adsorption coefficients ranging from 46 to 425 [10,12,22] indicate that it will display moderate to high mobility in soil [23]. Insufficient data are available to determine the importance or rate of biodegradation of 2-bromo-2-chloro-1,1,1-trifluoroethane in soil.

Aquatic Fate: If released to water, an estimated Henry's Law constant of 0.0313 atm-m^3/mol at 25 °C [14] for 2-bromo-2-chloro-1,1,1-tri-fluoroethane indicates that it will rapidly volatilize to the atmosphere. The estimated half-life for volatilization from a model river 1 m deep flowing at 1 m/sec with a wind speed of 3 m/sec is 4.1 hr [12,14]. Estimated bioconcentration factors ranging from 6 to 33 [10,12,22] indicate that it will not bioconcentrate in fish and aquatic organisms. Estimated soil adsorption coefficients ranging from 46 to 425 [10,12,22] indicate that it will not adsorb to sediment or suspended organic matter. Insufficient data are available to determine the importance or rate of biodegradation of 2-bromo-2-chloro-1,1,1-trifluoroethane in water.

Atmospheric Fate: If released to the atmosphere, 2-bromo-2-chloro-1,1,1-trifluoroethane will undergo a slow gas-phase reaction with photochemically produced hydroxyl radicals. An experimental rate constant of 6.0 x 10^{-14} cm^3/molecule-sec [4] for this process translates to an atmospheric half-life of 267 days [3]. The atmospheric lifetime of 2-bromo-2-chloro-1,1,1-trifluoroethane has been estimated as 0.7-2 yrs [4]. 2-Bromo-2-chloro-1,1,1-trifluoroethane is listed as a light sensitive compound [1,11], and it may undergo direct photochemical degradation in the atmosphere. The water solubility of 2-bromo-2-chloro-1,1,1-trifluoroethane, 3900 mg/L at 25 °C [22], indicates that it may undergo atmospheric removal by wet deposition processes; however, any removed is expected to rapidly re-volatilize to the atmosphere.

Biodegradation:

Abiotic Degradation: An experimental rate constant for the gas-phase reaction of 2-bromo-2-chloro-1,1,1-trifluoroethane with photochemically produced hydroxyl radicals of 6.0 x 10^{-14} cm^3/molecule-sec at 303 deg K [4] translates to an atmospheric half-life of 267 days using an atmospheric hydroxyl radical concn of 5 x 10^{+5} molec/cm^3 [3]. 2-Bromo-2-chloro-1,1,1-trifluoroethane has an estimated atmospheric lifetime of 0.7-2 yrs [4]. 2-Bromo-2-chloro-1,1,1-trifluoroethane is listed as a light sensitive compound [1,11], indicating that it may undergo direct photochemical decomposition.

Bioconcentration: Estimated bioconcentration factors ranging from 6 to 33 can be calculated for 2-bromo-2-chloro-1,1,1-trifluoroethane based on its experimental water solubility, 3900 mg/L at 25 °C [22], and its experimental log octanol/water partition coefficient, 2.30 [10], using appropriate regression equations [12]. These values indicate that 2-bromo-

2-chloro-1,1,1-trifluoroethane will not bioconcentrate in fish and aquatic organisms.

Soil Adsorption/Mobility: Estimated soil adsorption coefficients ranging from 46 to 425 can be calculated for 2-bromo-2-chloro-1,1,1-trifluoroethane based on its experimental water solubility, 3900 mg/L at 25 °C [22], and its experimental log octanol/water partition coefficient, 2.30 [10], using appropriate regression equations [12]. These values indicate that 2-bromo-2-chloro-1,1,1-trifluoroethane will display moderate to high mobility in soil [23].

Volatilization from Water/Soil: An estimated Henry's Law constant of 0.0313 atm-m^3/mol at 25 °C [14] indicates that 2-bromo-2-chloro-1,1,1-trifluoroethane will rapidly volatilize from water and moist soil. The estimated half-life for volatilization from a model river 1 m deep flowing at 1 m/sec with a wind speed of 3 m/sec is 4.1 hr [12]. The vapor pressure of 2-bromo-2-chloro-1,1,1-trifluoroethane, 302 mm Hg at 25 °C [7], indicates that it will rapidly volatilize from dry soil to the atmosphere.

Water Concentrations:

Effluent Concentrations:

Sediment/Soil Concentrations:

Atmospheric Concentrations: 2-Bromo-2-chloro-1,1,1-trifluoroethane was detected in 22 of 2013 air samples obtained during an air sampling campaign in France, 1981-85 [8].

Food Survey Values:

Plant Concentrations:

Fish/Seafood Concentrations:

Animal Concentrations:

Milk Concentrations:

2-Bromo-2-chloro-1,1,1-trifluoroethane

Other Environmental Concentrations:

Probable Routes of Human Exposure: Occupational exposure to 2-bromo-2-chloro-1,1,1-trifluoroethane may occur by inhalation or dermal contact during its production or use. Occupational exposure to 2-bromo-2-chloro-1,1,1-trifluoroethane may also occur during surgical procedures via the inhalation of waste anesthetic gases [9,18]. The general population may be exposed to 2-bromo-2-chloro-1,1,1-trifluoroethane during surgical procedures where it is used as an anesthetic.

Average Daily Intake:

Occupational Exposure: The time-weighted average concn of 2-bromo-2-chloro-1,1,1-trifluoroethane in personal air samples obtained for 13 veterinarians during 38 surgical procedures ranged from 0.08-9.19 ppm [18]. In a survey of 19 veterinary surgical procedures, 1979, the time-weighted average exposure ranged from 16.7 ppm to 30.9 ppm with peaks as high as 62.4 ppm [18]. The concn of 2-bromo-2-chloro-1,1,1-trifluoroethane in veterinarian offices ranged from 6.0-37.2 ppm [18,20]. At the Univ of California at Davis veterinary medical school, the concn of 2-bromo-2-chloro-1,1,1-trifluoroethane was 6.9 ppm in the small animal areas [6,13,15]. A USAF study conducted at 5 different veterinary surgical facilities found 2-bromo-2-chloro-1,1,1-trifluoroethane concns ranging from 0.06-37.20 ppm [6,13,15]. Of 156 samples collected at 11 different locations at the Univ of Georgia College of Veterinary medicine, 35% were above 2 ppm [6,13,15,18,20]. NIOSH (NOES 1981-1983) has statistically estimated that 94,685 workers are potentially exposed to 2-bromo-2-chloro-1,1,1-trifluoroethane in the US [16].

Body Burdens: 2-Bromo-2-chloro-1,1,1-trifluoroethane has been qualitatively detected in human blood samples of patients suspected of inhalation abuse of this compound [19].

REFERENCES

1. Aldrich; Catalog Handbook of Fine Chemicals Milwaukee, WI: Aldrich Chem Co pg. 188 (1994)
2. Astrologes G; Kirk-Othmer Encycl Chem Tech. 3rd 10: NY: Wiley 867 (1982)
3. Atkinson R; J Chem Phys Ref Data Monograph 1 (1989)

2-Bromo-2-chloro-1,1,1-trifluoroethane

4.   Brown AC et al; Atmos Environ 24A: 2499-511 (1990)
5.   Budavari S (ed); The Merck Index - Encyclopedia of Chemicals, Drugs, and Biologicals. Rahway, NJ: Merck and Co. Inc pg. 726 (1989)
6.   Creesen DW et al; J Am Vet Med Assoc 179: 787 (1981)
7.   Daubert TE, Danner RP; Physical & Thermodynamic Properties of Pure Chemicals NY: Hemisphere Pub Corp (1989)
8.   Ensminger A; Caniers de Notes Documentaires 131: 299-301 (1988)
9.   Gray WM et al; Comm Eur Communities Eur 10555: 89-92 (1987)
10. Hansch C et al; Explori ng SAR Hydrophobic, Electronic, and Steric Constants ACS Professional Reference Book. Heller SR, consult ed, Washington, DC: Amer Chem Soc pg. 3 (1995)
11. Lewis RJ; Hawley's Condensed Chemical Dictionary 12th ed. NY: Van Nostrand Reinhold Co pg. 583 (1993)
12. Lyman WJ et al; Handbook of Chemical Property Estimation Methods Washington, DC: Amer Chem Soc (1990)
13. Manley SV, McDonnell WN; J Am Vet Med Assoc 176: 515 (1980)
14. Meylan WM, Howard PH; Environ Toxicol Chem 10: 1283-93 (1991)
15. Milligan JE, Sablan JL; J Am Vet Med Assoc 177: 1021 (1980)
16. NIOSH; National Occupational Exposure Survey (NOES) (1983)
17. Petrick DM, Dougherty RB; Kirk-Othmer Encycl Chem Tech. 3rd NY: Wiley 23: 750 (1985)
18. Potts DL, Craft BF; Appl Ind Hyg 3: 132-38 (1988)
19. Ramsey JD, Flanagan RJ; J Chromatog 240: 423-44 (1982)
20. Ruby DL et al; Am Ind Hyg Assoc J 41: 229 (1980)
21. Siegmund G et al; p. 359 in Ullmann's Encycl Indust Chem A11 NY: VCH Publ (1985)
22. Suzuki T; J Comput-Aided Molec Design 5: 149-66 (1991)
23. Swann RL et al; Res Rev 85: 17-28 (1983)
24. Wolleneber H; p. 292 in Ullmann's Encycl of Indust Chem A2 NY: VCH Publ (1985)

2-Bromotoluene

SUBSTANCE IDENTIFICATION

Synonyms: o-Bromotoluene

Structure:

CAS Registry Number: 95-46-5

Molecular Formula: C_7H_7Br

SMILES Notation: c(c(ccc1)Br)(c1)C

CHEMICAL AND PHYSICAL PROPERTIES

Boiling Point: 181.7 °C

Melting Point: -27.8 °C

Molecular Weight: 171.04

Dissociation Constants:

Log Octanol/Water Partition Coefficient: 3.43 (estimated) [5]

Water Solubility: 51.3 mg/L (value based on solubility of the m-isomer) [12]

Vapor Pressure: 1 mm Hg at 24.4 °C [11]

Henry's Law Constant: 2.39×10^{-3} atm-m³/mol (estimated by the Group Method) [6]

ENVIRONMENTAL FATE/EXPOSURE POTENTIAL

Summary: 2-Bromotoluene may be released to the environment through evaporation from solvent uses or from wastes generated at sites of production and use. If released to the atmosphere, 2-bromotoluene should degrade by reaction with photochemically produced hydroxyl radicals; the half-life for this reaction in air with an average concn of hydroxyl radical can be estimated to be 6.2 days. If released to water, 2-bromotoluene can be removed from the water column by volatilization and adsorption. Volatilization half-lives of 1.7 hr and 14.9 days can be estimated for a model river (1 m deep) and a pond, respectively. The potential effect of adsorption to sediment is considered in estimating the volatilization half-life from the pond, but not from the model river. If released to soil, significant leaching is not expected to occur in most soil types. Evaporation from dry surfaces may occur. No data are available pertaining to biodegradation. Occupational exposure may occur through inhalation or dermal contact at sites where bromotoluene is manufactured, used as a solvent, or used as a chemical intermediate.

Natural Sources:

Artificial Sources: The 2-bromotoluene is used as a solvent (for fats, waxes or resins, and as a medium for carrying out reactions) and as chemical intermediate [8]. Use of 2-bromotoluene as a solvent may result in direct environmental release through evaporation. Environmental releases may also occur from sites of its production or use as a chemical intermediate.

Terrestrial Fate: Based on an estimated Koc of 910 [3], 2-bromotoluene is expected to have low soil mobility [9]; therefore, it should not leach significantly in most types of soil. 2-Bromotoluene has a reported vapor pressure of 1 mm Hg at 24.4 °C [11], thus making volatilization likely from dry surfaces. No data are available pertaining to chemical or biological degradation in soil systems.

Aquatic Fate: Volatilization may be the major process by which 2-bromotoluene is removed from the aquatic environment. Volatilization half-lives of 1.7 hr and 14.9 days can be estimated for a model river (1 m deep) and a pond, respectively [4,10]. The potential effect of adsorption to sediment is considered in estimating the volatilization half-life from the pond, but not from the model river. An estimated Koc of 910 [3] indicates that some partitioning from the water column to sediment and suspended material may occur; this would have the effect of decreasing the volatilization rate. Aquatic hydrolysis and direct photolysis do not appear to be important. Some bioaccumulation may be possible. No experimental data are available pertaining to biodegradation.

Atmospheric Fate: The vapor pressure of 2-bromotoluene is reported to be 1 mm Hg at 24.4 °C [11]. Based on this vapor pressure, 2-bromotoluene should exist almost entirely in the vapor phase in the ambient atmosphere [2]. Vapor phase 2-bromotoluene should degrade in the atmosphere by reaction with photochemically produced hydroxyl radicals; the half-life for this reaction in typical air can be estimated to be 6.2 days [1].

Biodegradation:

Abiotic Degradation: The rate constant for the vapor-phase reaction of 2-bromotoluene with photochemically produced hydroxyl radicals can be estimated to be 1.72×10^{-12} m^3/molecule-sec at 25 °C [1]; assuming a 12-hr daylight average atmospheric hydroxyl radical concn of 1.5×10^6 molecules/cm^3, the half-life for this reaction can be estimated to be approximately 6.2 days. 2-Bromotoluene does not contain any functional groups that are generally susceptible to aqueous hydrolysis [4], therefore, hydrolysis is not expected to be significant in water. Irradiation of 2-bromotoluene (in hexane solution) with artificial UV light of wavelengths above 290 nm resulted in a dehalogenation of only 1.03% over a two hour period [7]. Considering that the intensity of the light used was higher than available from sunlight, photodegradation may not be an important process.

Bioconcentration: Based upon an estimated log Kow of 3.43 [5], the bioconcentration factor (BCF) for 2-bromotoluene can be estimated to be 238 from a regression equation [4]. This BCF value indicates that some bioconcentration of 2-bromotoluene may occur in aquatic organisms.

Soil Adsorption/Mobility: Based upon an estimated log Kow of 3.43 [5], the Koc for 2- bromotoluene can be estimated to be approximately 910 from a regression equation that is applicable to methylated and halogenated benzenes [3]. This Koc value is indicative of low soil mobility for 2-bromotoluene [9].

Volatilization from Water/Soil: Based on a value of Henry's Law constant of 2.39×10^{-3} atm-m^3/mol at 25 °C [6], the volatilization of 2-bromotoluene from waters is probably significant, and may be rapid [4]. The volatilization half-life from a model river (1 meter deep flowing 1 m/sec with a wind speed of 5 m/sec) can be estimated to be 1.7 hr [4]; this estimate does not consider the effect of adsorption to sediment, which may be important. The volatilization half-life from a model pond (considering the effect of adsorption) can be estimated to be 14.9 days [10].

Water Concentrations:

Effluent Concentrations:

Sediment/Soil Concentrations:

Atmospheric Concentrations:

Food Survey Values:

Plant Concentrations:

Fish/Seafood Concentrations:

Animal Concentrations:

Milk Concentrations:

Other Environmental Concentrations:

Probable Routes of Human Exposure: Occupational exposure may occur through inhalation or dermal contact at sites where 2-bromotoluene is manufactured, used as a solvent or used as a chemical intermediate.

Average Daily Intake:

Occupational Exposure:

Body Burdens:

REFERENCES

1. Atkinson R; Inter J Chem Kinet 19: 799-828 (1987)
2. Eisenreich SJ et al; Environ Sci Technol 15: 30-8 (1981)
3. Karickhoff SW; In Environmental Exposure from Chemicals Vol I, Neely WB, Blau GE eds Boca Raton, FL: CRC Press p. 55 (1985)
4. Lyman WJ et al; Handbook of Chemical Property Estimation Methods. Environmental Behavior of Organic Compounds. Washington DC: American Chemical Society p. 5-4; 7-4 to 7-6; 15-15 to 15-32 (1990)
5. Meylan WM, Howard, PH; J Pharm. Sci. (1995)
6. Meylan WM, Howard PH; Environ Toxicol Chem 10:1283-1293 (1991)
7. Parlar H et al; Chemosphere 12: 93-6 (1983)
8. Stenger VA; Kirk-Othmer Encycl Chem Technol 4: 256 (1978)
9. Swann RL et al; Res Rev 85: 17-28 (1983)
10. USEPA; EXAMS II Computer Simulation, Athens, GA: USEPA (1987)
11. Weast RC (ed). CRC Handbook of Chemistry and Physics, 66th ed, Boca Raton, FL: CRC Press p D-203 (1986)
12. Yalkowsky SH et al; Arizona Database of Aqueous Solubility, Tucson, AZ: Univ of Arizona (1987)

3-Bromotoluene

SUBSTANCE IDENTIFICATION

Synonyms: m-Bromotoluene

Structure:

CAS Registry Number: 591-17-3

Molecular Formula: C_7H_7Br

SMILES Notation: c(cccc1Br)(c1)C

CHEMICAL AND PHYSICAL PROPERTIES

Boiling Point: 183.7 °C at 760 mm Hg

Melting Point: -39.8 °C

Molecular Weight: 171.04

Dissociation Constants:

Log Octanol/Water Partition Coefficient: 3.43 (estimated by the bond method) [5]

Water Solubility: 51.3 mg/L [11]

Vapor Pressure: 1 mm Hg at 14.8 °C

Henry's Law Constant: 4.39×10^{-3} atm-m^3/mol (estimated from the ratio of vapor pressure and water solubility)

ENVIRONMENTAL FATE/EXPOSURE POTENTIAL

Summary: 3-Bromotoluene may be released to the environment through evaporation from solvent uses or from sites of production and use. If released to the atmosphere, 3-bromotoluene should degrade by reaction with photochemically produced hydroxyl radicals; the half-life for this reaction in air with an average concn of hydroxyl radical can be estimated to be 3 days. If released to water, 3-bromotoluene can be removed from the water column by volatilization and adsorption. Volatilization half-lives of 1.5 hr and 14.9 days can be estimated for a model river (1 m deep) and a typical pond, respectively. The potential effect of adsorption to sediment is considered in the half-life from the pond, but not from the model river. If released to soil, significant leaching is not expected to occur in most soil types. Evaporation from dry surfaces may occur. No data are available pertaining to biodegradation. Occupational exposure may occur through inhalation or dermal contact at sites where bromotoluene is manufactured and used as a solvent or chemical intermediate.

Natural Sources:

Artificial Sources: 3-Bromotoluene is used as a solvent (for fats, waxes or resins and as a medium for carrying out reactions) and as a chemical intermediate [7]. Use of 3-bromotoluene as a solvent may result in direct environmental release through evaporation. Environmental releases may also be possible from sites of production or use as a chemical intermediate.

Terrestrial Fate: Based on an estimated Koc of 910 [3], 3-bromotoluene is expected to have low soil mobility; therefore, it should not leach significantly in most types of soil. 3-Bromotoluene has a reported vapor pressure of 1 mm Hg at 14.8 °C [10]. Based on this vapor pressure and an estimated value of Henry's Law constant of 4.39×10^{-3} atm-m^3/mol , volatilization may be important both from dry and wet soil. No data are available pertaining to chemical or biological degradation of 3-bromotoluene in soil.

Aquatic Fate: Volatilization may be the major process by which 3-bromotoluene is removed from the aquatic environments. Volatilization half-lives of 1.5 hr and 14.9 days can be estimated for a model river (1 m deep) and a typical pond, respectively [4,9]. The potential effect of adsorption to sediment is considered in the half-life from the pond, but not from the model river. An estimated Koc of 910 [3] indicates that some partitioning from the water column to sediment and suspended material may occur; this would have the effect of decreasing the volatilization rate. Aquatic hydrolysis and direct photolysis do not appear to be important [4,6]. Based on a bioconcentration factor estimated from K_{ow} [4], some bioconcentration of 3-bromotoluene in aquatic organisms may be possible. No experimental data are available pertaining to biodegradation.

Atmospheric Fate: The vapor pressure of 3-bromotoluene is reported to be 1 mm Hg at 14.8 °C [10]. Based on this vapor pressure, 3-bromotoluene should exist almost entirely in the vapor phase in the ambient atmosphere [2]. Vapor phase 3-bromotoluene should degrade in the atmosphere by reaction with photochemically produced hydroxyl radicals; the half-life for this reaction in typical air can be estimated to be 3 days [1].

Biodegradation:

Abiotic Degradation: The rate constant for the vapor-phase reaction of 3-bromotoluene with photochemically produced hydroxyl radicals in the atmosphere has been estimated to be 3.51×10^{-12} m^3/molecule-sec at 25 °C [1]; assuming an average 12-hr daylight hydroxyl radical concn of 1.5×10^6 molecules/cm^3, the half-life for this reaction can be estimated to be approximately 3 days [1]. 3-Bromotoluene does not contain any functional groups that are generally susceptible to aqueous hydrolysis [4]; therefore, hydrolysis is not expected to be significant in water. Irradiation of 3-bromotoluene (in hexane solution) with artificial UV light of wavelengths above 290 nm resulted in a dehalogenation of only 0.36% over a two hour period [6]. Considering that the intensity of the light used was higher than available from sunlight, photodegradation may not be an important process.

Bioconcentration: Based upon an estimated log Kow of 3.43 [5], the bioconcentration factor (BCF) for 3-bromotoluene can be estimated to be 238 from a regression equation [4]. This BCF value indicates that some bioconcentration of 3-bromotoluene may occur in aquatic organisms.

Soil Adsorption/Mobility: Based upon an estimated log Kow of 3.43 [5], the Koc for 3- bromotoluene can be estimated to be approximately 910 from a regression equation applicable to methylated and halogenated benzenes [3]. This Koc value is indicative of low soil mobility [8].

Volatilization from Water/Soil: Based upon a Henry's Law constant of 4.39×10^{-3} atm-m^3/mol for 3-bromotoluene derived from the ratio of its vapor pressure over water solubility, volatilization from waters is probably significant, and may be rapid [4]. The volatilization half-life from a model river (1 m deep flowing 1 m/sec with a wind speed of 5 m/sec) can be estimated to be 1.5 hr [4]; this estimate does not consider the effect of adsorption to sediment, which may slow down the rate of evaporation. The volatilization half-life from a model pond (considering the effect of adsorption) can be estimated to be 14.9 days [9].

Water Concentrations:

Effluent Concentrations:

Sediment/Soil Concentrations:

Atmospheric Concentrations:

Food Survey Values:

Plant Concentrations:

Fish/Seafood Concentrations:

Animal Concentrations:

Milk Concentrations:

Other Environmental Concentrations:

Probable Routes of Human Exposure: Occupational exposure may occur through inhalation or dermal contact at sites where 3-bromotoluene is manufactured, used as a solvent, or used as chemical intermediate.

Average Daily Intake:

3-Bromotoluene

Occupational Exposure:

Body Burdens:

REFERENCES

1. Atkinson R; Inter J Chem Kinet 19: 799-828 (1987)
2. Eisenreich SJ et al; Environ Sci Technol 15: 30-8 (1981)
3. Karickhoff SW; In: Environmental Exposure from Chemicals Vol 1, Neely WB, Blau GE eds Boca Raton, FL: CRC Press, p.55 (1985)
4. Lyman WJ et al; Handbook of Chemical Property Estimation Methods. Environmental Behavior of Organic Compounds. Washington DC: American Chemical Society p. 5-4; 7-4 to 7-6; 15-15 to 15-32 (1990)
5. Meylan WM, Howard, PH; J Pharm. Sci. (1995)
6. Parlar H et al; Chemosphere 12: 93-6 (1983)
7. Stenger VA; Kirk-Othmer Encycl Chem Technol 4: 256 (1978)
8. Swann RL et al; Res Rev 85: 16-28 (1983)
9. USEPA; EXAMS II Computer Simulation, Athens, GA: USEPA (1987)
10. Weast RC (ed). CRC Handbook of Chemistry and Physics, 66th ed, Boca Raton, FL: CRC Press p D-203 (1986)
11. Yalkowsky SH et al; Arizona Database of Aqueous Solubility, Tucson, AZ: Univ of Arizona (1987)

4-Bromotoluene

SUBSTANCE IDENTIFICATION

Synonyms: p-Bromotoluene

Structure:

$$Br-C_6H_4-CH_3$$

CAS Registry Number: 106-38-7

Molecular Formula: C_7H_7Br

SMILES Notation: c(ccc(c1)Br)(c1)C

CHEMICAL AND PHYSICAL PROPERTIES

Boiling Point: 184.5 °C at 760 mm Hg

Melting Point: 28.5 °C

Molecular Weight: 171.04

Dissociation Constants:

Log Octanol/Water Partition Coefficient: 3.42 [2]

Water Solubility: 110 mg/L at 25 °C [5]

Vapor Pressure: 1.15 mm Hg at 25 °C [5]

Henry's Law Constant: 2.35×10^{-3} atm-m^3/mol [5]

ENVIRONMENTAL FATE/EXPOSURE POTENTIAL

Summary: 4-Bromotoluene may be released to the environment through evaporation from solvent uses or from wastes generated at sites of production and use. If released to the atmosphere, 4-bromotoluene should

34

degrade by reaction with photochemically produced hydroxyl radicals; the half-life for this reaction in air with an average concn of hydroxyl radical can be estimated to be 9.4 days. If released to water, 4-bromotoluene can be removed from the water column by volatilization and adsorption. Volatilization half-lives of 4.3 hr and 14.9 days can be estimated for a model river (1 m deep) and an environmental pond, respectively. The potential effect of adsorption to sediment is considered in the half-life from the pond, but not from the model river. If released to soil, significant leaching is not expected to occur in most soil types. Evaporation from dry surfaces may occur. Data are insufficent to assess the importance of biodegradation in the environment. Occupational exposure may occur through inhalation or dermal contact at sites where bromotoluene is manufactured, used as a solvent or used as a chemical intermediate.

Natural Sources:

Artificial Sources: 4-Bromotoluene is used as a solvent (for fats, waxes or resins, and as a medium for carrying out reactions) and as a chemical intermediate [10]. Use of 4-bromotoluene as a solvent may result in direct environmental release through evaporation. Environmental releases may also be possible from wastes generated at sites of production or use as a chemical intermediate.

Terrestrial Fate: Based on an estimated Koc values of 1270-1860, 4-bromotoluene is expected to have low soil mobility; therefore, it should not leach significantly in most types of soil. 4-Bromotoluene has a reported vapor pressure of 1.15 mm Hg at 25 °C [5]; thus making volatilization likely. A single pure culture biodegradation study has observed biooxidation of 4-bromotoluene utilizing a bacterium isolated from soil; however, sufficient biodegradation data are not available to predict the significance of biodegradation in the environment. No data are available pertaining to chemical degradation in soil systems.

Aquatic Fate: Volatilization may be the major process by which 4-bromotoluene is removed from the aquatic environment. Volatilization half-lives of 4.3 hr and 14.9 days can be estimated for a model river (1 m deep) and an environmental pond, respectively [8]. The potential effect of adsorption to sediment is considered in the half-life from the pond, but not from the model river. Estimated Koc values of 1270-1860 [6] indicate that some partitioning from the water column to sediment and suspended material may occur; this would have the effect of decreasing the volatilization rate. Aquatic hydrolysis and direct photolysis do not appear to be important. Some bioaccumulation may be possible. No experimental data are available pertaining to aquatic biodegradation.

Atmospheric Fate: The vapor pressure of 4-bromotoluene is reported to be 1.15 mm Hg at 25 °C [5]. Based on this vapor pressure, 4-bromotoluene should existed almost entirely in the vapor phase in the ambient atmosphere [3]. Vapor phase 4-bromotoluene should degrade in the atmosphere by reaction with photochemically produced hydroxyl radicals; the half-life for this reaction in average air can be estimated to be 9.4 days [1].

Biodegradation: 4-Bromotoluene was oxidized to a dihydrodiol compound in a pure culture study utilizing the soil isolated bacterium *Pseudomonas putida* [4].

Abiotic Degradation: The rate constant for the vapor-phase reaction of 4-bromotoluene with photochemically produced hydroxyl radicals has been estimated to be 1.72×10^{-12} m^3/molecule-sec at 25 °C [1]; assuming an average atmospheric hydroxyl radical concn of $5 \times 10^{+5}$ molecules/cm^3, the half-life for this reaction can be estimated to be approximately 9.4 days [1]. 4-Bromotoluene does not contain any functional groups that are generally susceptible to aqueous environmental hydrolysis [8]; therefore, hydrolysis is not expected to be significant in water. Irradiation of 4-bromotoluene (in hexane solution) with artificial UV light above 290 nm resulted in a dehalogenation of only 1.0% over a two hour period [9].

Bioconcentration: Based upon a measured log Kow of 3.42 [2], the bioconcentration factor (BCF) for 4-bromotoluene can be estimated to be 350 from a recommended regression-derived equation [8]. Based upon a water solubility of 110 ppm [5], the BCF can be estimated to be 44 from a recommended regression-derived equation [8]. These BCF values may indicate that some bioconcentration may occur.

Soil Adsorption/Mobility: Based upon a measured log Kow of 3.4265 [2], the Koc for 4-bromotoluene can be estimated to be approximately 1300 from a regression-derived equation applicable to methylated and halogenated benzenes [6]. Based upon a water solubility of 110 ppm at 25

°C [5], the Koc value can be estimated to range from about 1270 to 1861 from appropriate regression-derived equations [8]. These Koc values are indicative of low soil mobility [11].

Volatilization from Water/Soil: Based upon a water solubility of 110 ppm and a vapor pressure of 1.15 mm Hg at 25 °C, the Henry's Law constant of 4-bromotoluene has been estimated to be 2.35×10^{-3} atm-m^3/mol [5]. This value of Henry's Law constant indicates that volatilization from environmental waters is probably significant, and may be rapid [8]. The volatilization half-life from a model river (1 m deep flowing 1 m/sec with a wind speed of 3 m/sec) has been estimated to be 4.3 hr [8]; this estimate does not consider the effect of adsorption to sediment, which may be important. The volatilization half-life predicted by a model that considers adsorption has been estimated to be 14.9 days [12].

Water Concentrations: DRINKING WATER: 4-Bromotoluene was qualitatively detected in drinking water concentrates collected from Cincinnati, OH, in Oct. 1978 [7].

Effluent Concentrations:

Sediment/Soil Concentrations:

Atmospheric Concentrations:

Food Survey Values:

Plant Concentrations:

Fish/Seafood Concentrations:

Animal Concentrations:

Milk Concentrations:

Other Environmental Concentrations:

4-Bromotoluene

Probable Routes of Human Exposure: Occupational exposure may occur through inhalation or dermal contact at sites where 4-bromotoluene is manufactured, used as a solvent or used as a chemical intermediate.

Average Daily Intake:

Occupational Exposure:

Body Burdens:

REFERENCES

1. Atkinson R; Inter J Chem Kinet 19: 799- 828 (1987)
2. Dunn, WJ II et al.; Quant. Struct-Act. Relat. Pharmacol. Chem. Biol. 2:156-63 (1983)
3. Eisenreich SJ et al; Environ Sci Technol 15: 30-8 (1981)
4. Gibson DT; p. 36-46 in Fate and Effects of Petroleum Hydrocarbons in Marine Organisms and Ecosystems, Wolfe DA ed NY: Pergamon Press (1977)
5. Hine J, Mookerjee PK; J Org Chem 40: 292-8 (1975)
6. Karickhoff SW; p. 55 in Environmental Exposure from Chemicals Vol I, Neely WB, Blau GE eds Boca Raton, FL: CRC Press (1985)
7. Lucas SV; GC/MS Analysis of Organics in Drinking Water Concentrates and Advanced Waste Treatment Concentrates: Vol 1. EPA-600/1-84-020a p. 45,153 (1984)
8. Lyman WJ et al; Handbook of Chemical Property Estimation Methods NY: McGraw-Hill p. 4-9, 5-4 to 5-10, 7-4 to 7-6, 15-15 to 15-29 (1982)
9. Parlar H et al; Chemosphere 12: 93-6 (1983)
10. Stenger VA; Kirk-Othmer Encycl Chem Technol 4: 256 (1978)
11. Swann RL et al; Res Rev 85: 16-28 (1983)
12. USEPA; EXAMS II Computer Simulation (1987)

n-Butane

Synonyms:

Structure:

H_3C CH_3

CAS Registry Number: 106-97-8

Molecular Formula: C_4H_{10}

SMILES Notation: C(CC)C

CHEMICAL AND PHYSICAL PROPERTIES

Boiling Point: -0.50 °C

Melting Point: -138.4 °C

Molecular Weight: 58.12

Dissociation Constants:

Log Octanol/Water Partition Coefficient: 2.89 [22]

Water Solubility: 61.4 mg/L water at 20 °C [75]

Vapor Pressure: 1856.4 mm Hg at 25 °C [16].

Henry's Law Constant: 9.47 x 10^{-1} atm-m³/mol at 25 °C (calculated from the ratio of vapor pressure of 760 mm Hg and water solubility)

ENVIRONMENTAL FATE/EXPOSURE POTENTIAL

Summary: n-Butane is a highly volatile constituent in the paraffin fraction of crude oil and natural gas. n-Butane is released to the environment via the manufacture, use, and disposal of many products associated with the petroleum and natural gas industries. Extensive data show release of n-butane into the environment from hazardous waste disposal sites, landfills, waste incinerators and the combustion of gasoline fueled engines. Photolysis, hydrolysis, and bioconcentration of n-butane are not expected to be important environmental fate processes. Biodegradation of n-butane may occur in soil and water; however, volatilization is expected to be the dominant fate process. To a lesser extent adsorption may be important. A Koc range of 450 to 900 indicates a low to medium mobility class in soil for n-butane. In aquatic systems, n-butane may partition from the water column to organic matter contained in sediments and suspended materials. A Henry's Law constant of 9.47×10^{-1} atm-m^3/mol at 25 °C suggests extremely rapid volatilization of n-butane from environmental waters. The volatilization half-lives from a model river and a model pond, the latter considers the effect of adsorption, have been estimated to be 2.2 hr and 2.6 days, respectively. n-Butane is expected to exist entirely in the vapor phase in ambient air. Reactions with photochemically produced hydroxyl radicals in the atmosphere have been shown to be important (average half-life of 6 days). Data also suggest the nighttime reactions with radical species and nitrogen oxides may contribute to the atmospheric transformation of n-butane. The most probable route of human exposure to n-butane is by inhalation. Extensive monitoring data indicate n-butane is a widely occurring atmospheric pollutant.

Natural Sources: n-Butane is a constituent in the paraffin fraction of crude oil and natural gas [72]. Raw natural gas contains an average of 0.30 mole-percent n-butane.

Artificial Sources: n-Butane is released to the environment via the manufacture, use and disposal of many products associated with the petroleum [5,56] and natural gas industries [5]. The combustion of gasoline is a major mechanism for the release of n-butane into the atmosphere [3,21,43,44,65,67,77]. Waste incinerators [12], hazardous waste disposal sites [35] and landfills [57,76] also release n-butane into the environment.

Terrestrial Fate: Photolysis or hydrolysis [38] of n-butane is not expected to be important in soils. The biodegradation of dissolved n-butane may occur in soils; however, primarily volatilization and to some extent adsorption are expected to be far more important fate processes. A calculated Koc range of 450 to 900 [38] indicates a medium to low mobility class for n-butane in soils [69]. Based upon an estimated Henry's Law constant of 9.47 x 10^{-1} atm-m^3/mol at 25 °C [38], n-butane is expected to rapidly volatilize from most surface soils [69].

Aquatic Fate: Photolysis or hydrolysis [38] of n-butane in aquatic systems is not expected to be important. The bioconcentration factor (log BCF) for n-butane has been estimated to range from 1.78 to 1.97 [38], suggesting bioconcentration is not an important factor in aquatic systems. Biodegradation of n-butane may occur in aquatic environments; however, primarily volatilization and to some extent adsorption are expected to be more important fate processes. An estimated range for Koc from 450 to 900 [38] indicates n-butane may partition from the water column to organic matter [69] contained in sediments and suspended materials. An estimated Henry's Law constant of 9.47 x 10^{-1} atm-m^3/mol at 25 °C suggests rapid volatilization of n-butane from environmental waters [38]. Based on this Henry's Law constant, the volatilization half-life from a model river has been estimated to be 2.2 hr [38]. The volatilization half-life from a model pond, which considers the effect of adsorption, can be estimated to be about 2.6 days [71].

Atmospheric Fate: Based on a vapor pressure of 1856 mm Hg at 25 °C [16], n-butane is expected to exist entirely in the vapor phase in ambient air [17]. n-Butane does not absorb UV light in the environmentally significant range, >290 nm [66], and probably will not undergo direct photolysis in the atmosphere. Vapor phase reactions with photochemically produced hydroxyl radicals in the atmosphere have been shown to be important. Rate constants for n-butane were measured [6,15,53] to be about 2.67 x 10^{-12} cm^3/molecule-sec at 25 °C [48], which corresponds to an atmospheric half-life of about 6 days at an atmospheric concentration of 5 x 10^{+5} hydroxyl radicals per cm^3. Experimental data showed that 7.7% of the n-butane fraction in a dark chamber reacted with nitrogen oxide to form the corresponding alkyl nitrate [7,9], suggesting nighttime reactions with radical species and nitrogen oxides may contribute to the atmospheric transformation of n-butane.

Biodegradation: Within 24 hr, n-butane was oxidized to its corresponding methyl ketone, 2-butanone [24,49], and the corresponding alcohol, 2-

butanol [24,50], by cell suspensions of over 20 methyltrophic organisms isolated from lake water and soil samples [24,49,50]. Incubation with natural flora in ground water in the presence of other components of high octane gasoline (100 μL/L) yielded 0% biodegradation after 192 hr at 13 °C. The initial concentration was 0.63 μL/L. After 192 hr, the concentration of n-butane contained in gasoline was reduced from 0.63 to 0.37 μL/L for both a sterile control and a mixed culture sample collected from ground water contaminated with gasoline [28]. *Mycobacterium crassa* and *M. phlei* grow on n-butane. In combination with various concentrations of oxygen, n-butane supports the growth of *Neurospora crassa*, as well as the germination of *N. ascrospores* and growth of *Escherichia coli* strains B and Sd4, thus rendering n-butane potentially biodegradable. The degradation of n-alkanes by microorganisms is similar to the degradation of fatty acids. The terminal methyl group is enzymatically oxidized by incorporation of a molecular oxygen by a monooxygenase producing a primary alcohol with further oxidation to an acid group, although involvement of a dioxygenase is also postulated. Once the fatty acid is produced, it is degraded into 2-carbon units via the β-oxidation pathway. Another pathway for n-alkane degradation that is encountered less often is the oxidation of both terminal carbons to form a dioic acid with subsequent β-oxidation. Subterminal oxidation of the 2-carbon atom is seen mainly in C3 to C6 alkanes. A dehydrogenation of the n-alkane may also occur, yielding an alkene that is then converted to an alcohol, although there is little evidence for this theory. Some microorganisms have been shown to have both terminal and subterminal oxidation, each having different rates of activity. In a study comparing growth on long and short chain alkanes by some bacteria, the initial oxidase had a broad specificity and would oxidize C1 to C8 alkanes. However, cells grown on C4 to C8 alkanes did not oxidize the shorter chain alkanes to a significant extent.

Abiotic Degradation: A detailed mechanism is presented for reactions occurring during irradiation of ppm concentrations of n-butane and oxides of nitrogen in air. A smog chamber solar simulator facility designed for providing data suitable for quantitative model validation was used to elucidate several unknown or uncertain kinetic parameters and details of the reaction mechanism. Products of the photo-oxidation included 2-butyl nitrate, butyraldehyde, 1-butyl nitrate, methyl nitrate, peroxyacetyl nitrate, propene oxide, propionaldehyde, formaldehyde, and acetaldehyde [13]. Alkanes are generally resistant to hydrolysis [6]. Based on data for isooctane and n-hexane, n-butane is not expected to absorb UV light in the environmentally significant range, >290 nm [66]. Therefore, n-butane probably will not undergo hydrolysis or direct photolysis in the environment. An air sample's n-butane concentration of 286 ppbC was reduced by 14% within 6 hr of irradiation by natural sunlight in downtown Los Angeles, CA [31]. The rate constants for the vapor phase reaction of n-butane with photochemically produced hydroxyl radicals were measured to be 2.58×10^{-12} [6], 2.67×10^{-12} [48], 2.72×10^{-12} [53] and 2.70×10^{-12} [15] cm^3/molecule-sec at 25 [1-3] and 22 [15] $^\circ$C, respectively, which correspond to atmospheric half-lives of about 6.2 [6], 6.0 [48], 5.9 [53], and 5.9 [15] days at an atmospheric concentration of $5 \times 10^{+5}$ hydroxyl radicals per cm^3. The photooxidation rates for n-butane in air and water with ozone and peroxy radicals were 3.1×10^{-14} cm^3/molecule-sec [8] and 3.0×10^{-4} 1/mol-sec [23] at 30 and 28 $^\circ$C, respectively. Neither reaction is expected to be environmentally important. Experimental data showed that 7.7% of the n-butane fraction in a dark chamber reacted with nitrogen oxide to form the corresponding alkyl nitrate [7,9], suggesting nighttime reactions with radical species and nitrogen oxides may contribute to the atmospheric transformation of n-butane.

Bioconcentration: Based upon a water solubility of 61.4 ppm [75] at 25 $^\circ$C and a log Kow of 2.89 [22], the bioconcentration factor (log BCF) for n-butane has been calculated, using recommended regression-derived equations, to be 1.78 and 1.97, respectively [38]. These bioconcentration factor values do not indicate that bioconcentration in aquatic organisms is important.

Soil Adsorption/Mobility: Based on a water solubility of 61.4 ppm [75] and a log Kow of 2.89 [22], the Koc of dissolved n-butane has been calculated, using various regression-derived equations, to range between 450 and 900 [38]. These Koc values indicate a medium to low soil mobility class for n-butane [69].

Volatilization from Water/Soil: Using a water solubility of 61.4 ppm [75] and a vapor pressure of 760 mm Hg at 25 $^\circ$C [16], the Henry's Law constant for n-butane has been calculated to be 9.47×10^{-1} atm-m^3/mol . This value of Henry's Law constant indicates extremely rapid volatilization from environmental waters [38]. The volatilization half-life from a model river (1 meter deep flowing 1 m/sec with a wind speed of 3 m/sec) has been estimated to be 2.2 hr [38]. The volatilization half-life from a model pond, which considers the effect of adsorption, has been estimated to be 2.6 days [71].

Water Concentrations: DRINKING WATER: n-Butane was listed as one of the many organic chemicals identified in drinking water in the US as of 1974 [1,30,32]. SURFACE WATER: n-Butane was identified in Delaware River water at Philadelphia, PA [18]. n-Butane was listed as a contaminant present in the waters of Lake Ontario [18]. The Inner Harbor Navigation Canal of Lake Pontchartrain at New Orleans, LA, was found to contain n-butane at an average concentration for 8 samples of 2.4 ppb on May 6, 1980 [39]. SEAWATER: All 8 near surface sea water samples from the intertropical Indian Ocean contained small amounts of n-butane [10]. GROUND WATER: One of 11 ground water monitoring wells near the Granby Landfill, CT, contained n-butane at a concentration of 20 ppb [57].

Effluent Concentrations: Flue gases from a waste incinerator at Babylon, Long Island, NY, were found to contain n-butane at concentrations generally less than 0.4 ppm [12]. Two of five hazardous waste sites listed on the National Priorities List emitted gaseous n-butane with a 75 to 100% frequency of occurrence [35]. Landfills also release n-butane into the environment [76]. One of 11 ground water monitoring wells near the Granby Landfill, CT, contained trace quantities of n-butane [57]. A Texaco refinery located in Tulsa, OK, was attributed with emissions to the surrounding atmosphere where the n-butane concentration was measured to be 175.2 and 342.7 ppbC for 2 min before and after 1:33 PM [5]. The n-butane content of the air downwind of a Mobil natural gas facility in Rio Blanco, CO, was 56.0 ppbC [5]. Underwater hydrocarbon vent discharges from offshore oil production platforms were found to contain n-butane concentrations in the vapor phase at 740 umol/L of gas [56]. n-Butane is a product of gasoline combustion [3,67,77]. Data from September 2, 1979, identified n-butane as a gaseous emission of the vehicle traffic through the Allegheny Mountain Tunnel of the Pennsylvania Turnpike [21]. The average exhaust from 67 gasoline-fueled vehicles was found to contain n-butane at a concentration of 2.2% by weight [44]. n-Butane from car exhaust ranged in concentration from 0.12 to 0.28 ppmV, with an average for 8 samples of 0.18 ppmV [43]. The average concentration of n-butane for the exhaust of 46 automobiles was 4.8, 23.2 and 4.9 % by weight of total hydrocarbon according to the federal test procedure, hot soak test and the

New York City cycle, respectively [65]. n-Butane has been measured as exhaust from diesel engines at 22 ppm. n-Butane was detected in flue gas of a municipal incinerator at levels < 0.4 ppm. n-Butane accounted for 5.3% of the emitted hydrocarbons in the exhaust gas of a diesel engine. n-Butane was found in combustion gas of household central heating at approximately 50 ppm at 7% carbon dioxide system on gasoil (3.3 g/kg gasoil at 6% carbon dioxide, 1.6 g/kg gasoil at 7% carbon dioxide). Automotive gasoline engine combustion produces n-butane at 4.31 to 5.02% by exhaust gas volume. According to a survey of 62 cars, 5.3% by volume of total exhaust hydrocarbons was n-butane. A study of 15 different fuels showed 4% n-butane by volume of total exhaust hydrocarbons. n-Butane accounted for 16.5 to 48.5% by volume of the total evaporated hydrocarbons from a gasoline fuel tank. n-Butane accounted for 9.1 to 23.0% by volume of the total evaporated hydrocarbons from a gasoline engine carburetor.

Sediment/Soil Concentrations: n-Butane was detected in 10 of 10 sediment samples from Walvis Bay of the Namibian shelf of SW Africa at concentrations of 2.2, 0.45, 2.2, 1.5, 0.31, 0.24, 0.01, 0.22, 0.52, and 0.27 ng/g [74]. Sediments from the Bering Sea contained n-butane gas at concentrations ranging from 4 to 43 nL/L [34].

Atmospheric Concentrations: URBAN: The average n-butane concentration for 2 samples per 4 sites in Tulsa, OK, was 102.0 ppbC, with a range of 16.5 to 342.7 ppbC [5]. The n-butane concentrations for 6 sites in Rio Blanco, CO, averaged 10.3 ppbC, with a range from 0.7 to 56.0 [5]. n-Butane was detected in 21 of 21 air samples from Houston, TX, ranging in concentration from 11.5 to 1604.1 ppm, with an average of 316.1 ppm [36]. According to the Total Exposure Assessment Methodology (TEAM) conducted in New Jersey, 4 of 12 air samples contained n-butane [73]. The arithmetic and geometric means were 16.6 and 11.3 ppbC, respectively, for the atmospheric n-butane content at urban locations in New England [4]. The average n-butane concentration in the air at the 6th floor of the Cooper Union Building in New York City, NY, was 43, 48 and 38 ppbC for 19, 12 and 10 samples taken at 6:00-9:00 AM, 9:00-11:00 AM and 1:00-3:00 PM, respectively, in July 1978 [2]. The average n-butane concentration in the air at the 82nd floor of the Empire State Building in New York City, NY, was 17, 27 and 19 ppbC for 18, 21 and 17 samples taken at 6:00-9:00 AM, 9:00-11:00 AM and 1:00-3:00 PM, respectively, in July 1978 [2]. At street level at the Empire State and World Trade Buildings in Manhattan, NY, the average n-butane concentration of 4 samples was 72 ppbC in July 1978 [2]. In 1975 the average n-butane concentration of 14 air samples taken between 05:30-08:30 and 12:30-15:30 at the World Trade Center in New York City, NY, was 36 and 30 ppbC, respectively [2]. In 1975 the average n-butane concentration of 11 and 8 air samples taken between 5:30-8:30 AM and 12:30-3:30 PM at the Interstate Sanitation Commission in New York City, NY, was 64 and 97 ppbC, respectively [2]. n-Butane was detected at an average concentration of 27.4 ug/m^3 for 5 samples collected at the 82nd floor of the World Trade Center in New York City between 5:00 AM-5:30 PM August 23, 1977 [4]. The ground level atmospheric concentration of n-butane at 13:25 was 29 ppb and 165 ppb at 08:00 for Huntington Park, CA [59]. At 1500 ft the n-butane concentration was 8 ppb at 07:43 and at 08:07 at a height of 2,200 ft the n-butane concentration was 3 ppb [59]. The n-butane concentration ranged from 21 to 70 ppb of volume at a downtown Los Angeles location for the Fall of 1981 [19]. The n-butane concentration at 1100 ft just east of Antioch, CA, was 13.5 ug/m^3, at 1000 ft near Pittsburgh, CA, was 21.0 ug/m^3, at 1100 ft over Carquinez Strait, CA, was 2.5 ug/m^3 and at 1000 ft over San Pablo Bay, CA, was 0.5 ug/m ^{3} [62]. According to the National Ambient Volatile Organic Compounds (VOCs) Database, the median urban atmospheric concentration of n-butane is 9.174 ppb of volume for 546 samples [64]. n-Butane was detected in the atmospheres of Pretoria, Johannesburg and Durban, South Africa [37]. n-Butane was identified in the ambient air of Sydney, Australia [42], ranging in concentration from 0.8 to 52.0 ppb of volume, with an average concentration of 9.5 ppb of volume [45]. n-Butane was detected at an average concentration of 121.61 ppbC in the atmosphere over the British Columbia Research Council Laboratory at the University of British Columbia [68]. The average n-butane concentration in the air over Tokyo, Japan, in 1980 and 1981 was 1.9 and 2.4 ppb for 66 and 192 samples, respectively [70]. At Deuselbach, Hunsruck in Germany, the atmospheric n-butane concentration was 0.54 ppb for October 23, 1983 [55]. n-Butane was detected in the atmospheres of 6 industrialized cities of the USSR, ranging in size of population from 0.4 to 4.5 million people [26,27]. The minimum, maximum and average n-butane concentration in the ambient air of Bombay, India, were 0.6, 145.0 and 21.2 ppb, respectively [41]. In September 1967, ground level air concentrations at Point Barrow, AK, were from 0.03 to 0.19 ppb. At downtown Los Angeles in 1967, the 10th percentile was 20 ppb, with an average of 46 ppb. The 90th percentile was 80 ppb. The average ground level air concentration in US urban air ranged

n-Butane

from 0.05 to 0.45 ppm. SUBURBAN: According to the National Ambient Volatile Organic Compounds (VOCs) Database, the median suburban atmospheric concentration of n-butane is 8.832 ppb of volume for 226 samples [64]. The n-butane concentration was 6.0, 2.5 and 2.0 ug/m^3 at 10, 15 and 40 mi downwind of Janesville, WI August 14, 1978 [63]. RURAL: At a rural site near Duren, Germany, the atmospheric n-butane concentration was 5.5 ppb for March 1984 [55]. The respective median, minimum and maximum atmospheric concentration of n-butane for 5 rural locations in NC ranged from 1.1 to 33.6, 0.2 to 14.8, and 2.5 to 47.5 ppb [60]. The atmospheric concentration of n-butane for Jones State Forest, TX, ranged from 12.0 to 49.6 ppb, with an average of 24.1 ppb for 10 samples [61]. According to the National Ambient Volatile Organic Compounds (VOCs) Database, the median rural atmospheric concentration of n-butane is 0.779 ppb of volume for 36 samples [64]. The arithmetic and geometric means were 1.1 and 1.3 ppbC, respectively, for the atmospheric n-butane content at rural locations in New England [14]. REMOTE: According to the National Ambient Volatile Organic Compounds (VOCs) Database, the median remote atmospheric concentration of n-butane is 0.510 ppb of volume for 7 samples [64]. For 9 samples collected over a 30 hour period, the average n-butane concentration in the Smokey Mountains, NC, was 4.8 ppbC, with a range from 3.6 to 7.6 ppbC [5]. On August 27, 1976, the average n-butane concentration for air over Lake Michigan at altitudes of 2000, 2500 and 3000 ft was 5.9 ppb of volume [40]. On August 28, 1976, the average n-butane concentration for air over Lake Michigan at altitudes of 1000 and 1500 ft was 1.5 ppb of volume [40]. The air over the Norwegian Artic had an average n-butane concentration for 5 samples from Bear Island, 2 from Hopen and 2 from Spitsbergen of less than 0.02 ppb of volume in July 1982 and 0.805 ppb of volume in the spring of 1983 [25]. All 27 air samples from the intertropical Indian Ocean contained n-butane at concentrations ranging from 0.03 to 0.70 ppb of volume [10]. SOURCE DOMINATED: According to the National Ambient Volatile Organic Compounds (VOCs) Database, the median source dominated atmospheric concentration of n-butane is 9.500 ppb of volume for 53 samples [64]. A Texaco refinery located in Tulsa, OK, was attributed with emissions to the surrounding atmosphere where the n-butane concentration was measured to be 175.2 and 342.7 ppbC for 2 min before and after 1:33 PM [5]. The n-butane content of the air downwind of a Mobil natural gas facility in Rio Blanco, CO, was 56.0 ppbC [5]. The arithmetic and geometric means were 7.3 and 3.4 ppbC, respectively, for the atmospheric n-butane content at polluted rural locations in New England [14]. n-Butane was detected in the ambient air over 1 of 5 National Priorites List (NPL) hazardous waste sites and one industrial/municipal landfill in New Jersey [35]. The frequency of occurrence for n-butane at both site was between 75 and 100% [35]. n-Butane was detected in the air of the Lincoln Tunnel at an average concentration of 198.1 ppbC [58]. Air over the roadway outside the tunnel had an average n-butane concentration of 16.42 ug/m^3 [58].

Food Survey Values:

Plant Concentrations:

Fish/Seafood Concentrations:

Animal Concentrations:

Milk Concentrations: n-Butane was detected in 6 of 12 samples of mothers' breast milk from the cities of Bayonne, NJ, Jersey City, NJ, Bridgeville, PA, and Baton Rouge, LA [51].

Other Environmental Concentrations: An air sample taken near an oil fire was found to contain n-butane and n-butene at a combined concentration of 1.63 mg/m^3 [52]. The background measurements utilized as a set of controls for air expired from humans contained n-butane in 8 of 20 samples collected over 18 mo [33].

Probable Routes of Human Exposure: The most probable route of human exposure to n-butane is by inhalation. Atmospheric workplace exposures have been documented [20,29,54]. n-Butane is a highly volatile compound and monitoring data indicates that it is a widely occurring atmospheric pollutant.

Average Daily Intake: According to the National Ambient Volatile Organic Compounds (VOCs) Database, the median urban atmospheric concentration of n-butane is 9.174 ppbV for 546 samples [64]. Based upon this figure and the value for average daily inhalation by a human adult of 20 m^3 of air, the average daily intake of n-butane via air is 183 mg.

n-Butane

Occupational Exposure: NIOSH (NOHS 1972-1974) has estimated that 71,296 workers are potentially exposed to n-butane in the US [47]. NIOSH (NOES 1981-1983) has estimated that 422,474 workers are potentially exposed to n-butane in the US [46]. A 1984 study showed n-butane was emitted from gasoline exposing outside operators at the refineries to an average air concentration of 3.437 mg/m^3; n-butane was detected in 54 of 56 samples [54]. Transport drivers were exposed to n-butane at atmospheric concentration of 9.701 mg/m^3 and n-butane was detected in 49 of 49 samples [54]. Gas station attendants were exposed to n-butane at atmospheric concentration of 21.605 mg/m^3 and n-butane was detected in 49 of 49 samples [54]. Attendants at a high volume service station in eastern PA were exposed to levels of n-butane ranging from 0.1 to 0.3 ppm for 18 of 18 air samples [29]. Workers at separate gasoline bulk handling facilities were exposed to vapors that contained n-butane at concentration of 33.7% by weight, 21.2% by weight and 38.1% by volume [20] of total hydrocarbons. Exposures to total hydrocarbons at one of the facilities exceeded 240 ppm for 5% of the sampling time [20].

Body Burdens: n-Butane was detected in 6 of 12 samples of mothers' breast milk from the cities of Bayonne, NJ, Jersey City, NJ, Bridgeville, PA and Baton Rouge, LA [51]. According to the Total Exposure Assessment Methodology (TEAM) conducted in New Jersey, 2 of 12 personal breath samples contained n-butane [73]. The background measurements utilized as a set of controls for air expired from humans were collected over 18 mo [33].

REFERENCES

1. Abrams EF et al; Identification of Organic Compounds in Effluents from Industrial Sources USEPA-560/3-75-002 (1975)
2. Altwicker ER et al; J Geophys Res 85: 7475-87 (1980)
3. Altwicker ER, Whitby RA; Atmos Environ 12: 1289-96 (1978)
4. Altwicker ER, Whitby RA; Sampling, Sample Prep and Measurement of Specific Non-methane Hydrocarbons 72 Ann Meet Air Pollut Contr Asssoc (1979)
5. Arnts RR, Meeks SA; Atmos Environ 15: 1643-51 (1981)
6. Atkinson R et al; Internat J Chem Kin 14: 781-8 (1982)
7. Atkinson R et al; Preprints Div Environ Chem 23: 173-6 (1983)
8. Atkinson R, Pitts JN Jr; J Phys Chem 78: 1780-4 (1974)
9. Atkinson R et al; J Phys Chem 86: 4563-9 (1982)
10. Bonsang B et al; J Atmos Chem 6: 3-20 (1988)

11. Burnham AK et al; AWWA J 65: 722-5 (1973)
12. Carotti AA, Kaiser ER; J Air Pollut Contr Assoc 22: 224-53 (1972)
13. Carter W PL et al; Int J Chem Kinet 11 (1): 45-102 (1979)
14. Colbeck I, Harrison RM; Atmos Environ 19: 1899-904 (1985)
15. Cox RA et al; Environ Sci Technol 15: 587-92 (1981)
16. Daubert TE, Danner RP; Data Compilation, Tables of Properties of Pure Cmpds, Design Inst for Phys Prop Data, Am Inst for Phys Prop Data, NY NY (1985)
17. Eisenreich SJ et al; Environ Sci Technol 15: 30-8 (1981)
18. Great Lakes Water Quality Board; Inventory Chem Subst Id Great Lakes Ecos p 195 (1983)
19. Grosjean D, Fung K; J Air Pollut Control Assoc 34: 537-43 (1984)
20. Halder CA et al; Am Ind Hyg Assoc J 47: 164-72 (1986)
21. Hampton CV et al; Environ Sci Technol 16: 287-98 (1982)
22. Hansch C, Leo AJ; Medchem Project Issue No 26. Claremont CA: Pomona College (1985)
23. Hendry DG et al; J Phys Chem Ref Data 3: 944-78 (1974)
24. Hou CT et al; Appl Environ Microbiol 46: 178-84 (1983)
25. Hov O et al; Geophys Res Lett 11: 425-8 (1984)
26. Ioffe BV et al; Environ Sci Technol 13: 864-8 (1979)
27. Ioffe BV et al; Dokl Akad Nauk Sssr 243: 1186-9 (1978)
28. Jamison VW et al; pp. 187-96 in Proc Int Biodeg Symp 3rd Sharpley JM and Kapalan AM (eds) Essex Eng (1976)
29. Kearney CA, Dunham DB; Am Ind Hyg Assoc J 47: 535-9 (1986)
30. Kool HJ et al; Crit Rev Env Control 12: 307-57 (1982)
31. Kopczynski SL et al; Environ Sci Technol 6: 342-7 (1972)
32. Kopfler FC et al; Adv Environ Sci Technol 8: 419-33 (1977)
33. Krotoszynski BK et al; J Anal Toxicol 3: 225-34 (1979)
34. Kvenvolden KA, Redden GD; Geochimica et Cosmochimica Acta 44: 1145-50 (1980)
35. LaRegina J et al; Environ Proc 5: 18-27 (1986)
36. Lonneman WA et al; Hydrocarbons in Houston Air USEPA-600/3-79/018 p. 44 (1979)
37. Louw CW et al; Atmos Environ 11: 703-17 (1977)
38. Lyman WJ et al; Handbook of Chemical Property Estimation Methods NY: McGraw-Hill pp. 4-9, 5-4, 5-10, 7-4 (1982)
39. McFall JA et al; Chemosphere 14: 1253-65 (1985)
40. Miller MM, Alkezweeny AJ; Ann NY Acad Sci 338: 219-32 (1980)
41. Mohan Rao AM, Panditt GG; Atmos Environ 2: 395-401 (1988)
42. Mulcahy MFR et al; Paper IV p 17 in Occurrence Contr Photochem Pollut, Proc Symp Workshop Sess (1976)
43. Neligan RE; Arch Environ Health 5: 581-91 (1962)
44. Nelson PF, Quigley SM; Atmos Environ 18: 79-87 (1984)
45. Nelson PF, Quigley SM; Environ Sci Technol 16: 650-5 (1982)
46. NIOSH; National Occupational Exposure Survey (NOES) (1989)
47. NIOSH; National Occupational Hazard Survey (NOHS) (1974)

48. Paraskevopoulos G, Nip WS; Can J Chem 58: 2146-9 (1980)
49. Patel RN et al; Appl Environ Microbiol 39: 727-33 (1980)
50. Patel RN et al; Appl Environ Microbiol 39: 720-6 (1980)
51. Pellizzari ED et al; Bull Environ Contam Toxicol 28: 322-8 (1982)
52. Perry R; Mass Spectroscopy in the Detection and Identification of Air Pollutants, Int Symp Ident Meas Environ Pollut p. 130-7 (1971)
53. Perry RA et al; J Chem Phys 64: 5314-6 (1976)
54. Rappaport SM et al; Appl Ind Hyg 2: 148-54 (1987)
55. Rudolph J, Khedim A; Int J Environ Anal Chem 290: 265-82 (1985)
56. Sauer TC Jr; Org Geochem 7: 1-16 (1981)
57. Sawhney BL, Raabe JA; Ground Water Contamination Mvmt Org Pollut in Granby Landfill, Bull 833 p 9 (1986)
58. Scheff PA et al; JAPCA 37A: 469-78 (1989)
59. Scott Research Labs Inc; Atmospheric Reaction Studies in the Los Angeles Basin, NTIS PB-194-058 p. 86 (1969)
60. Seila RL et al; Atmospheric Volatile Hydrocarbon Composition at Five Remote Sites in NW NC, USEPA-600/D-84-092 (1984)
61. Seila RL; Non-urban Hydrocarbons Concentration in Ambient Air No of Houston TX USEPA-500/3-79-010 p38 (1979)
62. Sexton K, Westberg H; Environ Sci Tech 14: 329-32 (1980)
63. Sexton K; Environ Sci Technol 17: 402-7 (1983)
64. Shah JJ, Heyerdahl EK; National Ambient VOC Database Update USEPA 600/3-88/010 (1988)
65. Sigsby JE et al; Environ Sci Technol 21: 466-75 (1987)
66. Silverstein RM, Bassler GC; Spectrometric Id of Org Cmpd, J Wiley and Sons Inc p 148-169 (1963)
67. Stump FD et al; Atmos Environ 23: 307-20 (1989)
68. Stump FD, Dropkin DL; Anal Chem 57: 2629-34 (1985)
69. Swann RL et al; Res Rev 85: 16-28 (1983)
70. Uno I et al; Atmos Environ 19: 1283-93 (1985)
71. USEPA; EXAMS II Computer Simulation (1987)
72. USEPA; Drinking water Criteria Document for Gasoline ECAO-CIN-D006, 8006-61-9 (1986)
73. Wallace LA et al; Environ Res 35: 293-319 (1984)
74. Whelan JK et al; Geochim Cosmochim Acta 44: 1767-85 (1980)
75. Yalkowsky SH et al; Arizona Data Base of Water Solubility (1987)
76. Young P, Parker A; Vapors Odors and Toxic Gases from Landfills ASTM Spec Tech Publ 851: 24-41 (1984)
77. Zweidinger RB et al; Environ Sci Tech 22: 956-62 (1988)

1,3-Butanediol

Synonyms:

Structure:

CAS Registry Number: 107-88-0

Molecular Formula: $C_4H_{10}O_2$

SMILES Notation: OCCC(O)C

CHEMICAL AND PHYSICAL PROPERTIES

Boiling Point: 207.5 °C at 760 mm Hg

Melting Point: Below 50 °C

Molecular Weight: 90.12

Dissociation Constants:

Log Octanol/Water Partition Coefficient:

Water Solubility: Miscible in water [19]

Vapor Pressure: 0.02 mm Hg at 25 °C [3]

Henry's Law Constant: 2.3 x 10^{-7} atm-m^3/mol [15]

ENVIRONMENTAL FATE/EXPOSURE POTENTIAL

Summary: 1,3-Butanediol can be released to the environment in waste effluents generated at sites of its commercial production. Its use in de-icing mixtures for removing ice from airplanes will release 1,3-butanediol directly to the surrounding environment. If released to the atmosphere, it will degrade in the vapor-phase by reaction with photochemically produced hydroxyl radicals (estimated half-life of 1.2 days). If released to soil or water, 1,3-butanediol will probably degrade via biodegradation. Leaching in soil is possible since 1,3-butanediol is miscible in water. Occupational exposure to 1,3-butanediol occurs through dermal contact and inhalation of vapor.

Natural Sources: Some esters of 1,3-butanediol occur in nature [7].

Artificial Sources: 1,3-Butanediol's use as an aircraft de-icing agent for melting ice from airplane surfaces [7] will release the compound directly to surrounding terrestrial surfaces. Waste streams (both gaseous and aqueous) generated at commercial sites of 1,3-butanediol manufacture probably contain 1,3-butanediol [9].

Terrestrial Fate: The dominant environmental fate process for 1,3-butanediol in soil is probably biodegradation. However, insufficient experimental data are available to predict biodegradation rates in the environment. Leaching in soil is possible since 1,3-butanediol is miscible in water.

Aquatic Fate: The dominant environmental fate process for 1,3-butanediol in water is probably biodegradation. However, insufficient experimental data are available to predict biodegradation rates in the environment. Aquatic hydrolysis, volatilization, adsorption to sediment, and bioconcentration are not expected to be environmentally important.

Atmospheric Fate: Based upon the vapor pressure, 1,3-butanediol is expected to exist almost entirely in the vapor phase in the ambient atmosphere [5]. It will degrade in an average ambient atmosphere by reaction with photochemically produced hydroxyl radicals (estimated half-life of 1.2 days) [1,16]. Physical removal from air via wet deposition is possible since 1,3-butanediol is miscible in water.

Biodegradation: Based primarily upon biodegradation screening studies with ethylene glycol (ethanediol) and propylene glycol (propanediol), the glycol chemical class is considered biodegradable by both acclimated and unacclimated soil, water and sewage microorganisms [14]. Although several pure culture screening studies have demonstrated that 1,3-butanediol can be biodegraded [4,6,8,18], insufficient data are available from mixed culture screening studies to experimentally demonstrate environmental biodegradability.

Abiotic Degradation: The rate constant for the vapor-phase reaction of 1,3-butanediol with photochemically produced hydroxyl radicals has been estimated to be 1.340×10^{-11} cm^3/molecule-sec at 25 °C, which corresponds to an atmospheric half-life of about 1.2 days at an atmospheric concn of $5 \times 10^{+5}$ hydroxyl radicals per cm^3 [15]. Glycols are generally resistant to aqueous environmental hydrolysis [13]; therefore, 1,3-butanediol is not expected to chemically hydrolyze in environmental waters. The rate constant for the reaction between photochemically produced hydroxyl radicals in water and 1,3-butanediol is $2.2 \times 10^{+9}$ 1/mol-sec [2]; assuming that the concn of hydroxyl radicals in brightly sunlit natural water is 1×10^{-17} M [13], the half-life would be about 365 days of continuous (24 hr/day) sunlight.

Bioconcentration: 1,3-Butanediol is miscible in water [19]; this suggests that bioconcentration in aquatic organisms will be not be important environmentally.

Soil Adsorption/Mobility: 1,3-Butanediol is miscible in water [19]; this suggests that 1,3-butanediol will be very mobile in soil and will probably leach from soil to ground water.

Volatilization from Water/Soil: The Henry's Law constant for 1,3-butanediol can be estimated to be 2.30×10^{-7} atm-m^3/mol using a structure estimation method [15]. This value of Henry's Law constant indicates that a compound is essentially nonvolatile from water [12].

Water Concentrations:

Effluent Concentrations: 1,3-Butanediol was qualitatively detected in water samples collected from an advanced waste treatment facility in Pomona, CA, on September 25, 1974 [10].

1,3-Butanediol

Sediment/Soil Concentrations:

Atmospheric Concentrations:

Food Survey Values:

Plant Concentrations:

Fish/Seafood Concentrations:

Animal Concentrations:

Milk Concentrations:

Other Environmental Concentrations:

Probable Routes of Human Exposure: Occupational exposure to 1,3-butanediol occurs through dermal contact and inhalation of vapor [11].

Average Daily Intake:

Occupational Exposure: NIOSH (NOES 1981-1983) has statistically estimated that 19,323 workers are potentially exposed to 1,3-butanediol in the US [17]. NIOSH (NOHS 1972-1974) has statistically estimated that 41,354 workers are potentially exposed to 1,3-butanediol in the US [17].

Body Burdens:

REFERENCES

1. Atkinson R; J Inter Chem Kinct 19: 799-828 (1987)
2. Buxton GV et al; J Phys Chem Ref Data 17: 704 (1988)
3. Daubert TE; Danner RP; Physical and Thermodynamic Properties of Pure Chemicals: Data Compilation, NY: Hemisphere Pub Corp (1989)
4. Daugherty LC; Lubrication Engin 36: 718-23 (1980)
5. Eisenreich SJ et al; Environ Sci Technol 15: 30-8 (1981)
6. Enomoto K et al; Inst Phys Chem Res 53: 637-42 (1975)
7. Fay RH; Kirk-Othmer Encycl Chem Technol 3rd ed. NY: John Wiley & Sons 3: 93 (1978)
8. Kersters K, Deley J; Biochim Biophys Acta 71: 311-31 (1963)

9.   Leipins R et al; Industrial Process Profiles for Environmental Use: Chpt 6 USEPA-600/2-77-023f (NTIS PB-281478) p. 6-241, 6-242 (1977)

10. Lucas SV; GC/MS Analysis of Organics in Drinking Water Concentrates and Advanced Waste Treatment Concentrates: Volume 1, USEPA-600/1-84-020A (NTIS PB85-128221) p. 49, 176 (1984)

11. Parmeggiani L; Encyl Occup Health & Safety 3rd ed Geneva, Switzerland: International Labour Office p. 973-4 (1983)

12. Lyman WJ et al; Handbook of Chemical Property Estimation Methods Washington, DC: Amer Chem Soc p. 7-4; 15-15 to 15-29 (1990)

13. Mill T et al; Sci 207: 886-7 (1980)

14. Miller LM; Investigation of Selected Potential Environmental Contaminants: Ethylene Glycol, Propylene Glycols and Butylene Glycols. USEPA-560/11-79-006 Washington, DC: USEPA p. 67 (1979)

15. Meylan W, Howard PH; Environ Toxicol Chem 10: 1283-93 (1991)

16. Meylan WM, Howard PH; Chemosphere 26:2293-2299 (1993)

17. NIOSH; National Occupational Exposure Survey (NOES) (1983)

18. NIOSH; National Occupational Hazard Survey (NOHS) (1974)

19. Riddick JA et al; Organic Solvents: Physical Properties and Methods of Purification. Techniques of Chem 4th ed. NY: Wiley-Interscience p. 268 (1986)

20. Tsukamura M; Amer Rev Resp Dis 94: 796-8 (1966)

n-Butyl Chloride

SUBSTANCE IDENTIFICATION

Synonyms:

Structure:

$$Cl\diagup\diagdown\diagup\diagdown\diagup CH_3$$

CAS Registry Number: 109-69-3

Molecular Formula: C_4H_9Cl

SMILES Notation: CCCCCl

CHEMICAL AND PHYSICAL PROPERTIES

Boiling Point: 78.5 °C at 760 mm Hg

Melting Point: -123.1 °C

Molecular Weight: 92.57

Dissociation Constants:

Log Octanol/Water Partition Coefficient: 2.64 [8]

Water Solubility: 1100 mg/L at 25 °C [16]

Vapor Pressure: 101 mm Hg at 25 °C [3]

Henry's Law Constant: 1.67×10^{-2} atm-m^3/mol [10]

ENVIRONMENTAL FATE/EXPOSURE POTENTIAL

Summary: Environmental emissions of n-butyl chloride arise from process and fugitive emissions and waste water from its production and use as an alkylating agent and evaporation from its use as a solvent. There is also evidence that it may be formed during chlorination of waste water. n-Butyl chloride has a high vapor pressure and Henry's Law constant. In addition it has a low adsorptivity to soil. Therefore releases to the land or water will partition, to a large extent, to the atmosphere. Limited data suggest that biodegradation is slow. It's rate of hydrolysis is unknown and estimates of its rate from analogous compounds indicate that hydrolysis could not compete with volatilization as a fate process except in ground water. If released in surface water, the volatilization half-life in a model river and pond are estimated to be 2.9 hr and 34 hr, respectively. In the atmosphere, n-butyl chloride will degrade by reaction with photochemically produced hydroxyl radicals in the atmosphere with a half-life of 7.0 days. Human exposure will be primarily occupational by inhalation and dermal contact.

Natural Sources:

Artificial Sources: n-Butyl chloride is used as a solvent, an alkylating agent, and an antihelmintic medicine [9]. It may be released to the environment in emissions or wastewater related to its manufacture and these uses. n-Butyl chloride was formed during the chlorination of leachate obtained from a simulated landfill used to study the codisposal of metal plating sludge with municipal solid waste [6].

Terrestrial Fate: n-Butyl chloride has a very high vapor pressure, indicating that volatilization from soil will be rapid and a primary removal process. Based on its water solubility, it has a low estimated adsorptivity to soil (Koc = 93-102) [11] and therefore would be mobile in soil. Biodegradation in soil is unknown.

Aquatic Fate: If released in water, n-butyl chloride will be lost rapidly by volatilization. Using its Henry's Law constant, its estimated volatilization half-life in a model river and model pond are 2.9 hr [11] and 34 hr [18], respectively. In ground water, where volatilization may not occur, n-butyl chloride may be lost by hydrolysis. Based upon hydrolysis rates for alkyl chlorides, the half-life is estimated to be between 6 hr and 38 days at neutral pH [12]. The half-life at higher pHs will be shorter. Photodegradation, adsorption to sediment, and bioconcentration in fish are not important aquatic fate processes.

n-Butyl Chloride

Atmospheric Fate: n-Butyl chloride will degrade via reaction with photochemically produced hydroxyl radicals in the atmosphere. The half-life of n-butyl chloride in the atmosphere is estimated to be 7.0 days [1]. n-Butyl chloride does not absorb radiation >290 nm and therefore it will not directly photolyze [2]. Due to its moderate water solubility, washout by rain may occur. However, n-butyl chloride removed in this manner will revolatilize into the atmosphere.

Biodegradation: Microbial enzymes and pure cultures have been reported that are capable of degrading n-butyl chloride under aerobic conditions [19]. Limited data from screening studies suggest that n-butyl chloride biodegrades slowly under aerobic conditions. When incubated with activated sludges from three municipal treatment plants, 2.6% of the n-butyl chloride (500 mg/L) was oxidized after 24 hr [5]. At the concentration used, n-butyl chloride was toxic to one of the three sludges [5]. Another screening test using sewage seed and much lower concentrations of n-butyl chloride (1 ppm) resulted in 10% of the theoretical BOD being consumed in 1.4 days [14].

Abiotic Degradation: Alkyl halides hydrolyze in water by neutral and base catalyzed reactions to give the corresponding alcohols; no acid catalyzed processes have been reported [12]. No rate for neutral hydrolysis of n-butyl chloride is available. Based on extrapolations of hydrolysis rates of analogous compounds determined at higher temperatures and generalizations concerning the relative reactivity of analogs, the hydrolysis half-life of n-butyl chloride at 25 °C is estimated to lie in the range of 6 hr to 38 days [12]. In the atmosphere, n-butyl chloride will react with photochemically produced hydroxyl radicals by H-atom extraction. The estimated rate of reaction is 2.29×10^{-12} cm^3/molecule-sec [1]. Assuming a hydroxyl radical concentration of $5 \times 10^{+5}$ per cm^3, the half-life of n-butyl chloride in the atmosphere is 7.0 days. n-Butyl chloride does not absorb radiation >290 nm and therefore it will not directly photolyze [2]. The half-life of n-butyl chloride in a eutrophic lake due to the photochemical production of hydroxyl radicals in full summer noon sunlight is 1000 hr [7]. Therefore, this reaction would not be a significant degradation route for n-butyl chloride in surface waters.

Bioconcentration: Using the log octanol/water partition coefficient for n-butyl chloride, one estimates a BCF of 60 using a recommended regression equation [11]. Therefore, n-butyl chloride will not bioconcentrate in fish and aquatic organisms.

Soil Adsorption/Mobility: Using the water solubility for n-butyl chloride, its Koc can be estimated to be 93 and 102 using two recommended regression equations [11]. These estimated values indicate that n-butyl chloride will have high mobility in soil [17].

Volatilization from Water/Soil: Based on its Henry's Law constant, the volatilization of n-butyl chloride from a model river 1 m deep, flowing at 1 m/sec with a 3 m/sec wind is 2.9 hr [11]. The volatilization half-life from a model pond is 34 hr [18]. Due to its high vapor pressure and Henry's Law constant and low adsorptivity to soil, n-butyl chloride should volatilize rapidly from both dry and moist soils.

Water Concentrations:

Effluent Concentrations: In a study of 63 industrial effluents that discharge into surface waters, butylchloride was found in one effluent at a concentration of <10 ug/L [15].

Sediment/Soil Concentrations:

Atmospheric Concentrations: SOURCE DOMINATED: n-Butyl chloride was identified, but not quantified, in one of 10 samples of ambient air in the Kanawha Valley, WV [4]. Various locations were sampled during three trips.

Food Survey Values:

Plant Concentrations:

Fish/Seafood Concentrations:

Animal Concentrations:

Milk Concentrations:

Other Environmental Concentrations:

Probable Routes of Human Exposure: Exposure to n-butyl chloride is primarily in the work place. Probable routes of occupational exposure are inhalation and dermal contact.

Average Daily Intake:

Occupational Exposure: NIOSH (NOES 1981-1983) has statistically estimated that 3,201 workers are exposed to n-butyl chloride [13].

Body Burdens:

REFERENCES

1. Atkinson R et al, Environ Sci Technol 19: 799-828 (1987)
2. Calvert JG, Pitts JN Jr; Photochemistry NY: Wiley pp. 427-30 (1966)
3. Daubert TE, Danner RP; Data Compilation Tables of Properties of Pure Compounds NY, NY: Amer Inst for Phys Prop Data (1989)
4. Erickson MD, Pellizzari ED; Analysis of Organic Air Pollutants in the Kanawha Valley, WV & the Shenandoah Valley VA USEPA-903/9-78-007 (1978)
5. Gerhold RM, Malaney GW; J Water Pollut Contr Fed 38: 562-79 (1966)
6. Gould JP et al; pp. 525-39 in Water Chlorination: Environ Impact Health Eff Vol 4 (1983)
7. Haag WR, Hoigne J; Chemosphere 14: 1659-71 (1985)
8. Hansch C, Leo AJ; Medchem Project Issue No 26 Claremont, CA: Pomona College (1985)
9. Hawley CG; The Condensed Chemical Dictionary 10th ed. NY: Van Nostrand Reinhold Co (1981)
10. Leighton DT Jr, Calo JM; J Chem Eng 26: 382-5 (1981)
11. Lyman WJ et al; Handbook of Chemical Property Estimation Methods NY: McGraw-Hill Chapt 4, 5, 15 (1982)
12. Mabey W, Mill T; J Phys Chem Ref Data 7: 383-415 (1978)
13. NIOSH; National Occupational Exposure Survey (NOES) (1989)
14. Okey RW, Bogan RH; J Water Pollut Control Fed 37: 692-712 (1965)
15. Perry DL et al; Identification of Organic Compound in Industrial Effluent Discharges USEPA-600/4-79-016, NTIS-PB-294794 (1979)
16. Riddick JA et al; Organic Solvents 4th ed; NY, NY: Wiley (1986)
17. Swann RL et al; Res Rev 85: 17-28 (1983)
18. USEPA; EXAMS II (1987)
19. Visscher K, Brinkman J; Haz Waste Haz Materials 6: 210-12 (1989)

Butyric Acid

SUBSTANCE IDENTIFICATION

Synonyms:

Structure:

CAS Registry Number: 107-92-6

Molecular Formula: $C_4H_8O_2$

SMILES Notation: O=C(O)CCC

CHEMICAL AND PHYSICAL PROPERTIES

Boiling Point: 163.7 °C

Melting Point: -7.9 °C

Molecular Weight: 88.11

Dissociation Constants: pKa = 4.82 [1]

Log Octanol/Water Partition Coefficient: 0.79 [18]

Water Solubility: infinite [37]

Vapor Pressure: 0.001 mm Hg at 25 °C [37]

Henry's Law Constant: 5.35×10^{-7} atm-m^3/mol at 25 °C [34]

ENVIRONMENTAL FATE/EXPOSURE POTENTIAL

Summary: Butyric acid is both a natural and a commercially produced organic compound. It may be released to the environment as a fugitive emission during its production and formulation, or in the effluent of commercial processes, sewage treatment plants, landfills, and in the exhaust of motor vehicles. If released to soil, butyric acid is expected to be relatively mobile, although adsorption may occur by attractive interactions with active sites in the soil. Butyric acid is not expected to significantly volatilize from either moist or dry soil to the atmosphere. If released to water, butyric acid will exist predominately in the dissociated form under environmental conditions. Butyric acid is expected to biodegrade rapidly under both aerobic and anaerobic conditions. Volatilization from water to the atmosphere is not expected to occur to any significant extent; the half-life for volatilization from a model river is 59 days. Butyric acid will not significantly adsorb to sediment and suspended organic matter, nor is it expected to significantly bioconcentrate in fish and aquatic organisms. If released to the atmosphere, butyric acid is expected to undergo a gas-phase reaction with photochemically produced hydroxyl radicals with a half-life of 8 days. Butyric acid may also undergo atmospheric removal by wet deposition. Occupational exposure to butyric acid may occur by inhalation or dermal contact during its production or use. Exposure to the general population may occur by inhalation or dermal contact if commercial products containing this compound are used in the home. Ingestion of butyric acid is a probable route of exposure due to its presence in foods.

Natural Sources: Butyric acid is found in vegetable oils and in animal fluids, such as sweat, tissue fluids and milk fat [36]. Free butyric acid is an important metabolite in the breakdown of carbohydrates, fats and proteins [36]. Butyric acid may arise from natural fermentive processes occurring in sediment [31]. It has also been detected as a volatile flavor component in the fruit of the deciduous palm Dalieb at a concn of 58 mg/kg pulp [19]. It is a component of the essential oils of citronella ceylon, eucalyptus globulus, araucaria cunninghamii, lippia scaberrima, monarda fistulosa, cajeput, heracleum giganteum, lavender, hedeoma pulegioides, valerian, nutmeg, hops, pastinaca sativa, and spanish anise [14].

Artificial Sources: Butyric acid is used in the synthesis of esters, which are then used to manufacture lacquers, plastics, perfumes and flavor ingredients [36,39,46]. Butyric acid is also used as a deliming agent, in disinfectants and emulsifying agents, for the decalcification of hides and for sweetening gasoline [39,46]. Butyric acid has also been detected in new motor oil [24]. Butyric acid may enter the environment as a fugitive emission during its production, transport, or formulation into other products. Butyric acid may also enter the environment in the effluent of energy related processes [35], sewage treatment plants [32], landfills [11,13,25] and in the exhaust of motor vehicles [24]. Butyric acid may enter the environment as a result of the biological breakdown of other organic compounds [22].

Terrestrial Fate: If released to soil, experimental studies indicate that butyric acid may be relatively mobile in soil [20,38]. Any adsorption of butyric acid to soil is driven by attractive interactions with active sites in the soil and not due to hydrophobic characteristics [40,45]. Butyric acid's low Henry's Law constant indicates that it will not significantly volatilize from either moist or dry soil, respectively.

Aquatic Fate: If released to water, butyric acid's pKa, 4.82 [37], indicates that the ionic form will predominate under environmental conditions. Butyric acid is expected to biodegrade rapidly under both aerobic [3,41,43] and anaerobic [5,26] conditions. Volatilization from water to the atmosphere is not expected to occur to any significant extent; the half-life for volatilization from a model river 1 m deep, flowing at 1 m/sec and a wind speed of 3 m/sec is 59 days [7,26]. An experimental Koc of 19.1 obtained using a lateritic muddy sand (1.3% organic carbon) [8] indicates that butyric acid will not significantly adsorb to sediment and suspended organic matter. A bioconcentration factor of 2.3 calculated from butyric acid's log octanol/water partition coefficient using an appropriate regression equation [7] indicates that it will not significantly bioconcentrate in fish and aquatic organisms.

Atmospheric Fate: If released to the atmosphere, butyric acid is expected to undergo a gas-phase reaction with photochemically produced hydroxyl radicals. An experimental rate constant of 2.4×10^{-12} cm/molecule-sec translates to a half-life of 8 days for this process [2,3]. Butyric acid is miscible with water and may be removed from the atmosphere by wet deposition.

Biodegradation: At an initial concentration of 100 mg/L, butyric acid displayed a 72% theoretical biochemical oxygen demand (BODT) after 5 hr when incubated with activated sludge [43]. Butyric acid at an initial concentration of 5 ppm displayed a BODT of 76.6% in fresh water and 72.4% in sea water after 5 days [41]. Butyric acid had a BODT of 17.4%, 23.8%, 26.2%, and 27.7% after 6, 12, 18, and 24 hr, respectively, when incubated with an activated sludge seed at an initial concentration of 500 ppm [29]. In a screening study, butyric acid displayed a 46%, 48%, and 58% BODT after 2, 10, and 30 days, respectively, using a sewage seed [9]. In a screening study using a sewage seed, butyric acid had a 5 day BODT of 72-78% and a 20 day BODT of 92-99% [15,16]. Several other screening studies with activated sludge inocula have shown that butyric acid is amenable to biodegradation under aerobic conditions [8,23,30]. In a screening study, methanogenic microbes raised on acetate were found to completely remove butyric acid after a 3 day lag period at a rate of 284 mg/L/day, initial concentration not provided [5]. In a laboratory experiment using a flow-through methanogenic digester with a sewage sludge seed, butyric acid was found to be amenable to biodegradation under anaerobic conditions [26].

Abiotic Degradation: Experimental rate constants for the gas-phase reaction of butyric acid with photochemically produced hydroxyl radicals of 2.00×10^{-12} cm³/molecule-sec [7], 1.80×10^{-12} cm³/molecule-sec [44] and 2.4×10^{-12} cm³/molecule-sec [2] at 25 °C have been reported. Using a value of 2.4×10^{-12} cm³/molecule-sec and an average atmospheric hydroxyl radical concentration of $5 \times 10^{+5}$ molec/cm³ [3], a half-life of 8 days is calculated for this process.

Bioconcentration: Based on an experimental log octanol/water partition coefficient of 0.79 [18], a bioconcentration factor of 2.3 can be calculated for butyric acid [27], indicating that it will not significantly bioconcentrate in fish and aquatic organisms.

Soil Adsorption/Mobility: Experimental Koc values for butyric acid on a clastic mud (3.5% organic carbon), a lateritic muddy sand (1.3% organic carbon) and a fine carbonate sand (0.17% organic carbon) were 19.1, 27.6 and 14.7, respectively [38]. The percent of butyric acid sorbed to a kalonite or montmorillonite clay at 22 °C was 14.0% and 19.9% after 48 hours, respectively, which increased to 31.4% and 24.2% after 144 hr [20]. In a field study in which 100 ppm butyric acid was injected underground, the retardation, relative to the linear ground-water velocity, was calculated to be 3% [40]. Butyric acid is listed as a compound displaying an L-type adsorption isotherm, indicating that specific binding sites may be involved [4]. Experimental studies indicate that adsorption of butyric acid is dominated by attractive forces between the compound and soil and not by hydrophobic interactions [42].

Volatilization from Water/Soil: An experimental Henry's Law constant of 5.35×10^{-7} atm-m^3/mol at 25 °C [34] indicates that butyric acid will not significantly volatilize from water or moist soil. Based on this value, the estimated half-life for volatilization from a model river 1 m deep, flowing at 1 m/sec with a wind speed of 3 m/sec is 59 days [27]. The vapor pressure of butyric acid [37] indicates that it will not significantly volatilize from dry soil to the atmosphere.

Water Concentrations: SURFACE WATER: The concn of butyric acid in the Ohio River, Little Miami River, and Tanners Creek ranged from 0.1-0.3 µg/L, 0.4-0.5 µg/L, and 0.5 µg/L, respectively [1]. GROUND WATER: Studies near a closed wood preserving facility in Pensacola, FL, found butyric acid concentrations in ground water ranging from 12.87 mg/L at 6 m depth to 0.17 mg/L at 18 m depth ca. 170 m from the plant site. At ca. 330 m from the site, butyric acid was not detected at any well tested (6-24 m depth) [17].

Effluent Concentrations: Butyric acid was detected in 1 of 7 aqueous effluent samples from energy-related processes at a concn of 90 ppb [35]. It was detected in both the primary and secondary effluent of sewage treatment plants at concentrations ranging from 17-1,540 µg/L, although the concn was always lower in the secondary effluent [32]. Butyric acid was measured in the automobile exhaust at a concn of 0.123 ppb [24]. The effluent from a landfill in Norman, OK, 1972, contained butyric acid at an estimated concn of 1.5 µg/L [11,12]. It was detected in solid waste leachates in the Netherlands, UK, Canada, France, and Spain, representing from 0.34% to 8.5% of the volatile fatty acid fraction [25]. It was also identified in the leachate from low-level radioactive waste disposal sites in KY and NY [13]. Butyric acid was detected in the leachate from a 1 year old simulated solid waste landfill at a concn of 48.8 g/L, representing 65% of the acid fraction [4], and it has been detected in the leachate from a Barcelona, Spain, sanitary landfill [1]. Australian oil shale retort water contained butyric acid at a concn of 174 mg/L [10].

Sediment/Soil Concentrations: Butyric acid was detected in the sediment of Loch Eil, Scotland, at a concn ranging from trace to 160 µg/g dry weight [31]. It was detected at a concn of 0.273 mg/g in the sediment of Lake Biwa, Japan, 1981 [28].

Atmospheric Concentrations: URBAN/SUBURBAN: The concn of butyric acid in Los Angeles, CA, July and September 1984, ranged from 0.014-0.083 ppb (8 samples) [24].

Food Survey Values: Butyric acid was identified as a volatile component of baked potatoes [6]. It has also been detected as a volatile flavor component in the fruit of the deciduous palm Dalieb at a concn of 58 mg/kg pulp [19].

Plant Concentrations:

Fish/Seafood Concentrations:

Animal Concentrations:

Milk Concentrations:

Other Environmental Concentrations:

Probable Routes of Human Exposure: The probable routes of exposure to butyric acid are by inhalation or dermal contact during its production or use. Exposure to the general population may occur by inhalation or dermal contact if commercial products containing this compound are used in the home. Ingestion of butyric acid is a probable route of exposure due to its presence in foods [6,19].

Average Daily Intake:

Occupational Exposure: NIOSH (NOES 1981-1983) has statistically estimated that 9,547 workers are exposed to butyric acid in the US [33]. Butyric acid has been detected as an emission during the welding of steel coated with protective paints [21].

Body Burdens:

REFERENCES

1. Albaiges J et al; Wat Res 20: 1153-59 (1986)
2. Atkinson R; Chem Rev 85: 69-201 (1985)
3. Atkinson R; Int J Chem Kinet 19: 799-828 (1987)
4. Burrows WD, Rowe RS; JWPCF 47: 921-3 (1975)
5. Chou WL et al; Biotechnol Bioeng Symp 8: 391-414 (1979)
6. Coleman EC et al; J Agric Food Chem 29: 42-8 (1981)
7. Daugaut P et al; Int J Chem Kinet 20: 331-8 (1988)
8. Dawson PSS, Jenkins SH; Sewage and Indust Waste 22: 490-507 (1950)
9. Dias FF , Alexander M: App Microbiol 22: 1114-8 (1971)
10. Dobson KR et al; Water Res 19: 849-56 (1985)
11. Dunlap WJ et al; Organic Pollutants Contributed to Ground Water from a Landfill. USEPA Off Res Dev USEPA-600/9-76-004 pp 96-110 (1976)
12. Dunlap WJ et al; pp. 453-7 in Identif Anal Org Pollut Water (1976)
13. Francis AJ et al; Nuc Tech 50: 158-63 (1980)
14. Furia; Handbook of Food Additives, 2nd ed, Vol 2, p. 78 (1980)
15. Gaffney PE, Heukelekian H; J Wat Pollut Contr Fed 33: 1169-83 (1961)
16. Gaffney PE, Heukelekian H; Sewage and Indust Waste 30:673-9 (1958)
17. Goerlitz DF et al; Environ Sci Technol 19: 955-61 (1985)
18. Hansch C, Leo AJ; Medchem Project Issue No 26. Claremont,CA: Pomona College (1985)
19. Harper DB et al; J Sci Food Agric 37: 685-8 (1986)
20. Hemphill L et al; Proc 18th Indust Waste Conf 18: 204-17 (1964)
21. Henriks-Eckerman M et al; Am Ind Hyg Assoc J 51: 241-4 (1990)
22. Hoviuos JC et al; Anaerobic Treatment of Synthetic Organic Wastes Washington, DC USEPA 12020 DIS (1972)
23. Ishikawa T et al; Wat Res 13: 681-5 (1979)
24. Kawamura K et al; Environ Sci Tech 19: 1082-6 (1985)
25. Lema JM et al; Water Air Soil Pollut 40: 223-50 (1988)
26. Lin C et al; Water Res 20: 385-94 (1986)
27. Lyman WJ et al; Handbook of Chemical Property Estimation Methods NY: McGraw-Hill pp. 5-1 to 5-30, 15-15 to 15-29 (1982)
28. Maeda H, Kawai A; Bull Jap Soc Sci Fish 52: 1205-8 (1986)
29. Malaney GW, Gerhold RM; J Wat Pollut Control Fed 41: R18-33 (1969)
30. McKinney RE et al; Sewage and Ind Waste 28: 547-57 (1956)
31. Miller D et al; Marine Biol 50: 375-83 (1979)
32. Murtauch JJ, Bunch RL; JWPCF 37: 410-5 (1965)
33. NIOSH; National Occupational Exposure Survey (NOES) (1989)
34. Nirmalakhandan NN, Speece RE; Env Sci Tech 22: 1349-57 (1988)
35. Pellizzari ED et al; ASTM Spec Tech Publ STP 686: 256-7 (1979)

36. Reimenschneider W; Ullmann's Encycl Indust Tech 5th ed. Gerhartz W et al Eds. VCH Publishers A5: 235-48 (1986)
37. Riddick JA et al; Organic Solvents 4th ed, pp. 365-7 NY, NY: Wiley (1986)
38. Sansone FJ et al; Geochim Cosmochim Acta 51: 1889-96 (1987)
39. Sax NI, Lewis RJSR; Hawley's Condensed Chemical Dictionary 11th ed. NY,NY: Van Nostrand Reinhold Co p. 193 (1987)
40. Sutton PA, Barker JF; Ground Water 23: 10-6 (1985)
41. Takemoto S et al; Suishitsu Odaku Kenkyu 4: 80-90 (1981)
42. Ulrich H et al; Env Sci Tech 22: 37-41 (1988)
43. Urano K, Katz Z; J Haz Mat 13: 135-45 (1986)
44. Wallington TJ et al.; J Phys Chem 92: 5024-8 (1988)
45. Weber JB, Miller CT pp. 305-33 in Reactions and Movement of Organic Chemicals in Soils SSSA Spec Publ No. 22 (1989)
46. Windholz M; The Merck Index 10th ed Rahway, NJ: Merck & Co Inc p. 1562 (1983)

Chloroacetic Acid

SUBSTANCE IDENTIFICATION

Synonyms:

Structure:

CAS Registry Number: 79-11-8

Molecular Formula: $C_2H_3ClO_2$

SMILES Notation: O=C(O)CCl

CHEMICAL AND PHYSICAL PROPERTIES

Boiling Point: 189 °C

Melting Point: α, 63 °C; β, 55-56 °C; γ, 50 °C

Molecular Weight: 94.50

Dissociation Constants: 2.86 [9]

Log Octanol/Water Partition Coefficient: 0.22 [4]

Water Solubility: Very soluble in water

Vapor Pressure: 1 mm Hg at 43.0 °C

Henry's Law Constant:

ENVIRONMENTAL FATE/EXPOSURE POTENTIAL

Summary: Chloroacetic acid may enter the environment in emissions and wastewater from its production and use as a chemical intermediate primarily in the manufacture of chlorophenoxy herbicides and carboxymethyl cellulose. Such release of the chemical would be limited to industrial settings. If released into surface water, chloroacetic acid would biodegrade (73% in 8-10 days). It would not adsorb appreciably to sediment or bioconcentrate in fish. If spilled on land it would biodegrade and leach into the ground water. Its fate in ground water is unknown. If released into the air, probably as an aerosol, it will gravitationally settle out and undergo slow photodegradation.

Natural Sources:

Artificial Sources: Emissions and wastewater from its production and use as a chemical intermediate in the manufacture of 2,4-dichlorophenoxyacetic acid, 2,4,5-trichlorophenoxyacetic acid, carboxymethyl cellulose and many other chemicals [7]. The chemical itself has been used as a pre-emergent herbicide and defoliant [8], and these applications, if still in use, would constitute an emission source and ground contaminant of a more general nature.

Terrestrial Fate: When released on soil, chloroacetic acid will leach into the ground and biodegrade. While no rates of biodegradation in soil were found in the literature, the aqueous biodegradation literature suggests that it is a relatively rapid process.

Aquatic Fate: When released into water, chloroacetic acid will be mineralized (73% in 8-10 days). It will not adsorb appreciably to sediment.

Atmospheric Fate: If chloroacetic acid is used as a pesticide, it could possibly be released to the atmosphere during spraying and will generally be associated with aerosols and sprays. The aerosol will be subject to gravitational settling and undergo slow photodechlorination.

Biodegradation: Chloroacetic acid is degraded in laboratory biodegradation tests using sewage or acclimated sludge inocula with greater than 70-90% degradation being reported in 5-10 days [2,5,10,11]. Degradation is increased by acclimation [5,10] and involves dechlorination [5]. Mineralization occurs in river water with 73% of the chemical being converted to carbon dioxide in 8-10 days at 29 °C [1]. Degradation occurs in soil; however, under acidic conditions at low temperature chloroacetic acid is comparatively resistant [6].

Abiotic Degradation: Chloroacetic acid does not absorb UV radiation above 290 nm appreciably [3] and would therefore not directly photolyze. It very slowly photodechlorinates in air-saturated solutions with <0.4% being converted to free chloride when irradiated for 11 hr in a laboratory photoreactor [3]. The rate decreases after a few hours. Direct photodechlorination is much lower in the absence of oxygen [3]. The presence of sensitizers such as p-cresol and tryptophan that generate superoxide radicals increase the rate of photodechlorination by up to 16-fold [3]. Hydrolysis was negligible during the course of these experiments [3].

Bioconcentration: Chloroacetic acid has a very low log octanol/water partition coefficient, 0.22 [4], and therefore would not be expected to bioconcentrate in fish.

Soil Adsorption/Mobility: Chloroacetic acid has a very low log octanol/water partition coefficient, 0.22 [4], and therefore would not be expected to adsorb appreciably to soil.

Volatilization from Water/Soil: Chloroacetic acid has a pKa of 2.86 [9] and will be completely ionized at environmental pHs. Evaporation from water will therefore not be a significant loss process.

Water Concentrations:

Effluent Concentrations:

Sediment/Soil Concentrations:

Atmospheric Concentrations:

Food Survey Values:

Plant Concentrations:

Fish/Seafood Concentrations:

Chloroacetic Acid

Animal Concentrations:

Milk Concentrations:

Other Environmental Concentrations:

Probable Routes of Human Exposure:

Average Daily Intake:

Occupational Exposure:

Body Burdens:

REFERENCES

1. Boethling RS, Alexander M; Appl Environ Microbiol 37: 1211-6 (1979)
2. Dias FF, Alexander M; Appl Microbiol 22: 1114-8 (1971)
3. Draper WM, Crosby DG; J Agric Food Chem 31:734-7 (1983)
4. Hansch C, Leo AJ; Medchem Project Issue no 19; Pomona College Claremont CA (1981)
5. Jacobson SN, Alexander M; Appl Environ Microbiol 42: 1062-6 (1981)
6. Jensen HL; Tidsskr Planteavl 63: 470-99 (1959)
7. Kirk-Othmer Encyclopedia of Chemical Technology; 3rd ed Wiley New York 1:142-4 (1978)
8. Pesticide Manual British Crop Protection Manual; p.104 (1977)
9. Sergeant EP, Dempsey B; Ionization constants of organic acids in aqueous solutions; Pergamon New York p.989 (1979)
10. Thom NS, Agg AR; Proc Roy Soc London B 189: 347-57 (1975)
11. Zahn R, Wellens H; Z Wasser Abwasser Forschung 13: 1-7 (1980)

Chloroacetonitrile

SUBSTANCE IDENTIFICATION

Synonyms:

Structure:

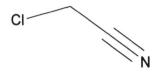

CAS Registry Number: 107-14-2

Molecular Formula: C_2H_2ClN

SMILES Notation: N#CCCl

CHEMICAL AND PHYSICAL PROPERTIES

Boiling Point:

Melting Point:

Molecular Weight: 75.50

Dissociation Constants:

Log Octanol/Water Partition Coefficient: 0.45 [5]

Water Solubility: 73,700 mg/L at 25 °C (estimated) [4]

Vapor Pressure: 15 mm Hg at 30 °C [14]

Henry's Law Constant: 1.08×10^{-5} atm-m³/mol (estimated) [8]

ENVIRONMENTAL FATE/EXPOSURE POTENTIAL

Summary: Chloroacetonitrile may be released to the environment in wastewater streams that have undergone chlorination treatment; chloroacetonitriles can be formed through chlorination of aquatic humic materials. If released to water, chloroacetonitrile can be transported to air through volatilization. Volatilization half-lives of 16 hr and 7.3 days have been estimated for a model river and environmental pond, respectively. If released to soil, chloroacetonitrile may leach readily based upon an estimated Koc value of 40. Surface evaporation may occur from soil surfaces. Insufficient data are available to predict the relative importance of biodegradation in soil or water. If released to the atmosphere, chloroacetonitrile will degrade slowly by reaction with photochemically produced hydroxyl radicals (half-life of 360 days). Physical removal from air via wet deposition is possible since it is relatively soluble in water; physical removal may be important since the degradation rate in air is slow. Occupational exposure to chloroacetonitrile occurs through inhalation of vapor and dermal contact. Exposure may occur through ingestion of contaminated drinking water.

Natural Sources:

Artificial Sources: Chloroacetonitrile has been detected in the wastewater effluent from the electronics industry [2]. Chloroacetonitrile's major uses as an organic intermediate in the manufacture of the insecticide fenoxycarb and the cardiovascular drug guanethidine [10] could result in its release in various manufacturing waste streams. Chloroacetonitriles have been detected in water treatment facilities that use chlorination treatment [11]; the presence of chloroacetonitriles in chlorinated natural water is the result of formation through chlorination of aquatic humic materials [11].

Terrestrial Fate: Using the log octanol-water partition coefficient, a Koc value of 42 can be estimated from a regression-derived equation [7]; according to a suggested classification scheme [12], this estimated Koc suggests that chloroacetonitrile will leach readily in soil. Insufficient data are available to predict the relative importance of biodegradation. Chloroacetonitrile's vapor pressure of 15 mm Hg at 30 °C [14] suggests that it will evaporate from dry surfaces.

Aquatic Fate: Volatilization may be an important transport process for chloroacetonitrile in water. Volatilization half-lives of 16 hr and 7.3 days have been estimated for a model river (1 m deep) and environmental pond (2 m deep), respectively [7,13]. Using the log octanol/water partition coefficient of 0.45 [5], Koc and BCF values of 42 and 1 can be estimated from regression-derived equations [7], which suggest that adsorption to sediment and bioconcentration are not important fate processes. Insufficient data are available to predict the relative importance of biodegradation.

Atmospheric Fate: Based upon the vapor pressure, chloroacetonitrile is expected to exist almost entirely in the vapor-phase in the ambient atmosphere [3]. It will degrade slowly in the ambient atmosphere by reaction with photochemically produced hydroxyl radicals (estimated half-life of about 360 days) [1]. Chloroacetonitrile's estimated solubility in water suggests that physical removal from air via wet deposition is possible; physical removal may be important since the degradation rate is slow.

Biodegradation:

Abiotic Degradation: The rate constant for the vapor-phase reaction of chloroacetonitrile with photochemically produced hydroxyl radicals has been estimated to be 4.46×10^{-14} cm^3/molecule-sec at 25 °C, which corresponds to an atmospheric half-life of about 360 days at an atmospheric concn of $5 \times 10^{+5}$ hydroxyl radicals per cm^3 [1].

Bioconcentration: Based upon the octanol/water coefficient, the BCF for chloroacetonitrile can be estimated to be about 1 from a regression-derived equation [7]. This BCF value suggests that bioconcentration in aquatic organisms is not environmentally important.

Soil Adsorption/Mobility: Based upon the octanol/water coefficient, the Koc for chloroacetonitrile can be estimated to be about 42 from a regression-derived equation [7]. According to a suggested classification scheme [12], this estimated Koc suggests that chloroacetonitrile is highly mobile in soil.

Volatilization from Water/Soil: The Henry's Law constant indicates that volatilization from environmental waters is possibly significant, but may not be rapid [7]. Based on the Henry's Law constant, the volatilization half-life from a model river (1 m deep flowing 1 m/sec with a wind velocity of 3 m/sec) can be estimated to be about 3 days [7]. Volatilization half-life from a model environmental pond (2 m deep) can be estimated to be about 33 days [13]. Chloroacetonitrile has a relatively high vapor pressure, which suggests that evaporation from dry surfaces could occur.

Water Concentrations: DRINKING WATER: Chloroacetonitrile was qualitatively detected in drinking water samples collected in Miami, FL, on February 3, 1976 [6].

Effluent Concentrations: Chloroacctonitrilc has bccn qualitativcly detected in a wastewater effluent collected from the electronics industry (source and sampling date not reported) [2].

Sediment/Soil Concentrations:

Atmospheric Concentrations:

Food Survey Values:

Plant Concentrations:

Fish/Seafood Concentrations:

Animal Concentrations:

Milk Concentrations:

Other Environmental Concentrations:

Probable Routes of Human Exposure: Occupational exposure to nitriles, such as chloroacetonitrile, can occur through inhalation of vapor and dermal contact [9]. Since chloroacetonitrile has been detected in drinking water samples [6], exposure could occur through consumption of water; the presence of chloroacetonitrile in drinking water may result from chlorination treatments [11].

Average Daily Intake:

Occupational Exposure:

Body Burdens:

REFERENCES

1. Atkinson R; J Inter Chem Kinet 19: 799-828 (1987)
2. Bursey JT, Pellizzari ED; Analysis of Industrial Wastewater for Organic Pollutants in Consent Decree Survey. Contract No. 68-03-2867. Athens, GA: USEPA Environ Res Lab pp. 79, 90, 92, 118 (1982)
3. Eisenreich SJ et al; Environ Sci Technol 15: 30-8 (1981)
4. GEMS; Graphical Exposure Modeling System. PCCHEM. USEPA (1987)
5. Hansch C, Leo AJ; Medchem Project Issue No 26. Claremont CA: Pomona College (1985)
6. Lucas SV; GC/MS Analysis of Organics in Drinking Water Concentrates and Advanced Waste Treatment Concentrates: Volume 1. USEPA-600/1-84-020A (NTIS PB85-128221) pp. 45, 145 (1984)
7. Lyman WJ et al; Handbook of Chemical Property Estimation Methods. Washington, DC: Amer Chem Soc p. 4-9, 5-4. 15-15 to 15-29 (1990)
8. Meylan WM, Howard PH; Environ Toxicol Chem 10: 1283-93 (1991)
9. Parmeggiani L; Encycl Occup Health & Safety 3rd ed. Geneva, Switzerland: International Labour Office pp. 1445-7 (1983)
10. Pollack P et al; p. 368 in Ullmann's Encycl Industr Chem A17 NY: VCH Publ (1991)
11. Singer PC, Chang SD; J Amer Water Works Assoc 81: 61-5 (1989)
12. Swann RL et al; Res Rev 85: 23 (1983)
13. USEPA; EXAMS II Computer Simulation (1987)
14. Weast RC; Handbook of Chemistry and Physics 66th ed. Boca Raton, FL: CRC Press p. C-204 (1985)

1-Chloro-2-propanone

SUBSTANCE IDENTIFICATION

Synonyms: 1-Chloroacetone

Structure:

CAS Registry Number: 78-95-5

Molecular Formula: C_3H_5ClO

SMILES Notation: O=C(CCl)C

CHEMICAL AND PHYSICAL PROPERTIES

Boiling Point: 119.7 °C

Melting Point: -44.5 °C

Molecular Weight: 92.53

Dissociation Constants:

Log Octanol/Water Partition Coefficient: 0.02 (estimated) [8]

Water Solubility: 100,000 mg/L [6]

Vapor Pressure: 23 mm Hg at 25 °C (estimated) [5]

Henry's Law Constant: 1.2 x 10^{-5} atm-m³/mol (estimated) [7]

ENVIRONMENTAL FATE/EXPOSURE POTENTIAL

Summary: 1-Chloro-2-propanone (chloroacetone) may be released to the environment in connection with its use as a chemical intermediate. Little is known about its fate in the environment. From its physical/chemical properties, one can predict that it will not be reactive in air, but may be removed by wet deposition. It will hydrolyze in water and possibly in moist soil. Photolysis in the environment, adsorption to soil/sediment or bioconcentration in aquatic organisms will not be important.

Natural Sources:

Artificial Sources: Possible release in wastewater is connected with its use as a chemical intermediate in the manufacture of perfumes, antioxidants, drugs, couplers for color photography, etc. [4].

Terrestrial Fate: Photolysis of chloroacetone in soil may not be important because it lacks chromophoric groups that can absorb sunlight. It may hydrolyze in moist soil because hydrolysis is important in water [1]. Due to its high water solubility, chloroacetone may readily leach from soil. From its estimated vapor pressure of 23 mm Hg at 25 °C [5] and Henry's Law constant value, it is estimated that volatilization may be important from dry soil but not from wet soil.

Aquatic Fate: Volatilization from water and adsorption to sediment may not be important transport processes for chloroacetone. Based on an estimation method [5], the volatilization half-live of 30 days can be estimated for chloroacetone from a model river (1 m deep) at a current and wind speed of 1 m/sec and 3 m/sec, respectively. Based upon its high water solubility [6], the partitioning of chloroacetone from the water column to sediment and suspended materials will not be important. It will hydrolyze in water with the production of hydrochloric acid [1]. Insufficient data are available to predict the importance of biodegradation. Because of its high water solubility [6], bioconcentration in aquatic organisms will not be important.

Atmospheric Fate: Based upon an estimated vapor pressure of 23 mm Hg at 25 °C [5], chloroacetone is expected to exist almost entirely in the vapor phase in the ambient atmosphere [3]. Its reaction with photochemically produced hydroxyl radicals in the atmosphere will be slow (estimated half-life of 44 days) [2]; however, its removal from the atmosphere by wet deposition may be important because of high water solubility [6].

1-Chloro-2-propanone

Biodegradation:

Abiotic Degradation: No information could be found on the abiotic degradation of chloroacetone; however, the rate constant for the vapor phase reaction of chloroacetone with hydroxyl radicals in the atmosphere is estimated to be 3.68×10^{-13} m^3/molecule-sec by an estimation method [2]. This rate constant corresponds to a half-life of 44 days at an atmospheric hydroxyl radical concentration of 5×10^5 per m^3. Since this chemical does not contain a chromophore, which absorbs light at wavelengths >290 nm [5], it would not be expected to directly photolyze. It will hydrolyze, releasing HCl, which is related to its irritation as a lachrymator [1].

Bioconcentration: Chloroacetone would not be expected to bioconcentrate because of its very low estimated log Kow (0.02) [8].

Soil Adsorption/Mobility The adsorption of chloroacetone to soil or sediment would not be important due to its high solubility in water (1g/10g water) [5,6].

Volatilization from Water/Soil: Since chloroacetone has an estimated vapor pressure of 23 mm Hg at 25 °C [5], it would be expected to volatilize from dry soil. The estimated Henry's Law constant of 1.2×10^5 atm-m^3/mol [7] indicates that volatilization from water and moist soil may not be rapid [5]. Based on an estimation method [7], the volatilization half-life of chloroacetone from a river of depth 1 m flowing at a current velocity of 1 m/sec and a wind speed of 3 m/sec would be about 30 days.

Water Concentrations:

Effluent Concentrations:

Sediment/Soil Concentrations:

Atmospheric Concentrations: RURAL/REMOTE: Jones State Forest, TX, 1.0-2.6 ppb [9].

Food Survey Values:

Plant Concentrations:

81

1-Chloro-2-propanone

Fish/Seafood Concentrations:

Animal Concentrations:

Milk Concentrations:

Other Environmental Concentrations:

Probable Routes of Human Exposure:

Average Daily Intake:

Probable Exposure:

Body Burdens:

REFERENCES

1. Allinger NL et al; Organic Chemistry; Worth Publ New York NY pp.500 (1973)
2. Atkinson RA; J Internat Chem Kinet 19: 799-828 (1987)
3. Eisenreich SJ et al; Environ Sci Technol 15: 30-8 (1981)
4. Hawley GG; The Condense Chem Dictionary 10th ed NY: Von Nostrand Reinhold p. 32 (1981)
5. Lyman WJ et al; Handbook of Chemical Property Estimation Methods. Environmental Behavior of Organic Compounds, Washington, DC: American Chemical Society. pp.4-1 to 4-33, 8-1 to 8-43, 14-1 to 14-20, 15-1 to 15-34 (1990)
6. Merck Index; 10th ed p 297 (1983)
7. Meylan WM, Howard PH; Environ Toxicol Chem 10: 283-93 (1991)
8. Meylan WM, Howard PH; J Pharm Sci 84: 83-92 (1995)
9. Seila RL; Non-urban Hydrocarbon Concentrations in Ambient Air North of Houston TX p.38 USEPA-500/3-79-010 (1979)

2-Chloro-1,1,1,2-tetrafluoroethane

SUBSTANCE IDENTIFICATION

Synonyms:

Structure:

CAS Registry Number: 2837-89-0

Molecular Formula: C_2HClF_4

SMILES Notation: ClC(F)C(F)(F)F

CHEMICAL AND PHYSICAL PROPERTIES

Boiling Point: -9.93 °C (estimated)

Melting Point: -130.31 °C (estimated)

Molecular Weight: 136.48

Dissociation Constants:

Log Octanol/Water Partition Coefficient: 1.867 [7] (estimated)

Water Solubility: 404 mg/L at 25 °C [6] (estimated from Kow).

Vapor Pressure: 2600 mm Hg at 25 °C [10] (estimated)

Henry's Law Constant: 0.540 atm-m^3/mol at 25 °C [8]

ENVIRONMENTAL FATE/EXPOSURE POTENTIAL

Summary: 2-Chloro-1,1,1,2-tetrafluoroethane is an anthropogenic compound that holds promise as an alternative to chlorofluorocarbons (CFCs). It may be released to the environment as a fugitive emission during its production or use. If released to soil, 2-chloro-1,1,1,2-tetrafluoroethane will rapidly volatilize from both moist and dry soil to the atmosphere. It will display moderate mobility in soil. If released to water, 2-chloro-1,1,1,2-tetrafluoroethane will rapidly volatilize to the atmosphere. The estimated half-life for volatilization from a model river is 3.4 hr. 2-Chloro-1,1,1,2-tetrafluoroethane will not bioconcentrate in fish and aquatic organisms nor will it adsorb to sediment or suspended organic matter. If released to the atmosphere, 2-chloro-1,1,1,2-tetrafluoroethane will undergo a slow gas-phase reaction with photochemically produced hydroxyl radicals with an estimated half-life of 1573 days. The atmospheric lifetime of 2-chloro-1,1,1,2-tetrafluoroethane has been estimated to range from 5.3-10 years. 2-Chloro-1,1,1,2-tetrafluoroethane may undergo atmospheric removal by wet deposition processes; however, any removed is expected to rapidly re-volatilize to the atmosphere. Occupational exposure to 2-chloro-1,1,1,2-tetrafluoroethane may occur by inhalation or dermal contact during its production or use.

Natural Sources: 2-Chloro-1,1,1,2-tetrafluoroethane is of anthropogenic origin and it is not known to be produced by natural sources.

Artificial Sources: 2-Chloro-1,1,1,2-tetrafluoroethane is used in the production of hexafluoropropylene via a high-temperature co-pyrolysis with difluoromethane [4]. It may also be used as a replacement for chlorofluorocarbons (CFCs) [5]; if so, it may be released to the environment as a fugitive emission during its production and use.

Terrestrial Fate: If released to soil, the estimated vapor pressure indicates that it will rapidly volatilize from dry soil to the atmosphere. Estimated soil adsorption coefficients ranging from 154-200 [6,9] indicate that it will display moderate mobility in soil [11]. The estimated Henry's Law constant indicates that it will also rapidly volatilize from moist soil to the atmosphere.

Aquatic Fate: If released to water, the estimated Henry's Law constant indicates that it will rapidly volatilize to the atmosphere. The estimated half-life for volatilization from a model river 1 m deep flowing at 1 m/sec with a wind speed of 3 m/sec is 3.4 hr [6]. An estimated bioconcentration factor of 15 [6] indicates that 2-chloro-1,1,1,2-tetrafluoroethane will not bioconcentrate in fish and aquatic organisms. Estimated soil adsorption coefficients ranging from 154 to 200 [6,9] indicate that it will not adsorb to sediment or suspended organic matter.

Atmospheric Fate: If released to the atmosphere, 2-chloro-1,1,1,2-tetrafluoroethane will undergo a slow gas-phase reaction with photochemically produced hydroxyl radicals. The recommended rate constant for this process of 1.02×10^{-14} cm^3/molecule-sec [1] translates to an atmospheric half-life of 1573 days using an average atmospheric hydroxyl radical concn of $5 \times 10^{+5}$ molec/cm^3 [1]. The atmospheric lifetime of 2-chloro-1,1,1,2-tetrafluoroethane, calculated using both one- and two-dimensional models, ranges from 5.3 to 10 yrs [4]. The estimated water solubility of 2-chloro-1,1,1,2-tetrafluoroethane indicates that it may undergo atmospheric removal by wet deposition processes; however, any removed is expected to rapidly re-volatilize to the atmosphere.

Biodegradation:

Abiotic Degradation: Experimental rate constants for the gas-phase reaction of 2-chloro-1,1,1,2-tetrafluoroethane with photochemically produced hydroxyl radicals of 1.24×10^{-14} cm^3/molecule-sec at 296 K [2], 4.33×10^{-15} cm^3/molecule-sec at 250 K [2], 9.4×10^{-15} cm^3/molecule-sec at 296 K [2], and 1.23×10^{-14} cm^3/molecule-sec at 298 K [3] have been reported. The recommended value of 1.02×10^{-14} cm^3/molecule-sec [1] translates to an atmospheric half-life of 1573 days using an average atmospheric hydroxyl radical concn of $5 \times 10^{+5}$ molec/cm^3 [1]. The atmospheric lifetime of 2-chloro-1,1,1,2-tetrafluoroethane, calculated using both one- and two- dimensional models, ranges from 5.3 to 10 yrs [4].

Bioconcentration: An estimated bioconcentration factor of 15 can be calculated for 2-chloro-1,1,1,2-tetrafluoroethane based on its estimated log octanol/water partition coefficient using appropriate regression equations [6]. This value indicates that 2-chloro-1,1,1,2-tetrafluoroethane will not bioconcentrate in fish and aquatic organisms.

Soil Adsorption/Mobility: Estimated soil adsorption coefficients ranging from 154 to 200 can be calculated for 2-chloro-1,1,1,2-tetrafluoroethane based on a structure estimation method [9] or on the estimated log octanol/water partition coefficient using appropriate regression equations

[6]. These values indicate that 2-chloro-1,1,1,2-tetrafluoroethane will display moderate mobility in soil [11].

Volatilization from Water/Soil: The estimated Henry's Law constant indicates that 2-chloro-1,1,1,2-tetrafluoroethane will rapidly volatilize from water and moist soil to the atmosphere. The estimated half-life for volatilization from a model river 1 m deep flowing at 1 m/sec with a wind speed of 3 m/sec is 3.4 hr [6]. The estimated vapor pressure of 2-chloro-1,1,1,2-tetrafluoroethane indicates that it will rapidly volatilize from dry soil to the atmosphere.

Water Concentrations

Effluent Concentrations:

Sediment/Soil Concentrations:

Atmospheric Concentrations:

Food Survey Values:

Plant Concentrations:

Fish/Seafood Concentrations:

Animal Concentrations:

Milk Concentrations:

Other Environmental Concentrations:

Probable Routes of Human Exposure: Occupational exposure to 2-chloro-1,1,1,2-tetrafluoroethane may occur by inhalation or dermal contact during its production or use.

Average Daily Intake:

Occupational Exposure:

2-Chloro-1,1,1,2-tetrafluoroethane

Body Burdens:

REFERENCES

1. Atkinson R; J Chem Phys Ref Data Monograph 1 (1989)
2. Atkinson R; Chem Rev 85: 69-201 (1985)
3. Cohen N, Benson SW; J Phys Chem 91: 171-5 (1987)
4. Fisher DA et al; Nature 344: 508-12 (1990)
5. Gangal SV; Kirk-Othmer Encyl Chem Tech 3rd 11: NY, NY Wiley 25 (1980)
6. Lyman WJ et al; Handbook of Chemical Property Estimation Methods NY: McGraw-Hill (1982)
7. Meylan W, Howard PH; J Pharm Sci 84: 83-92 (1995)
8. Meylan WM, Howard PH; Environ Toxicol Chem 10: 1283-93 (1991)
9. Meylan WM et al; Environ Sci Technol 26:1560-1567 (1992)
10. SRC; MPBPVP Program from Syracuse Research Corp. (1995)
11. Swann RL et al; Res Rev 85: 17-28 (1983)

Cumene

SUBSTANCE IDENTIFICATION

Synonyms: Isopropylbenzene

Structure:

CAS Registry Number: 98-82-8

Molecular Formula: C_9H_{12}

SMILES Notation: c(cccc1)(c1)C(C)C

CHEMICAL AND PHYSICAL PROPERTIES

Boiling Point: 152.4 °C

Melting Point: -96.0 °C

Molecular Weight: 120.19

Dissociation Constants:

Log Octanol/Water Partition Coefficient: 3.66 [28]

Water Solubility: 73 mg/L [84]

Vapor Pressure: 4.50 mm Hg at 25 °C [17]

Henry's Law Constant: 0.0097 atm-m^3/mol (estimated from vapor pressure and water solubility data)

Cumene

ENVIRONMENTAL FATE/EXPOSURE POTENTIAL

Summary: Cumene is released to the environment as a result of its production and processing, from petroleum refining and the evaporation and combustion of petroleum products and by the use of a variety of products containing cumene. A variety of natural products contain cumene. When released to soil, cumene is expected to biodegrade and may volatilize from the soil surface. Cumene is expected to strongly adsorb to soils and is not expected to leach to ground water. When released to water, cumene is expected to volatilize with an estimated half-life of 5-14 days and to biodegrade rapidly. Compared to these processes, aqueous photooxidation by hydroxyl radicals (estimated half-life 0.7 years) and peroxy radicals (estimated half-life 2.2 years) are expected to be relatively slow, and so are not expected to be significant fate processes of cumene. Adsorption to sediments may occur based on the high soil-sorption coefficient of cumene. Bioconcentration is not expected to be significant. When released to the atmosphere, vapor phase cumene will react with photochemically generated hydroxyl radicals with an estimated half-life of 25 hr in polluted atmospheres and 49 hr in normal atmospheres. The reaction of vapor phase cumene with ozone has an estimated half-life of 3 years and the half-life of direct photolysis was estimated to be 1500 years. Cumene is a contaminant of air, sediments and surface, drinking and ground water and a natural constituent of a variety of foods and vegetation. Human exposure to cumene is expected to result primarily from inspiration of air contaminated with cumene, but lower exposures may result from ingestion of food and water.

Natural Sources: Cumene occurs in a variety of natural substances including essential oils from plants [73], marsh grasses [53,54], and a variety of foodstuffs [14,18,29,30,43,65,70,74,83].

Artificial Sources: About 98% of cumene produced in the US is used to produce acetone and phenol [11]. Cumene is released by manufacturing and processing plants and during the transport of cumene. Cumene is also a constituent of crude oil and finished fuels [2]. It is, therefore, released to the environment by oil spills and the incomplete combustion of fossil fuels by land transportation vehicles. It is also released during the transportation and distribution of motor fuels and by evaporative loss from gasoline stations. Cigarette tobacco also releases cumene during consumption [34]. Cumene release from all these sources was estimated to be 21 million pounds annually [32]. Other, unquantifiable anthropogenic cumene releases include operations involving vulcanization of rubber [13], building materials [55], jet engine exhaust [36], outboard motor operation [56], solvent uses [40], paint manufacture [6], pharmaceutical production [8], and textile plants [24]. Cumene is also released to the environment from leather tanning, iron and steel manufacturing, paving and roofing, paint and ink formulation, printing and publishing, ore mining, coal mining, organics and plastics manufacturing, pesticide manufacturing, electroplating and pulp and paper production [67].

Terrestrial Fate: Cumene is expected to biodegrade in soil and may volatilize from moist and dry soil surfaces based on a vapor pressure of 4.50 mm Hg at 25 °C [17] and a Henry's Law constant of 0.0116 atm-m^3/mol (estimated from vapor pressure and water solubility data). Based on an estimated Koc value of 3.45 [44], cumene is expected to strongly adsorb to soils and is not expected to leach to ground water [38].

Aquatic Fate: Half-lives of 5-14 days were estimated for the volatilization of cumene from water, depending on the turbulence, wind speed, current velocity, depth, diffusion and temperature [22]. Cumene is expected to biodegrade fairly rapidly in water. Photooxidation by alkylperoxy radicals and hydroxy radicals is expected to occur with half-lives of 2.2 and 0.7 years [50,51], respectively, so photooxidation in water is not expected to be a significant fate process relative to biodegradation. Bioconcentration is not expected to be significant.

Atmospheric Fate: A half-life of about 25 hr for the reaction between cumene and hydroxyl radicals was estimated for polluted atmospheres; for average atmospheres, a half-life of about 49 hr was estimated [32]. The rate of reaction of cumene with ozone is considerably slower and a half-life of about 3 years was estimated [32]. Since cumene has an absorption maximum at 258 nm in cyclohexane [64], and has been theoretically estimated to have a direct photolysis half-life of about 1500 years [61], direct photolysis is not expected to be significant in the atmosphere. The predominant atmospheric fate of vapor phase cumene is expected to be its reaction with photochemically generated hydroxyl radicals.

Biodegradation: Cumene was added to uncontaminated ground water and this solution was continuously percolated through a column containing homogeneous sand [35]. No additional nutrients were added to the system and it was assumed to be aerobic throughout. Biodegradation occurred following 5 days of acclimation and the cumene degraded to nondetectable levels within 48 hr [35]. Cumene biodegradation in batch reactors proceeded following about 144 hr of acclimation, and after 120 hr, cumene was nondetectable [80]. In unacclimated cultures, the half-life of cumene was 206 hr (62 hr after acclimation) [80]. Mixed cultures from contaminated and uncontaminated estuarine sediments were capable of degrading cumene with the higher rate observed in the culture from the contaminated sediment [80]. Cumene incubated with mixed cultures taken from various depths in the Atlantic Ocean at 15 °C were all capable of biodegrading cumene [81]. When cumene was incubated with an activated sludge acclimated to benzene, the theoretical BOD was reduced by 37.8% after 192 hr [47]. A 20-day biochemical oxygen demand study was conducted using unacclimated, settled, domestic wastewater as the inoculum [62]. After 10 and 20 days, the theoretical BOD was 62% and 70%, respectively [62]. Activated sludge acclimated to aniline degraded cumene following an acclimation period of about 30 hr [46]. Activated sludge samples from three different communities were able to degrade 50 mg/L cumene [49]. Incubation with a pure culture of *Pseudomonas putida* resulted in the degradation of cumene to an orthodihydroxy compound in which the isopropyl side chain of cumene was intact [23]. Incubation of cumene with *Pseudomonas desmolytica* and *Pseudomonas convexa* resulted in the formation of (+)-2-hydroxy-7-methyl-6-oxo-octanoic acid via 3-isopropylcatechol [33].

Abiotic Degradation: An absolute rate constant for the reaction of cumene with hydroxyl radicals of 7.79 x 10^{-12} cm³/molecule-sec was determined at 200 torr and 298 K using flash photolysis-resonance fluorescence [63]. Using a hydroxyl radical concentration of 1 x 10^{+6} molecule/cm³ for polluted atmospheres [32], a half-life of about 25 hr for the reaction between cumene and hydroxyl radicals was estimated [32]. Using a hydroxyl radical concentration of 5 x 10^{+5} molecule/cm³ for average atmospheres [32], a half-life of about 49 hr was estimated [32]. The major product of the reaction of it with hydroxyl radicals is likely to be isopropylphenols with minor compounds resulting from side chain attack [32] for aromatic hydrocarbons in general; a rate constant for the reaction with ozone is 1 x 10^{-20} cm³/molecule-sec [4]. Using an ozone concentration of 7 x 10^{+11} molecule/cm³ [4], a half-life for the reaction of cumene with ozone of about 3 years was estimated [32]. Since cumene has an absorption maximum at 258 nm in cyclohexane [64], and has been theoretically estimated to have a direct photolysis half-life of about 1500 years [61], direct photolysis is not expected to be significant in the atmosphere. Rate constants for the aqueous photooxidation of cumene by alkylperoxy and hydroxyl radicals were 10 1/M-sec and $3 \times 10^{+9}$ 1/M-sec, respectively [50]. Using alkylperoxy and hydroxyl radical concentrations of 1×10^9 M and 1×10^{-17} M [51], respectively, half-lives for the reaction of cumene with these two species were calculated to be about 2.2 and 0.7 years, respectively [32].

Bioconcentration: A bioconcentration factor (BCF) of about 35.5 was measured in goldfish that were exposed to cumene at 1 mg/L [59]. A BCF of this magnitude suggests that cumene will not significantly bioconcentrate in fish.

Soil Adsorption/Mobility: Based on a log octanol/water partition coefficient of 3.66 [28], a log soil-sorption coefficient (Koc) of 3.45 was estimated [44]. A Koc of this magnitude suggests that cumene will be strongly adsorbed to soil [38] and will, therefore, resist leaching to ground water.

Volatilization from Water/Soil: The water solubility of cumene is 73 mg/L [84] and the vapor pressure is 4.50 mm Hg at 25 °C [17]. These values were used to estimate a Henry's Law constant of 0.0097 atm-m^3/mol [44]. Using the Henry's Law constant and a model river 1 m in depth with a current of 1 m/sec and a wind speed of 1 m/sec, the volatilization half-life is 3.3 hr [44].

Water Concentrations: SURFACE WATER: River Lee, UK - <0.1 and >0.1 ug/L at two sampling points [78]. Cumene was detected but not quantified in surface water samples from Narraganset Bay, RI [79], and Japan [1]. GROUND WATER: Hoe Creek, WV (near underground coal gasification site) - 3 samples, 100% pos; 19, 27, and 59 ug/L, avg 35 ug/L [71]. United States (all 50 states) and Puerto Rico - <0.5 ug/L [82]. Great Ouse Basin, UK (near a gasoline storage tank) - 5 samples, 100% pos, 0.01-30 ug/L, avg 9.8 ug/L [75]. Cumene was detected but not quantified in ground water from Ames, IA [10], New York state [9], Melbourne, Australia (near a dump site) [69], and Milan, Italy (near underground storage tanks) [6]. DRINKING WATER: US cities [13] - 10 samples, 7.7%

pos, 0.01 ug/l [37]. Cincinnati, OH - 0.014 ug/L [15]. United States (all 50 states) and Puerto Rico - <0.5 ug/L [82]. Cumene was detected but not quantified in drinking water from New York state [9] and in tap water from Japan [68].

Effluent Concentrations: WATER: Timber products - 36-6319 ug/L, 228 ug/L median; leather tanning - 192 ug/L; iron and steel manufacturing - 17-18 ug/L, 18 ug/L median; petroleum refining - 13-1316 ug/L, 91 ug/L median; paving and roofing - 48 ug/L; paint and ink - 8-2621 ug/L, 168 ug/L median; printing and publishing - 25-739 ug/L, 41 ug/L median; ore mining - 42 ug/L; coal mining - 4-1646 ug/L, 50 ug/L median; organics and plastics - 4.9-17,933 ug/L, 85 ug/L median; inorganic chemicals - 14-606 ug/L, 109 ug/L median; textile mills - 112 ug/L; plastics and synthetics - 1.2- 57 ug/L, 4 ug/L median; pulp and paper - 8.4-341 ug/L, 47 ug/L median, rubber processing - 32-867 ug/L, 449 ug/L median; auto and other laundries - 35-3925 ug/L, 329 ug/L median; pesticide manufacturing - 217-1753 ug/L, 857 ug/L median; pharmaceuticals - 24 ug/L; plastics manufacturing - 112-1576 ug/L, 384 ug/L median; electroplating - 3.8 ug/L; oil and gas extraction - 3.1-334 ug/L, 6.8 ug/L median; organic chemicals - 21-328 ug/L, 63 ug/L median; mechanical products - 44-1766 ug/L, 259 ug/L median; transportation equipment - 21 ug/L [67].

Sediment/Soil Concentrations: SEDIMENT: Washington state (Puget Sound and Strait of Juan de Fuca) - 23 samples, 16% pos, 0.02-19 ug/g, 2.3 ug/g [8]. Cumene was detected but not quantified in Puget Sound, WA [48].

Atmospheric Concentrations: Los Angeles, CA - 17 samples, 94% pos, ND-9.8 ug/m^3 [26]; 136 samples, 100% pos, 144 ug/m^3 max, 14.7 ug/m^3 mean [41]; 10 samples, 80% pos, <2.45-36 ug/m^3, 16.66 ug/m^3 mean [57]. Houston, TX - 21 samples, 88% pos, ND-24.89 ug/m^3, 12.15 ug/m^3 mean [42]. Lake Michigan (1000-3000 ft altitude) - 2 samples, 100% pos, 0.49 ug/ m^3 [52]. Jones State Forest, TX (near Houston) - 15 samples, 100% pos, 0.108-9.8 ug/m^3, 2.45 ug/m^3 mean [66]. Smokey Mountains National Park, TN (near campfires) - 9 samples, 44% pos, <0.049-0.392 ug/m^3, 0.245 ug/m^3 mean [3]. Deer Park, TX (Shell Oil Refinery) - 29.4 ug/m^3, downwind, 53.9 ug/m^3, upwind [60]. Delft (the Netherlands) - <0.49-1.96 ug/m^3 [5]. Cumene was detected but not quantified in air samples taken in Elizabeth, Newark, Batsoto, and South Amboy, NY [7], Pullman, WA [58],

Allegheny Mountain Tunnel, PA [27], Leningrad, USSR [31], and Gatwick, UK [76].

Food Survey Values: Trace quantities of cumene have been detected in papaya [20], Sapodilla fruit [45], and Australian honey [25]. Cumene has been detected but not quantified in fried chicken [74], tomatoes [65], Concord grapes [70], cooked rice [83], oat groats [29], baked potatoes [14], Beaufort cheese [18], fried bacon [30], dried legumes (beans, split peas and lentils) [43], southern pea seeds [19], and Zinfandel wine [70].

Plant Concentrations: Cumene occurred in a variety of marsh grass species at 6.0×10^{-5} to 5.0×10^{-4} of the total fresh plant [53,54]. Cumene has been detected but not quantified in curly parsley [77], and oakmoss [21,72].

Fish/Seafood Concentrations:

Animal Concentrations:

Milk Concentrations:

Other Environmental Concentrations:

Probable Routes of Human Exposure: The most probable route of human exposure to cumene is exposure to air contaminated with the chemical from evaporation of petroleum products i.e., in individuals engaged in pumping gas or due to combustion of petroleum products or tobacco [32]. Additional exposure may result from food consumption. Little exposure is expected to result from water intake.

Average Daily Intake:

Occupational Exposure: Work area monitoring samples from cumene producers and processors were as follows. Distillation - 0.0001-3.35 ppm, 0.45 ppm mean; oxidation - 0.0001-5.58 ppm, 0.93 ppm mean; laboratory - 0.34-0.44 ppm, 0.39 ppm mean; repair - 0.16-2.50 ppm, 1.33 ppm mean; recovery - 0.001-1.20 ppm, 0.31 ppm mean; cumene unit - 0.078-0.620 ppm, 0.189 ppm mean [12]. Gasoline delivery truck drivers are exposed to air containing from <0.01-0.04 ppm cumene [2]. Cumene levels were

Cumene

60-250 ug/m^3 in shoe factory air and 2-200 ug/m^3 in the vulcanization area and ND-10 ug/m^3 in the extrusion area of a tire retreading plant [13]. No information was available to indicate whether or not these values are typical of the tire retreading industry.

Body Burdens: Cumene has been detected at 0.13 ug/hr [16] and detected but not quantified [39] in human expired air from nonsmoking individuals.

REFERENCES

1. Akiyama T et al; J UOEH 2: 285-300 (1980)
2. American Petroleum Institute; Letter to TSCA Interagency Testing Committee USEPA (1984)
3. Arnts RR, Meels SA; Biogenic Hydrocarbon Contribution to the Ambient Air of Selected Areas; Tulsa, Great Smokey Mountains, Rio Blanco County, CO USEPA-600/3-80-023 (1980)
4. Atkinson R, Carter WPL; Chem Rev 84: 437-70 (1984)
5. Bos R et al; Sci Total Environ 7: 269-81 (1977)
6. Botta D et al; Ground Water Pollut by Org Solvents and Their Microbial Degradation Products, Comm Eur Communities (Rep) EUR, EUR 8518, Anal Org Micropollut Water pp 261-75 (1984)
7. Bozzelli JW et al; Analysis of Selected Toxic and Carcinogenic Substances in Ambient Air in New Jersey, Office of Cancer and Toxic Substances Research, New Jersey Department of Environ Protection (1980)
8. Brown JM et al; Investigation of Petroleum in the Marine Environs of the Strait of Juan de Fuca and Northern Puget Sound USEPA-600/7-79-164 (1979)
9. Burmaster DE; Environ 24: 6-36 (1982)
10. Burnham AK et al; Anal Chem 44: 139-42 (1972)
11. Chemical Economics Handbook, Cumene, Stanford Research Institute International Menlo Park, CA (1984)
12. Chemical Manufacturers Association; Cumene Program Panel: Industrial Hygiene survey (1985)
13. Cocheo V et al; Amer Ind Hyg Assoc J 44: 521-7 (1983)
14. Coleman EC et al; J Agric Food Chem 29: 42-8 (1981)
15. Coleman WE et al; Arch Environ Contam Toxicol 13: 171-8 (1984)
16. Conkle JP et al; Arch Environ Health 30: 290-5 (1975)
17. Daubert TE, Danner RP; Data Compilation Tables of Properties of Pure Compounds. Amer Inst of Chem Engineers p. 450 (1985)
18. Dumont JP, Adda J; J Agric Food Chem 26: 364-7 (1978)
19. Fisher GS et al; J Agric Food Chem 27: 7-11 (1977)
20. Flath RA, Forrey RR; J Agric Food Chem 25: 103-9 (1977)
21. Gavin J et al; Helv Chim Acta 61: 352-7 (1978)
22. GEMS; Graphical Exposure Modeling System. EXAMS II (1986)

23. Gibson DT; Science 161: 1093-7 (1968)
24. Gordon AW, Gordon M; Trans Kentucky Acad Sci 42: 149-57 (1981)
25. Graddon AD et al; J Agric Food Chem 27: 832-7 (1979)
26. Grosjean D, Fung K; J Air Pollut Cont Assoc 34: 537-43 (1984)
27. Hampton CV et al; Environ Sci Technol 16: 287-98 (1982)
28. Hansch C et al; Exploring QSAR: Hydrophobic, Electronic, and Steric Constants. ACS Professional Reference Book, Amer Chem Soc, Washington, DC (1995)
29. Heydanek MG, McGorrin RJ; J Agric Food Chem 29: 950-4 (1981)
30. Ho CT et al; J Agric Food Chem 31: 336-42 (1983)
31. Ioffe BV et al; Environ Sci Technol 13: 864-8 (1979)
32. Jackson J et al; Test Rule Support Document Cumene Syracuse Res Corp pp 170 SRC-TR-85-098 (1985)
33. Jigami Y et al; Agric Biol Chem 39: 1781-8 (1975)
34. Johnstone RAW et al; Nature 195: 1267-69 (1962)
35. Kappeler T, Wuhrmann K; Water Res 12: 335-42 (1978)
36. Katzman H, Libby WF; Atmos Environ 9: 839-42 (1975)
37. Keith LH et al; pp 329-63 in Identification and Analysis of Org Pollut in Water Keith LH ed Ann Arbor, MI (1976)
38. Kenaga EE; Ecotox Env Safety 4: 26-38 (1980)
39. Krotoszynski BK, O'Neill HJ; J Environ Sci Health Part A Environ Sci Eng 17: 855-83 (1982)
40. Levy A; Advances in Chem Ser 124 American Chem Soc pp 70-94 (1973)
41. Lonneman WA et al; Environ Sci Technol 2: 1017-20 (1968)
42. Lonneman WA et al; Hydrocarbons in Houston Air, USEPA-600/3-79-018 (1979)
43. Lovegren NV et al; J Agric Food Chem 27: 851-3 (1979)
44. Lyman WJ et al; Handbook of Chemical Property Estimation Methods. Environmental Behavior of Organic Compounds. McGraw-Hill NY p 4-9 (1990)
45. Macleod AJ, Gonzales de Troconis N; J Agric Food Chem 30: 515-7 (1982)
46. Malaney GW; J Water Pollut Cont Fed 32: 1300-11 (1960)
47. Malaney GW, McKinney RE; Water Sewage Works 113: 302-9 (1966)
48. Malins DC et al; Environ Sci Technol 18: 705-13 (1984)
49. Marion CV, Malaney GW; Proc Indus Waste Conf 18: 297-308 (1964)
50. Mill T et al; Science 207: 886-7 (1980)
51. Mill T et al; Test Protocols for Evaluating the Fate of Organic Chemicals in Air and Water, Final Report, Office of Research and Development USEPA (1980)
52. Miller DF, Alkezweeny AJ; Ann NY Acad Sci 338: 219-32 (1980)
53. Mody NV et al; Phytochem 13: 1175-8 (1974)
54. Mody NV et al; Phytochem 14: 599-601 (1975)
55. Moelhave L; Proc First Int Indoor Clim Symp pp 89-110 (1979)
56. Montz WE Jr et al; Arch Environ Contam Toxicol 11: 561-65 (1982)
57. Neligan RE et al; pp 118-21 in ACS Nat Meeting: 118-21 (1965)
58. Nutmagul W et al; Anal Chem 55: 2160-64 (1983)
59. Ogata M et al; Bull Environ Contam Toxicol 33: 561-7 (1984)
60. Oldham RG et al; in Proc Spec Conf Cont Specific (Toxic) Pollut, Frederick ER ed, Air Pollution Control Association Pittsburgh, PA (1979)

Cumene

61. Parlar H et al; Fresenius Z Anal Chem 315: 605-9 (1983)
62. Price KS et al; J Water Pollut Cont Fed 46: 63-77 (1974)
63. Ravishankara AR et al; Int J Chem Kinetics 10: 783-804 (1978)
64. Sadtler; UV (1960)
65. Schormueller J, Kochmann HJ; Z Lebensm Unters Forsch 141: 1-9 (1969)
66. Seila RL; Non-urban Hydrocarbon Concentrations in Ambient Air North of Houston, TX USEPA-600/3-79-010 (1979)
67. Shackelford WM et al; Anal Chim Acta 146: 15-27 (1983)
68. Shiraishi H et al; Environ Sci Technol 19: 585-9 (1985)
69. Stepan S et al; Aust Water Res Coun Conf Ser 1: 415-24 (1981)
70. Stern DJ et al; J Agric Food Chem 15: 1100-3 (1967)
71. Steurmer DH et al; Environ Sci Technol 16: 582-87 (1982)
72. Tabacchi R, Nicollier G; Int Cong Essen Oils Oct 7-11, 1977 Kyoto, Japan (1979)
73. Tajuddin SAS et al; Indian Perfum 27: 56-9 (1983)
74. Tang J et al; J Agric Food Chem 31: 1287-92 (1983)
75. Tester DJ, Harker RJ; Water Pollut Cont 80: 614-31 (1981)
76. Tsani-Bazaca E et al; Chemosphere 11: 11-23 (1982)
77. Vernon F, Richard HMJ; Lebensum-Wiss Technol 16: 32-5 (1983)
78. Waggot A; pp 55-99 in Chem Water Reuse Vol. 2 Cooper WJ ed Ann Arbor, MI (1981)
79. Wakeham SG et al; Can J Fish Aquatic Sci 40: 304-21 (1983)
80. Walker JD, Colwell RR; J Gen Appl Microbiol 21: 27-39 (1975)
81. Walker JD et al; Mar Biol 34: 1-9 (1976)
82. Westrick JJ et al; J Am Water Works Assoc 76: 52-9 (1984)
83. Yajima I et al; Agric Biol Chem 42: 1229-33 (1978)
84. Yalkowsky SH, Dannenfelser RM; The AQUASOL Database of Aqueous Solubility. Fifth ed, Tucson, AZ: Univ AZ, College of Pharmacy (1992)

Cyclododecane

SUBSTANCE IDENTIFICATION

Synonyms:

Structure:

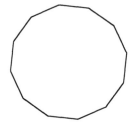

CAS Registry Number: 294-62-2

Molecular Formula: $C_{12}H_{24}$

SMILES Notation: C(CCCCCCCCC1)C1

CHEMICAL AND PHYSICAL PROPERTIES

Boiling Point:

Melting Point:

Molecular Weight: 168.13

Dissociation Constants:

Log Octanol/Water Partition Coefficient: 6.12 (estimated) [10]

Water Solubility: 0.005 mg/L (estimated from the ratio of vapor pressure and the Henry's Law constant)

Vapor Pressure: 0.0295 mm Hg at 25 °C (extrapolated) [2]

Henry's Law Constant: 1.40 atm-m^3/mol [9]

ENVIRONMENTAL FATE/EXPOSURE POTENTIAL

Summary: Cyclododecane may be released to the environment in wastewater streams and fugitive emissions generated at sites of its industrial production and use. If released to the atmosphere, it will degrade by reaction with photochemically produced hydroxyl radicals (estimated half-life of 23 hr). If released to soil, it is not expected to leach based upon an estimated Koc of 6500. If released to water, cyclododecane may volatilize and partition to sediment. Insufficient data are available to predict the importance of biodegradation in soil or water. Occupational exposure to cyclododecane can occur through dermal contact. Cyclododecane has been identified as a constituent of expired human breath.

Natural Sources:

Artificial Sources: The major use of cyclododecane is as an intermediate for the production of chemicals used to make polyamides, polyesters, synthetic lubricating oils, and nylon 12 [5]; it is also used as a high-purity solvent [7]. It may be released to the environment in wastewater streams and fugitive emissions generated at sites of its industrial production and use.

Terrestrial Fate: Based upon an estimated Koc value of 6500 [12], cyclododecane is expected to be relatively immobile in soil. Insufficient data are available to predict the rates or importance of soil degradation processes.

Aquatic Fate: Volatilization and adsorption to sediment may be important transport processes in water for cyclododecane. Volatilization half-lives of 3.8 and 45 hr can be estimated for a model river (1 m deep) and model pond (2 m deep), respectively, when the effects of adsorption are not present [8,15]; in the presence of adsorption (Koc of 6500), the volatilization half-life from the pond increases to 43 days [15]. Based upon an estimated Koc of 6500 [12], cyclododecane may partition from the water column to sediment and suspended material. Insufficient data are available to predict the importance of biodegradation. An estimated BCF of 74,000 indicates that bioconcentration in aquatic organisms may be important [8].

Atmospheric Fate: Based upon an extrapolated vapor pressure of 0.0295 mm Hg at 25 °C [2], cyclododecane is expected to exist almost entirely in the vapor-phase in the ambient atmosphere [3]. It will degrade in the ambient atmosphere by reaction with photochemically produced hydroxyl radicals (estimated half-life of 23 hr) [1].

Cyclododecane

Biodegradation:

Abiotic Degradation: The rate constant for the vapor-phase reaction of cyclododecane with photochemically produced hydroxyl radicals has been estimated to be 1.67×10^{-11} cm^3/molecule-sec at 25 °C, which corresponds to an atmospheric half-life of about 23 hr at an atmospheric concn of $5 \times 10^{+5}$ hydroxyl radicals per cm^3 [1]. Alkanes are generally resistant to aqueous environmental hydrolysis [8]; therefore, cyclododecane is not expected to hydrolyze in the environment.

Bioconcentration: Based upon the estimated log Kow, the BCF for cyclododecane can be estimated to be 74,000 from a recommended regression-derived equation [8]. This estimated BCF indicates that bioconcentration in aquatic organisms may be important environmentally.

Soil Adsorption/Mobility: Using a structure estimation method based on molecular connectivity indexes, the Koc for cyclododecane can be estimated to be 6500 [12]. According to a suggested classification scheme [14], this estimated Koc indicates that cyclododecane will be relatively immobile in soil [14].

Volatilization from Water/Soil: The Henry's Law constant for cyclododecane can be estimated to be 1.40 atm-m^3/mol using a structure-activity estimation method [9]. This value of Henry's Law constant indicates rapid volatilization from water [8]. Based on this Henry's Law constant, the volatilization half-life from a model river (1 m deep flowing 1 m/sec with a wind velocity of 3 m/sec) can be estimated to be about 3.8 hr [8]. The volatilization half-life from a model environmental pond can be estimated to be about 45 hr if the effect of adsorption to sediment is ignored [15]; if the effect of adsorption to sediment is predicted from an estimated Koc of 6500 [12], the estimated volatilization half-life from the pond increases to about 43 days [15].

Water Concentrations: SURFACE WATER: Cyclododecane has been detected in all five of the Great Lakes (Erie, Ontario, Huron, Superior, Michigan) aquatic ecosystems [4]; concns, sampling dates, and sample types (water, whole water or sediment) were not reported.

Effluent Concentrations:

Cyclododecane

Sediment/Soil Concentrations: Cyclododecane was qualitatively detected in sediments collected from Tobin Lake (collection date not reported) near the Saskatchewan-Manitoba border in Canada [13].

Atmospheric Concentrations:

Food Survey Values:

Plant Concentrations:

Fish/Seafood Concentrations:

Animal Concentrations:

Milk Concentrations:

Other Environmental Concentrations:

Probable Routes of Human Exposure: Occupational exposure to cyclododecane can occur through dermal contact.

Average Daily Intake:

Occupational Exposure: NIOSH (NOES 1981-1983) has statistically estimated that 28 workers are potentially exposed to cyclododecane in the US [11].

Body Burdens: The expired air of 62 nonsmoking humans was examined for the presence of environmental pollutants and other chemical constituents [6]; cyclododecane was detected in the expired air of all three subject groups (diabetic, pre-diabetic and control) [6].

REFERENCES

1. Atkinson R; J Inter Chem Kinet 19: 799-828 (1987)
2. Boublik T et al; The Vapor Pressures of Pure Substances (Physical Science Data 17). NY: Elsevier Sci Publ p. 854 (1984)
3. Eisenreich SJ et al; Environ Sci Technol 15: 30-8 (1981)

Cyclododecane

4. Great Lakes Water Quality Board; An Inventory of Chemical Substances Identified in the Great Lakes Ecosystem. Volume 1 - Summary. Report to the Great Lakes Water Quality Board. Windsor Ontario, Canada p. 12 (1983)
5. Griesbaum K; Ullmann's Encycl Industr Chem 5th ed NY: VCH Publishers A13: 238 (1989)
6. Krotoszynski BK, O'Neill HJ; J Environ Sci Health A17: 855-83 (1982)
7. Kuney JH; Chemcyclopedia 1991. Washington DC: Amer Chem Soc p. 58 (1991)
8. Lyman WJ et al; Handbook of Chemical Property Estimation Methods Washington, DC: Amer Chem Soc p. 5-4, 7-4, 15-15 to 15-29 (1990)
9. Meylan W, Howard PH; Environ Toxicol Chem 10:1283-93 (1991)
10. Meylan WM, Howard PH; J Pharm Sci 84: 83-92 (1995)
11. NIOSH; National Occupational Exposure Survey (NOES) (1989)
12. Sabljic A; Environ Sci Technol 21: 358-66 (1987)
13. Samoiloff MR et al; Environ Sci Technol 17: 329-34 (1983)
14. Swann RL et al; Res Rev 85: 23 (1983)
15. USEPA; EXAMS II Computer Simulation (1987)

p-Cymene

Synonyms: 1-Methyl-4-isopropylbenzene

Structure:

CAS Registry Number: 99-87-6

Molecular Formula: $C_{10}H_{14}$

SMILES Notation: c(ccc(c1)C(c1)C(C)C

CHEMICAL AND PHYSICAL PROPERTIES

Boiling Point: 177.10 °C

Melting Point: -67.94 °C

Molecular Weight: 134.22

Dissociation Constants:

Log Octanol/Water Partition Coefficient: 4.10 [2]

Water Solubility: 23.4 mg/L at 25 °C [3]

Vapor Pressure: 1 mm Hg at 17.3 °C, 1.46 mm Hg at 25 °C [8]

Henry's Law Constant: 0.011 atm-m³/mol at 25 °C (estimated from vapor pressure and water solubility)

p-Cymene

ENVIRONMENTAL FATE/EXPOSURE POTENTIAL

Summary: p-Cymene is released to the environment from natural sources such as volatile plant emissions. It is also released from anthropogenic sources such as motor vehicle exhaust, solvent evaporation, and industrial wastewaters. If released to the atmosphere, p-cymene will degrade in the vapor-phase by reaction with photochemically produced hydroxyl radicals (estimated half-life of 25.5 hr). If released to soil or water, p-cymene will probably biodegrade. The results of limited biodegradation screening studies suggest that p-cymene can biodegrade aerobically in the environment. Biodegradation is the only identifiable degradation process in soil or water. Limited leaching in soil is predicted. Volatilization will be a major removal mechanism from water in the absence of strong adsorption to sediment; however, strong adsorption will greatly reduce the volatilization rate. Occupational exposure to p-cymene occurs through dermal contact and inhalation of vapor. The general population is exposed to p-cymene through inhalation and consumption of foods that contain p-cymene as a natural constituent.

Natural Sources: p-Cymene is emitted to the biosphere from natural sources such as California black sage and "disturbed" eucalyptus foliage [1]; it is found in the gum terpentines of scotch pine and loblolly pine [1]. p-Cymene is a component of pimento oils from West Indies; barry oil contains 0.29% p-cymene, leaf oil contains 0.78%.

Artificial Sources: p-Cymene is produced as a byproduct in the manufacture of sulphite paper pulp [22], and it has been detected in wastewaters released from pulp manufacturing plants [2,3,4]. p-Cymene's use as a solvent and thinner for lacquers and varnishes [29] will release the compound directly to air through evaporation. p-Cymene has been identified as a gaseous exhaust product from motor vehicles [11].

Terrestrial Fate: The results of limited biodegradation screening studies suggest that p-cymene can biodegrade aerobically in the environment. Biodegradation is the only identifiable degradation process in soil. Estimated Koc values of 770 and 4050 suggest low soil mobility and limited leaching [15].

p-Cymene

Aquatic Fate: The results of limited biodegradation screening studies suggest that p-cymene can biodegrade aerobically in the environment. Biodegradation is the only identifiable degradation process in water. Estimated Koc values of 770 and 4050 suggest that some partitioning from the water column to sediment and suspended material may occur [27]. In the absence of adsorption, volatilization will be an important transport process. Volatilization half-lives (which exclude adsorption) of 3.5 and 41 hr can be estimated for a model river and model environmental pond, respectively [15,27]; if maximum predictable adsorption is included in the pond simulation, the volatilization half-life increases to 31 days [27].

Atmospheric Fate: Based upon a vapor pressure of 1.46 mm Hg at 25 °C [8], p-cymene is expected to exist almost entirely in the vapor-phase in the ambient atmosphere [10]. The dominant degradation process in the atmosphere is the vapor-phase reaction with photochemically produced hydroxyl radicals, which has an estimated half-life of 25.5 hr [7].

Biodegradation: A batch system die-away test using artificial seawater, a 10-day incubation period, and an inoculum of coastal water from the North Sea found p-cymene to undergo moderate biooxidation (actual rates not reported) [28]. p-Cymene was not degraded during batch system, anaerobic degradation tests using methanogenic bacteria [4]; at a concentration of 100 mg/L, it did not affect methanogenic gas production, but did cause a 14 day lag period in gas production at 1000 mg/L [4]. Pure culture biodegradation studies have identified cumic acid as a metabolite of p-cymene [21,30]. Both mono- and polynuclear aromatic hydrocarbons can be oxidized by different microorganisms. p-Cymene is converted to cumic acid.

Abiotic Degradation: The rate constant for the vapor-phase reaction of p-cymene with photochemically produced hydroxyl radicals has been experimentally determined to be 1.51×10^{-11} cm^3/molecule-sec at 22 °C, which corresponds to an atmospheric half-life of about 25.5 hr at an atmospheric concentration of $5 \times 10^{+5}$ hydroxyl radicals per cm^3 [7]; this is the dominant atmospheric degradation process [7]. The experimental rate constant for reaction with atmospheric nitrate radicals is 9.9×10^{-16} cm^3/molecule-sec, which corresponds to an atmospheric lifetime of 1.3 years [7]. The atmospheric lifetime with respect to reaction with ozone is in excess of 330 days [7]. Alkylated benzenes are generally resistant to aqueous environmental hydrolysis [15]; therefore, p-cymene is not expected to chemically hydrolyze in environmental waters.

Bioconcentration: Based upon a water solubility of 23.4 mg/L at 25 °C [3], the BCF for p-cymene can be estimated to be 104 from a regression-derived equation [15]. Based upon a measured log Kow of 4.10 [2], the BCF for p-cymene can be estimated to be 770 from a regression-derived equation [15]. These BCF values suggest that bioconcentration in aquatic organisms is not an important fate process for p-cymene.

Soil Adsorption/Mobility: Based upon a water solubility of 23.4 mg/L at 25 °C [3], the Koc for p-cymene can be estimated to be 770 from a regression-derived equation [15]. Based upon a measured log Kow of 4.10 [2], the Koc for p-cymene can be estimated to be 4050 from a regression-derived equation [15]. These Koc values suggest that p-cymene has low soil mobility [25].

Volatilization from Water/Soil: Based upon a vapor pressure of 1.46 mm Hg [8] and a water solubility of 23.4 mg/L at 25 °C [3], the Henry's Law constant for p-cymene can be estimated to be 0.011 atm-m^3/mol . This value of Henry's Law constant indicates that volatilization from environmental waters can be rapid [15]. Using this Henry's Law constant, the volatilization half-life from a model river (1 m deep flowing 1 m/sec with a wind velocity of 3 m/sec) can be estimated to be about 3.5 hr [15]. Volatilization half-life from a model environmental pond can be estimated to be about 41 hr [27]. However, both of these half-life estimates neglect the potentially important effects of adsorption to sediment and suspended materials; when maximum adsorption effects are included in the model pond simulation, the volatilization half-life increases to 31 days [27].

Water Concentrations: DRINKING WATER: p-Cymene was qualitatively detected in drinking water samples collected from Philadelphia, PA, on February 10, 1976, and Cincinnati, OH, on November 17, 1978 [14]. SURFACE WATER: Surface water samples collected from the lower 25 km of the Brazos River and at the river/ocean mixing in Texas (at the Gulf of Mexico) between June 25, 1981, and August 5, 1982, contained p-cymene levels of 0.001-0.01 ug/L [17].

Effluent Concentrations: A p-cymene concentration of 15.4 ug/L was detected in the effluent from a large, 5-yr-old community septic tank located near Tacoma, WA [9]. p-Cymene levels as high as 80 ug/L were detected in wastewaters from two pulp mills in WA and OR [29]. p-Cymene has been qualitatively detected in various wastewaters from the following industries: iron and steel manufacturing, petroleum refining, nonferrous metals, coal mining, pulp and paper, rubber processing, auto and other laundries, electronics, and mechanical products [5]. p-Cymene was qualitatively detected in water samples collected from an advanced waste treatment facility in Pomona, CA, on September 25, 1974 [14].

Sediment/Soil Concentrations:

Atmospheric Concentrations: URBAN/SUBURBAN: Ambient air samples collected in Long Beach, CA, and Inglewood, CA (data not reported) contained p-cymene levels of 0.001 and 0.005-0.008 ppm, respectively [19].

Food Survey Values: A p-cymene concentration of 0.1 ug/g was detected in fresh, ripe mangos [16]. p-Cymene has been qualitatively detected as a volatile constituent of nectarines [26], orange essence [18], chickpea flour [23], and roasted filbert nuts [13].

Plant Concentrations: p-Cymene has been identified in plants such as California black sage and eucalyptus foliage [1]; it is found in the gum terpentines of scotch pine and loblolly pine [1].

Fish/Seafood Concentrations:

Animal Concentrations:

Milk Concentrations:

Other Environmental Concentrations:

Probable Routes of Human Exposure: Occupational exposure to p-cymene occurs through dermal contact and inhalation of vapor [22]; p-cymene is a primary skin irritant; contact with the liquid can cause dryness, defatting and erythema [22]. The general population is exposed to p-cymene through inhalation and oral consumption of foods that contain p-cymene as a natural constituent.

Average Daily Intake:

Occupational Exposure: NIOSH (NOES 1981-1983) has statistically estimated that 74,141 workers are potentially exposed to p-cymene in the US [20]. Air samples collected inside a tire retreading factory in Italy contained p-cymene levels of 1-450 ug/m³ [6].

Body Burdens:

REFERENCES

1. Arnts RR, Gay BW Jr; Photochemistry of Some Naturally Emitted Hydrocarbons. USEPA-600/3-79-081 Research Triangle Park, NC: USEPA p. 6 (1979)
2. Banerjee S, Howard PH; Environ Sci Technol 22: 839-41 (1988)
3. Banerjee S et al; Environ Sci Technol 14: 1227-9 (1980)
4. Benjamin MM et al; Water Res 18: 601-607 (1984)
5. Bursey JT, Pellizzari ED; Analysis of Industrial Wastewater for Organic Pollutants in Consent Decree Survey. Contract No. 68-03-2867. Athens, GA: USEPA Environ Res Lab p. 79, 89 (1982)
6. Cocheo V et al; Amer Ind Hyg Assoc J 44: 521-7 (1983)
7. Corchnoy SB, Atkinson R; Environ Sci Technol 24: 1497-1502 (1990)
8. Daubert TE; Danner RP; Physical and Thermodynamic Properties of Pure Chemicals: Data Compilation, NY: Hemisphere Pub Corp (1989)
9. DeWalle FB et al; Determination of Toxic Chemicals in Effluent from Household Septic Tanks USEPA/600/S2-85/050 (1985)
10. Eisenreich SJ et al; Environ Sci Technol 15: 30-8 (1981)
11. Hampton CV et al; Environ Sci Technol 16: 287-98 (1982)
12. Hrutfiord BF et al; Tappi 58: 98-100 (1975)
13. Kinlin TE et al; J Agric Food Chem 20: 1021 (1972)
14. Lucas SV GC/MS Analysis of Organics in Drinking Water Concentrates and Advanced Waste Treatment Concentrates: Vol 1 USEPA-600/1-84-020A (NTIS PB85-128221) p. 46-47, 156, 167 (1984)
15. Lyman WJ et al; Handbook of Chemical Property Estimation Methods Washington, DC: Amer Chem Soc p. 4-9, 5-4, 7-4 15-15 to 15-29 (1990)
16. MacLeod AJ, Snyder CH; J Agric Food Chem 36: 137-9 (1988)
17. McDonald TJ et al; Chemosphere 17: 123-36 (1988)
18. Moshonas MG, Shaw PE; J Agric Food Chem 38: 2181-84 (1990)
19. Neligan RE et al; The Gas Chromatographic Determination of Aromatic Hydrocarbons in the Atmosphere. ACS Natl Mtg p. 118-21 (1965)
20. NIOSH; National Occupational Exposure Survey (NOES) (1983)

p-Cymene

21. Omori T, Yamada K; Agric Biol Chem 33: 979-85 (1969)
22. Parmeggiani L; Encyl Occup Health & Safety 3rd ed Geneva, Switzerland: International Labour Office p. 1074-5 (1983)
23. Rembold H et al; J Agric Food Chem 37: 659-62 (1989)
24. Suntio LR et al; Chemosphere 17: 1249-90 (1988)
25. Swann RL et al; Res Rev 85: 23 (1983)
26. Takeoka GR et al; J Agric Food Chem 36: 553-60 (1988)
27. USEPA; EXAMS II Computer Simulation (1987)
28. Van Der Linden AC; Dev Biodegrad Hydrocarbons 1: 165-200 (1978)
29. Wilson D, Hrutfiord B; Pulp Paper Can 76: 91-3 (1975)
30. Yamada K et al; Agric Biol Chem 29: 943-8 (1965)

1,2-Diaminoethane

SUBSTANCE IDENTIFICATION

Synonyms: Ethylenediamine

Structure:

CAS Registry Number: 107-15-3

Molecular Formula: $C_2H_8N_2$

SMILES Notation: NCCN

CHEMICAL AND PHYSICAL PROPERTIES

Boiling Point: 116-117 °C

Melting Point: 8.5 °C

Molecular Weight: 60.10

Dissociation Constants: pKa = 10.712; Ka = 1.94×10^{-11} at 0 °C (Step 1 in reaction) [11]

Log Octanol/Water Partition Coefficient: -2.04 [8]

Water Solubility: Very soluble [11]

Vapor Pressure: 10.7 mm Hg at 20 °C [10]

Henry's Law Constant: 7.08×10^{-8} at 25 °C (dimensionless) [9]

ENVIRONMENTAL FATE/EXPOSURE POTENTIAL

Summary: 1,2-Diaminoethane is produced in large quantities and large amounts of the chemical will be released as emissions and in wastewater

110

during its production and use as a chemical intermediate. Despite its wide use, no monitoring data or information concerning concentrations in effluents could be located. If released to soil 1,2-diaminoethane will have very high mobility. Volatilization of 1,2-diaminoethane may be important from moist and dry soil surfaces. Based on several screening studies, biodegradation should be the most important degradative process, but no experimental rates for soil are available. 1,2-Diaminoethane may form stable complexes with metal ions in soil, but again, experimental data are lacking. If released into water, 1,2-diaminoethane may adsorb to suspended solids and sediment. Little volatilization is expected from water surfaces; the estimated half-life from a model river is 45 years. 1,2-Diaminoethane will not bioconcentrate in aquatic organisms according to a suggested classification scheme. 1,2-Diaminoethane should biodegrade in water, based on several screening studies. However, no data are available with which to estimate half-lives in natural waters. Direct photolysis will not be significant; however, complexation with transition metals in the water or reaction with humic materials is likely. If released to the atmosphere, 1,2-diaminoethane will exist primarily in the vapor phase. Vapor-phase 1,2-diaminoethane will react with photochemically produced hydroxyl radicals; the half-life for this reaction in air is estimated to be about 6 hr. Another reaction that may occur in the atmosphere is with carbon dioxide to form an insoluble carbonate. Due to its high water solubility, washout by rain will also be an important removal process. Human exposure to 1,2-diaminoethane is primarily occupational via dermal contact and inhalation with the vapor and aerosol.

Natural Sources:

Artificial Sources: 1,2-Diaminoethane is a degradation product of the agricultural fungicide maneb [19]. 1,2-Diaminoethane is produced in large quantities, 64 million lbs in 1985 [24], and will be released to the atmosphere or in wastewater during its production and use (% of production) in the manufacture of chelating agents (30%), polyamides (15%), carbamate fungicides (10%), and aminoethylethanolamine (10%) [5]. Other uses that may lead to releases are as a solvent, emulsifying agent, textile lubricant and antifreeze inhibitor [10].

Terrestrial Fate: Based on a recommended classification scheme [22], an estimated Koc value of 1.85, determined from an experimental log Kow [8] and a recommended regression-derived equation [12], indicates that 1,2-diaminoethane has very high mobility in soil. Volatilization of 1,2-diaminoethane may be important from moist soil surfaces given an experimental Henry's Law constant of 7.08 x 10⁻⁸ [9], and from dry soil surfaces based on an experimental vapor pressure of 10.7 mm Hg [10]. Based on screening studies [14,17,18,23], biodegradation should be the most important degradative process, but no experimental rates for soil are available. 1,2-Diaminoethane may form stable complexes with metal ions in soil, but again, experimental data are lacking [21].

Aquatic Fate: Based on a recommended classification scheme [22], an estimated Koc value of 1.85, determined from an experimental log Kow [8] and a recommended regression-derived equation [12], indicates that 1,2-diaminoethane may adsorb to suspended solids and sediment in the water. Little volatilization is expected from water surfaces based on an experimental dimensionless Henry's Law constant of 7.08 x 10⁻⁸ [9]. The estimated half-life from a model river is 45 years [12]. An estimated BCF value of 0.001 [12], from an experimental log Kow [8], suggests that 1,2-diaminoethane will not bioconcentrate in aquatic organisms according to a suggested classification scheme [7]. 1,2-Diaminoethane should biodegrade in water, based on several screening studies [14,17,18,23]. No data are available with which to estimate half-lives in natural waters. Direct photolysis will not be significant [4]. However, complexation with transition metals [21] in the water or reaction with humic materials [20] is likely.

Atmospheric Fate: According to a recommended classification scheme [2], an experimental vapor pressure of 10.7 mm Hg at 25 °C [10] indicates that 1,2-diaminoethane will exist primarily in the vapor phase in the ambient atmosphere. Vapor-phase 1,2-diaminoethane will react with photochemically produced hydroxyl radicals; the half-life for this reaction in air is estimated to be about 6 hr [13]. Another reaction that may occur in the atmosphere is with carbon dioxide to form an insoluble carbonate [3]. Due to its high water solubility, washout by rain will also be an important removal process.

Biodegradation: Four screening studies were located for 1,2-diaminoethane. Two of these screening studies that used nonacclimated sewage seed in standard dilution tests with fresh and salt water obtained disparate results. In one, 24 and 47% of theoretical BOD was removed after 5 and 20 days, respectively, with freshwater, and 2 and 16% in salt water [18]. In the second study 0.6 and 17% BOD was removed in 5 days using fresh and salt water, respectively [23]. The concentration of the 1,2-diaminoethane was the same, 7-10 ppm, in both studies [18,23]. With acclimated sewage seed, 36 and 70% BOD was removed in 5 and 20 days, respectively [18]. Using acclimated activated sludge, 97.5% COD was removed in 5 days in one study [17] and 67% BOD was removed in a second study [14].

Abiotic Degradation: 1,2-Diaminoethane is a strong base and undergoes the typical reactions of primary amines [20]. Many of these reactions may occur in the environment, but there is little documentation as to what reactions take place in the environment and at what rates. It may readily combine with CO_2 from the atmosphere to form a nonvolatile carbonate, and one is cautioned to protect the chemical from undue exposure to the atmosphere [3]. No experimental data could be found containing information on the rate of this reaction. The ability of 1,2-diaminoethane to form five-membered rings by coordination of the two unshared pairs of electrons on the nitrogen atoms with metallic ions leads to complexes that are much more stable than those produced with other amines [21]. Metallic ions in soils or natural waters may therefore combine with 1,2-diaminoethane but no information could be found on reactions with soil components. Humic acids that occur in natural waters contain aldehyde groups that could also potentially react with the amine groups to form adducts, but again data in natural systems are lacking [20]. Reactions of aliphatic amines with photochemically produced hydroxyl radicals are rapid; however, no experimental data are available for 1,2-diaminoethane [1]. 1,2-Diaminoethane does not contain any chromophores, which absorb radiation > 290 nm, so direct photolysis will not be significant [4].

Bioconcentration: An estimated BCF value of 0.001 was calculated for 1,2-diaminoethane, using an experimental log Kow of -2.04 [8] and a recommended regression-derived equation [12]. According to a recommended classification scheme [7], the BCF value suggests that bioconcentration in aquatic organisms will not be an important fate process.

Soil Adsorption/Mobility: The Koc of 1,2-diaminoethane is estimated as approximately 1.85, using an experimental log Kow of -2.04 [8] and a regression-derived equation [12]. According to a recommended classification scheme [22], this estimated Koc value suggests that 1,2-diaminoethane has very high soil mobility.

Volatilization from Water/Soil: 1,2-Diaminoethane is miscible in water and forms hydrates [10]. It has an extremely low Henry's Law constant (7.08×10^{-8} at 25 °C [9]) and little evaporation would be expected despite its moderately high vapor pressure (10.7 mm Hg at 20 °C [10]). Using the reported dimensionless Henry's Law constant, a half-life of 45 yr was estimated for a model river 1 m deep, flowing at 1 m/sec with a wind velocity of 3 m/sec [12]. Some evaporation from soil and surfaces may occur due to its moderate vapor pressure [10] and low adsorption to soil [12].

Water Concentrations:

Effluent Concentrations:

Sediment/Soil Concentrations:

Atmospheric Concentrations:

Food Survey Values:

Plant Concentrations:

Fish/Seafood Concentrations:

Animal Concentrations:

Milk Concentrations:

Other Environmental Concentrations:

Probable Routes of Human Exposure: Skin sensitization was observed in a number of instances because of the use of 1,2-diaminoethane as a stabilizer in pharmaceutical skin creams [6]. Sensitization is less likely in industrial exposures because the contact is less intimate and because damaged skin is not usually involved [6]. Exposure to 1,2-diaminoethane may be by inhalation of vapor, percutaneous absorption, ingestion and skin and eye contact [19]. Exposure to 1,2-diaminoethane will be primarily occupational via inhalation and dermal contact.

Average Daily Intake:

Occupational Exposure: NIOSH (NOES 1981-1983) has statistically estimated that 9,033 workers are potentially exposed to 1,2-diaminoethane in the US [15]. Since this survey excludes exposure to trade name chemicals and plastics that may contain the plasticizer, levels of occupational exposure should be considerably higher. NIOSH (NOHS 1972-1974) has statistically estimated that 15,923 workers are exposed to 1,2-diaminoethane in the US [16].

Body Burdens:

REFERENCES

1. Atkinson R; Chem Rev 85: 69-201 (1985)
2. Bidleman TF; Environ Sci Technol 22:361-7 (1988)
3. Budavari S (ed); The Merck Encyclopedia of Chemicals, Drugs, and Biologicals. Rahway, NJ: Merck and Co. Inc pg. 598 (1989)
4. Calvert JG, Pitts JNJR; Photochemistry. New York, NY: John Wiley and Sons pg. 264 (1966)
5. Chem Marketing Reporter; Chemical Profiles 1/21/85 (1985)
6. Clayton GD, Clayton FE (eds); Patty's Industrial Hygiene and Toxicology, Volume III-Theory and Rationale of Industrial Hygiene Practice 2nd ed 3A The Work Environment New York, NY: John Wiley and Sons pg. 3163 (1985)
7. Franke C et al; Chemosphere 29:1501-14 (1994)
8. Hansch C et al; Exploring QSAR: Hydrophobic, Electronic, and Steric Constants. ACS Professional Reference Book. Heller SR (consult ed) Washington, DC: Amer Chem Soc pg. 5 (1995)
9. Hine J, Mookerjee PK; J Org Chem 40: 292-8 (1975)
10. Lewis RJ Sr (ed); Hawley's Condensed Chemical Dictionary 12th ed New York, NY: Van Nostrand Reinhold Co. pg 486 (1993)
11. Lide DR (ed) CRC Handbook of Chemistry and Physics 75th ed Boca Raton, FL: CRC Press, Inc pg. 8-45 (1994)
12. Lyman WJ et al; Handbook of Chemical Property Estimation Methods. Washington DC: Amer Chem Soc pp. 5-4,5-10 (1990)
13. Meylan WM, Howard PH; Chemosphere 26:2293-9 (1993)
14. Mills EJ, Stack VT Jr.; Sew Indust Wastes 27: 1061-4 (1955)

1,2-Diaminoethane

15. NIOSH; National Occupational Exposure Survey (1985)
16. NIOSH; National Occupational Hazard Survey (1975)
17. Pitter P; Water Res 10: 231-5 (1976)
18. Price KS et al; J Water Pollut Contr Fed 46: 63-77 (1974)
19. Sittig M; Handbook of Toxic and Hazardous Chemicals and Carcinogens 1985 2nd ed Park Ridge, NJ: Noyes Data Corporation pg. 420 (1985)
20. Spitz RD; Encyclopedia of Chem Technol 7: 580-602 (1979)
21. Stumm W, Morgan JJ; Aquatic Chemistry 2nd ed pp. 356-63 1981)
22. Swann RE et al; Res Rev 85:23 (1983)
23. Takemoto S et al; Suishitsu Okaku Kenkyu 4: 80-90 (1981)
24. USITC; United States Internatl Trade Commission, USITC 1892 (1986)

1,2-Dibromo-1,1-dichloroethane

SUBSTANCE IDENTIFICATION

Synonyms:

Structure:

CAS Registry Number: 75-81-0

Molecular Formula: $C_2H_2Br_2Cl_2$

SMILES Notation: BrCC(Br)(Cl)Cl

CHEMICAL AND PHYSICAL PROPERTIES

Boiling Point: 178.3 °C at 760 mm Hg

Melting Point: -66.8 °C

Molecular Weight: 256.75

Dissociation Constants:

Log Octanol/Water Partition Coefficient: 3.11 (estimated) [11]

Water Solubility: 700 mg/L at 25 °C (estimate based on the water solubility of its structural isomer) [6]

Vapor Pressure: 0.90 mm Hg at 25 °C [2]

Henry's Law Constant: 1.61x 10^{-4} atm-m^3/mol at 25 °C (estimated) [13]

ENVIRONMENTAL FATE/EXPOSURE POTENTIAL

Summary: Release of 1,2-dibromo-1,1-dichloroethane to the environment from anthropogenic sources is expected to be minimal since this compound is produced in very small quantities and is available only as a research chemical. If released to soil, 1,2-dibromo-1,1-dichloroethane may volatilize fairly rapidly or it may leach readily through soil. This compound may be susceptible to chemical hydrolysis. If released to water, volatilization is probably an important, if not the dominant, removal mechanism (half-life 13 hr from a model river). This compound may also be susceptible to chemical hydrolysis. If released to the atmosphere, 1,2-dibromo-1,1-dichloroethane will react with photochemically generated hydroxyl radicals (half-life 889 days) or it will diffuse slowly into the stratosphere, a process that may take decades, where it will slowly photolyze or react with atomic oxygen. Long distance transport from its emission sources before ultimate removal from the stratosphere is expected. The general population is not likely to be exposed to 1,2-dibromo-1,1-dichloroethane.

Natural Sources:

Artificial Sources: Release of 1,2-dibromo-1,1-dichloroethane to the environment from anthropogenic sources is expected to be minimal since this compound is available only as a research chemical [14,16].

Terrestrial Fate: If released to soil, 1,2-dibromo-1,1-dichloroethane may rapidly volatilize from soil surfaces or it may leach through soil possibly into ground water. This compound may be susceptible to chemical hydrolysis.

Aquatic Fate: If released to water, volatilization may be an important, if not the dominant, removal mechanism for 1,2-dibromo-1,1-dichloroethane (half-life from a model river 13 hr). This compound may also be susceptible to chemical hydrolysis. Bioaccumulation in aquatic organisms and adsorption to suspended solids and sediments in water are not expected to be important fate processes.

Atmospheric Fate: Based on the vapor pressure, 1,2-dibromo-1,1-dichloroethane is expected to exist almost entirely in the vapor phase in the atmosphere [5]. 1,2-Dibromo-1,1-dichloroethane will be partially removed by reaction with photochemically generated hydroxyl radicals (half-life 889 days) and partially removed by slow diffusion into the stratosphere, a process that may take decades [4]. In the stratosphere this compound may photolyze or react with singlet oxygen. While some 1,2-dibromo-1,1-

dichloroethane may be lost from the atmosphere by scavaging by rain, any loss will be returned to the atmosphere by volatilization. Due to its persistence this compound is expected to be transported long distances from its emission sources before ultimately being removed from the atmosphere.

Biodegradation:

Abiotic Degradation: Alkyl halides are potentially susceptible to chemical hydrolysis under environmental conditions [8]. The half-life for 1,2-dibromo-1,1-dichloroethane vapor reacting with photochemically generated hydroxyl radicals in the atmosphere has been estimated to be 889 days based on an estimated reaction rate constant of 1.80×10^{-14} cm^3/molecule-sec at 25 °C [12] and an average ambient hydroxyl concentration of $5 \times 10^{+5}$ molecules/cm^3 [1]. In the stratosphere, this compound will either photolyze slowly to release halogen atoms, which in turn participate in the catalytic removal of stratospheric ozone, or it will slowly react with singlet oxygen [3,9].

Bioconcentration: The estimated water solubility and log Kow of 1,2-dibromo-1,1-dichloroethane suggest that this compound would not bioconcentrate significantly in aquatic organisms [8].

Soil Adsorption/Mobility: A soil adsorption coefficient (Koc) of 96 was estimated using a molecular topology and quantitative structure activity relationship [10]. This estimated Koc value suggests that this compound would have high mobility in soil and that adsorption to suspended solids and sediments in water would be insignificant [15].

Volatilization from Water/Soil: The value of Henry's Law constant suggests that volatilization may be an important removal process from water and moist soil surfaces [8]. Based on the Henry's Law constant the volatilization half-life from a model river 1 m deep, flowing 1 m/sec with a wind speed of 3 m/sec, has been estimated to be 13 hr [8]. The relatively high vapor pressure of 1,2-dibromo-1,1-dichloroethane suggests that this compound would volatilize fairly rapidly from dry soil surfaces.

Water Concentrations: DRINKING WATER: Dibromodichloroethane (isomer not specified) was detected in finished drinking water samples from

2 out of 3 New Orleans area water plants during August 1974 [7].

Effluent Concentrations:

Sediment/Soil Concentrations:

Atmospheric Concentrations:

Food Survey Values:

Plant Concentrations:

Fish/Seafood Concentrations:

Animal Concentrations:

Milk Concentrations:

Other Environmental Concentrations:

Probable Routes of Human Exposure: Exposure of the general population to 1,2-dibromo-1,1-dichloroethane is not likely since this compound is prepared in laboratory quantities for research use only and has no commercial applications [14,16].

Average Daily Intake:

Occupational Exposure:

Body Burdens:

REFERENCES

1. Atkinson R; Inter J Chem Kinet 19: 799-828 (1987)
2. Boublik T et al; The Vapor Pressures of Pure Substances p. 95 NY: Elsevier (1984)
3. Chou CC et al; J Phys Chem 82: 1-7 (1978)
4. Dilling WL; Environmental Risk Analysis for Chemicals; Conway RA ed NY: Van Nostrand Reinhold Co pp. 154-97 (1982)
5. Eisenreich SJ et al; Environ Sci Tech 15: 30-8 (1981)

6. Horvath AL; Halogenated Hydrocarbons: Solubility-Miscibility with Water NY: Marcel Dekker p. 547 (1982)

7. Keith LH et al; pp. 329-73 in Ident Anal Organic Pollut Water; Keith LH et al eds; Ann Arbor, MI: Ann Arbor Press (1976)

8. Lyman WJ et al; Handbook of Chemical Property Estimation Methods NY: McGraw-Hill p. 7-5 (1982)

9. Makide T et al; Chem Lett 4: 355-8 (1979)

10. Meylan WM et al; Environ Sci Technol 26:1560-1567 (1992)

11. Meylan W, Howard PH; J Pharm Sci 84: 83-92 (1995)

12. Meylan WM, Howard PH; Chemosphere 26:2293-2299 (1993)

13. Meylan W, Howard PH; Environ Toxicol Chem 10: 1283-93 (1991)

14. SRI; 1987 Directory of Chemical Producers Menlo Park, CA: SRI International (1987)

15. Swann RL et al; Res Rev 85: 17-28 (1983)

16. USEPA; TSCA Inventory non-confidential production data (1988)

1,2-Dichloro-1,1-difluoroethane

SUBSTANCE IDENTIFICATION

Synonyms:

Structure:

CAS Registry Number: 1649-08-7

Molecular Formula: $C_2H_2Cl_2F_2$

SMILES Notation: FC(F)(Cl)CCl

CHEMICAL AND PHYSICAL PROPERTIES

Boiling Point: 57.95 °C (estimated) [9]

Melting Point: -88.62 °C (estimated) [9]

Molecular Weight: 134.94

Dissociation Constants:

Log Octanol/Water Partition Coefficient: 2.31 (estimated) [7]

Water Solubility: 170 mg/L at 25 °C (estimated) [10]

Vapor Pressure: 214 mm Hg at 25 °C (estimated) [9]

Henry's Law Constant: 0.048 atm-m³/mol at 25 °C [8]

ENVIRONMENTAL FATE/EXPOSURE POTENTIAL

Summary: 1,2-Dichloro-1,1-difluoroethane is an anthropogenic compound that may be released to the environment as a fugitive emission during its production or use. If released to soil, 1,2-dichloro-1,1-difluoroethane will rapidly volatilize from both moist and dry soil to the atmosphere. It will display moderate to high mobility in soil. If released to water, 1,2-dichloro-1,1-difluoroethane will rapidly volatilize to the atmosphere. The estimated half-life for volatilization from a model river is 3.4 hr. 1,2-Dichloro-1,1-difluoroethane will not bioconcentrate in fish and aquatic organisms nor will it adsorb to sediment or suspended organic matter. If released to the atmosphere, 1,2-dichloro-1,1-difluoroethane will undergo a slow gas-phase reaction with photochemically produced hydroxyl radicals with an estimated half-life of 617 days. 1,2-Dichloro-1,1-difluoroethane may also undergo atmospheric removal by wet deposition processes; however, any removed by this process is expected to rapidly revolatilize to the atmosphere. Occupational exposure to 1,2-dichloro-1,1-difluoroethane may occur by inhalation or dermal contact during its production or use.

Natural Sources: 1,2-Dichloro-1,1-difluoroethane is of anthropogenic origin, and it is not known to be produced by natural sources.

Artificial Sources: 1,2-Dichloro-1,1-difluoroethane is an anthropogenic compound that is a potential CFC replacement; if so it may be released to the environment as a fugitive emission during its production and use.

Terrestrial Fate: If released to soil, the estimated vapor pressure for 1,2-dichloro-1,1-difluoroethane indicates that it will rapidly volatilize from dry soil to the atmosphere. Estimated soil adsorption coefficients ranging from 97-430 [4,6] indicate that it will display moderate to high mobility in soil [11]. The estimated Henry's Law constant indicates that it will also rapidly volatilize from moist soil to the atmosphere.

Aquatic Fate: If released to water, the estimated Henry's Law constant for 1,2-dichloro-1,1-difluoroethane indicates that it will rapidly volatilize to the atmosphere. The estimated half-life for volatilization from a model river 1 m deep flowing at 1 m/sec with a wind speed of 3 m/sec is 3.4 hr [4]. An estimated bioconcentration factor of 34 [4] from the water solubility or Kow indicates that 1,2-dichloro-1,1-difluoroethane will not bioconcentrate in fish and aquatic organisms. Estimated soil adsorption coefficients ranging from 97-430 [4,6] indicate that it will not adsorb to sediment or suspended organic matter.

1,2-Dichloro-1,1-difluoroethane

Atmospheric Fate: If released to the atmosphere, 1,2-dichloro-1,1-difluoroethane will undergo a slow gas-phase reaction with photochemically produced hydroxyl radicals. The recommended rate constant for this process of 2.6×10^{-14} cm^3/molecule-sec [1] translates to an atmospheric half-life of 617 days using an average atmospheric hydroxyl radical concn of $5 \times 10^{+5}$ molec/cm^3 [1]. The estimated water solubility of 1,2-dichloro-1,1-difluoroethane indicates that it may undergo atmospheric removal by wet deposition processes; however, any removal by this process is expected to rapidly revolatilize to the atmosphere.

Biodegradation:

Abiotic Degradation: Experimental rate constants for the gas-phase reaction of 1,2-dichloro-1,1-difluoroethane with photochemically produced hydroxyl radicals of 2.6×10^{-14} cm^3/molecule-sec [2,3] and 1.1×10^{-14} cm^3/molecule-sec [5] have been reported. The recommended value of 2.6×10^{-14} cm^3/molecule-sec [1] translates to an atmospheric half-life of 617 days using an average atmospheric hydroxyl radical concn of $5 \times 10^{+5}$ molec/cm^3 [1].

Bioconcentration: An estimated bioconcentration factor of 34 can be calculated for 1,2-dichloro-1,1-difluoroethane based on its estimated log octanol/water partition coefficient and estimated water solubility using appropriate regression equations [4]. This value indicates that 1,2-dichloro-1,1-difluoroethane will not bioconcentrate in fish and aquatic organisms.

Soil Adsorption/Mobility: Estimated soil adsorption coefficients ranging from 97-430 can be calculated for 1,2-dichloro-1,1-difluoroethane based on its estimated log octanol/water partition coefficient and estimated water solubility using appropriate regression equations [4] and a structure estimation method [6]. These values indicate that 1,2-dichloro-1,1-difluoroethane will display moderate to high mobility in soil [11].

Volatilization from Water/Soil: The estimated Henry's Law constant indicates that 1,2-dichloro-1,1-difluoroethane will rapidly volatilize from water and moist soil to the atmosphere. The estimated half-life for volatilization from a model river 1 m deep flowing at 1 m/sec with a wind speed of 3 m/sec is 3.4 hr [4]. The estimated vapor pressure of 1,2-dichloro-1,1-difluoroethane indicates that it will rapidly volatilize from dry soil to the atmosphere.

Water Concentrations:

Effluent Concentrations:

Sediment/Soil Concentrations:

Atmospheric Concentrations:

Food Survey Values:

Plant Concentrations:

Fish/Seafood Concentrations:

Animal Concentrations:

Milk Concentrations:

Other Environmental Concentrations:

Probable Routes of Human Exposure: Occupational exposure to 1,2-dichloro-1,1-difluoroethane may occur by inhalation or dermal contact during its production or use.

Average Daily Intake:

Occupational Exposure:

Body Burdens:

REFERENCES

1.   Atkinson R; J Chem Phys Ref Data Monograph 1 (1989)
2.   Cohen N, Benson SW; J Phys Chem 91: 162-70 (1987)
3.   Cohen N, Benson SW; J Phys Chem 91: 171-5 (1987)

4. Lyman WJ et al; Handbook of Chemical Property Estimation Methods NY: McGraw-Hill Chapt 4, 5 & 15 (1982)
5. Makide Y, Rowland FS; Proc Natl Acad Sci USA 78: 5933-7 (1981)
6. Meylan WM et al; Environ Sci Technol 26:1560-1567 (1992)
7. Meylan W, Howard PH; J Pharm Sci 84: 83-92 (1995)
8. Meylan WM, Howard PH; Environ Toxicol Chem 10: 1283-93 (1991)
9. SRC; MPBPVP Program, Syracuse Research Corp. (1995)
10. SRC; WS/KOW Program, Syracuse Research Corp. (1995)
11. Swann RL et al; Res Rev 85: 17-28 (1983)

Dichlorodifluoromethane

SUBSTANCE IDENTIFICATION

Synonyms: Freon 12

Structure:

CAS Registry Number: 75-71-8

Molecular Formula: CCl_2F_2

SMILES Notation: C(Cl)(Cl)(F)(F)

CHEMICAL AND PHYSICAL PROPERTIES

Boiling Point: -29.8 °C

Melting Point: -158 °C

Molecular Weight: 120.91

Dissociation Constants:

Log Octanol/Water Partition Coefficient: 2.16 [18]

Water Solubility: 280 mg/L at 25 °C, 1 atm [25]

Vapor pressure: 4250 mm Hg at 20 °C [25]

Henry's Law Constant: 0.225 atm-m^3/mol at 20 °C [27]

ENVIRONMENTAL FATE/EXPOSURE POTENTIAL

Summary: Dichlorodifluoromethane (Freon 12) is produced in large quantities and will be released to the environment as emissions during its production, storage, transport, and use as a refrigerant, foam blowing agent, solvent, and chemical intermediate in the production of fluoropolymers. All of the dichlorodifluoromethane that is produced is eventually lost as emissions and the levels of this chemical have been building up in the atmosphere. If released on land, dichlorodifluoromethane will leach into the ground and volatilize from the soil surface. No degradative processes are known to occur in the soil. Dichlorodifluoromethane is also stable in water and the only removal process will be volatilization. It can enter water bodies from the atmosphere and the concentration of dichlorodifluoromethane in the surface water rapidly reaches equilibrium with the concentration in the air. Ocean currents also carry the chemical long distances and many kilometers below the surface. Dichlorodifluoromethane is extremely stable in the troposphere and will disperse over the globe and diffuse slowly into the stratosphere where it will be lost by photolysis. In this process, chlorine atoms are released that attack ozone. The realization that certain chlorofluorocarbons can accumulate in the upper atmosphere and deplete the earth's ozone layer has resulted in restrictions on the uses of these chemicals. This is particularly true in the case of dichlorodifluoromethane since it is used in such large quantities and is so stable and effective in destroying ozone. General population exposure occurs by inhalation of dichloro-fluoromethane in ambient air. Occupational exposure may occur via inhalation or dermal contact.

Natural Sources: There are no known natural sources of dichlorodifluoromethane [3].

Artificial Sources: Dichlorodifluoromethane is produced in large quantities and will be released to the environment as emissions during its production, storage, transport, and use. As of 1981 the distribution of commercial use (percentage) was: refrigerant (46%), foam blowing agent (20%), solvent (16%), fluoropolymers (7%), other (4%), export (7%) [4]. Solvent use is mainly in the aerospace and electronic industries [30]. All of the dichlorodifluoromethane that is produced is eventually lost as emissions. It is estimated that 3.3% of the dichlorodifluoromethane produced is lost from plant vents and during packaging, a loss which is immediate [13]. Losses from aerosols occur, on the average, 6 months after sale; losses from domestic refrigerators and freezers have an average life of 12 yr with a 2% loss during filling; industrial refrigerator charges have an average lifetime of 4 yr; all of the dichlorodifluoromethane used in closed cell foams are lost within 2 yr with 75% being lost during the first year [13]. The annual world production and release of dichlorodifluoromethane in 1982 was 443.7 and 422.8 million kg, respectively [13]. Cumulative production and release estimates up until the end of 1974 were 4698.5 and 4286.2 million kg [13]. By the end of 1982 these figures had increased to 8196.0 and 7520.2 million kg [13]. The realization that certain chlorofluorocarbons can accumulate in the upper atmosphere and deplete the earth's ozone layer has had a major impact on chemicals like dichlorodifluoromethane, which are used in large quantities and have the stability to reach the stratosphere. Uses such as propellants in aerosols, which had accounted for about 75% of the release of dichlorodifluoromethane and trichlorofluoromethane, the chemicals of greatest concern (refrigerants and foams accounted for about 14 and 12%, respectively), were banned in the US after December 15, 1978 [30]. Previously dichlorodifluoromethane was the principal propellant for nonfood aerosols [30] and 60% of dichlorodifluoromethane and trichlorofluoromethane production went into aerosols [30].

Terrestrial Fate: If released on land, dichlorodifluoromethane will leach into the ground and volatilize from the soil surface. No degradative processes are known to occur in the soil.

Aquatic Fate: Dichlorodifluoromethane is persistent in natural waters [32]. It primarily enters bodies of water from the atmosphere. Since it is extremely stable in the water as well as in the atmosphere and does not adsorb appreciably to sediment, the only removal process will be volatilization. The concentration of dichlorodifluoromethane in surface water would be expected to reach solubility equilibrium with the atmosphere relatively rapidly [3]. Ocean surface water measurements in the Eastern Pacific show that a rough equilibrium with air concentration exists [29]. Accordingly, the chemical has been used as a tracer for ocean circulation and mixing. Model calculations have resulted in the time scale for deep convection mixing in the Greenland and Norwegian Seas to be about 40 yr and the lateral mixing between the depths of these seas as 20-30 yr [3].

Atmospheric Fate: Dichlorodifluoromethane is extremely stable in the troposphere and will disperse over the globe and diffuse slowly into the stratosphere where it will be photolyzed by the short wavelength UV radiation. It will also be removed from the atmosphere by dry deposition. This can be estimated assuming a deposition velocity of 0.1 cm/sec to be

0.76 million lbs/yr in the US [16]. However, the chemical does not degrade in water or soil and the chemical will volatilize back into the atmosphere. The combined stability, deposition, and volatilization will result in a more or less equal distribution of dichlorodifluoromethane in bodies of water all over the earth as well as in the atmosphere.

Biodegradation: No evidence of dichlorodifluoromethane biodegradation was found in a microcosm designed to simulate Narragansett Bay in a month-long experiment [1].

Abiotic Degradation: Being fully halogenated, dichlorodifluoromethane does not react with photochemically produced hydroxyl radicals in the atmosphere and accordingly no evidence of reaction has been reported [17]. The only decomposition reactions available are UV photolysis and attack by singlet oxygen atoms [11]. The rate of uncatalyzed hydrolysis of dichlorodifluoromethane was too low to be detected (<0.005 g/L of water-yr) [10]. When catalyzed by the presence of steel, the hydrolysis rates are detectable but still quite low [10], for dichlorodifluoromethane absorption is negligible above 221 nm [5] and the maximum photolysis rate occurs at an altitude of 32 km [36]. The photolysis of chlorofluorocarbons produces chlorine atoms that are responsible for the destruction of earth's protective ozone layer [36]. The more heavily chlorinated molecules are more destructive because they photolyze at lower altitudes and produce more chlorine atoms in the process [36]. Dichlorodifluoromethane also reacts with singlet oxygen atoms (O(1D)) in the atmosphere, but the half-life for this reaction is over 300 yr [20], assuming an oxygen concentration of 5 x 10^{-1} radicals/cm^3 [14]. When adsorbed on silica gel and irradiated with sunlight for effectively 170 hr, 9.1% of the dichlorodifluoromethane was mineralized to CO_2 [12]. When the irradiation was performed using a high-pressure Hg lamp through a pyrex filter (wavelength >300 nm) for 144 hr, 1.4% mineralization occurred [12].

Bioconcentration: Using the log octanol/water for dichlorodifluoromethane, one can estimate a BCF of 26 in aquatic organisms [19]. Therefore, dichlorodifluoromethane would not bioconcentrate in fish. No evidence of tissue storage of dichlorodifluoromethane was found in higher animals [1].

Soil Adsorption/Mobility: Using the water solubility for dichlorodifluoro-methane, one can estimate a Koc of 200 using a recommended regression equation [19]. Therefore, dichlorodifluoromethane would not adsorb appreciably to soil.

Volatilization from Water/Soil: Using the Henry's Law constant for di-chlorodifluoromethane, one can estimate that its volatilization half-life from a model river 1 m deep flowing 1 m/sec with a 3 m/sec wind is 3.2 hr [26]. The resistance to volatilization will lie almost entirely in the liquid phase [27]. The mass transfer coefficient of dichlorodifluoromethane relative to oxygen is 0.66 [26]. Combining this with oxygen reaeration coefficients from various bodies of water, one can calculate volatilization half-lives of 4.6-9.5 days, 2.7-26 hr, and 3.4-10.9 days for ponds, rivers, and lakes, respectively [19]. Its half-life in the water column of a laboratory microcosm designed to simulate Narragansett Bay was 9.9 days [35]. Since dichlorodifluoromethane is a gas and does not adsorb appreciably to soil, evaporation from soil would be rapid.

Water Concentrations: GROUND WATER: Dichlorodifluoromethane was detected, but not quantified in ground water under 6 of 13 municipal landfills in Minnesota with suspected leakage but not in the ground water under 7 other municipal landfills [28]. SURFACE WATER: Of the 696 stations reporting dichlorodifluoromethane in ambient waters in EPA's STORET database, 1.0% contained detectable levels of the chemical [31]. Dichlorodifluoromethane was detected, but not quantified, in the open waters of Lake Erie [15]. The average concentration of dichlorodifluoro-methane in surface water at two stations in the Greenland and Norwegian Seas was 218 and 187 ppb, respectively [3]. The concentration decreased with depth over many km in the Greenland Sea and was relatively constant in the upper 0.4 km in the Norwegian Sea before decreasing [3]. The deep water concentrations of dichlorodifluoromethane in these seas were 37 and 18 ppb [3].

Effluent Concentrations: Of the 1,144 stations reporting dichlorodifluoro-methane in effluents in EPA's STORET database, 1.6% contained detectable levels of the chemical [31]. No dichlorodifluoromethane was found in the effluent of a large community septic tank (detection limit 0.7 ppb) [9]. In the National Urban Runoff Program in which samples of runoff were collected from 19 cities (51 catchments) in the US, no dichlorodifluoromethane was found in any samples [6]. Leachate from 2 of 6 municipal landfills tested in Minnesota had detectable quantities of dichlorodifluoromethane [28]. Leachate from 1 out of 5 landfills in Wisconsin contained 180 ppb of the chemical [28]. Dichlorodifluoromethane was emitted from a simulated landfill composed of municipal refuse and wastewater sludges [34].

Sediment/Soil Concentrations: None of the 204 stations in EPA's STORET database reported any dichlorodifluoromethane in sediment [31].

Atmospheric Concentrations: RURAL/REMOTE: Sites in the US (431 samples) - 120 ppt, median, 230 ppt, max [2]. The average concentration of dichlorodifluoromethane at two stations over the Greenland and Norwegian Seas was 343 ppt [3]. Adrigole, Ireland 338 ppt [3]. Rural and urban areas of China and rural areas of Oregon 320 ppt, median; urban areas of Oregon - 460 ppt [24]. Similarly, median concentrations of dichlorodifluoromethane in rural areas of Eastern Washington and the Caucasus Mountains of the USSR in 1979 were 289 and 313 ppt, respectively [7]. The respective average concentrations in the morning vs the nighttime in Eastern Washington were 310 and 318 ppt [7]. Site and diurnal variations are consistent with the micrometeorology of the site and local sources [7]. RURAL/REMOTE: The average concentrations of dichlorodifluoromethane in the Northern and Southern hemispheres show slight differences [33]. In 1976, their values were 218 and 204 ppt [33]. Concentrations of dichlorodifluoromethane at the South Pole have increased from 195 to 284 ppt between 1976 and 1980, while over the same period of time the increase in the Pacific Northwest was from 228 to 322 ppt [8]. The annual rate of increase in these areas is 7%. As one goes up in altitude, the concentrations of dichlorodifluoromethane in the atmosphere at northern mid-latitudes decrease from 350 ppt at the earth's surface to 250, 100, and 20 ppt at 10, 20, and 30 km [11]. URBAN/SUBURBAN: Sites in the US (903 samples) - 0.38 ppb median, 4.9 ppb max [2]. Various sites in the US had dichlorodifluoromethane levels ranging from 0.37 to 4.8 ppb while sites in continental Europe and Japan had levels of 0.410-11.4 ppb [32]. In 1975, the average concentration of dichlorodifluoromethane was 0.23 ppb and rising at an annual rate of 0.0157 ppb [16].

Food Survey Values:

Plant Concentrations:

Fish/Seafood Concentrations: None of the 47 stations in EPA's STORET database reported any dichlorodifluoromethane in biota [31].

Animal Concentrations:

Milk Concentrations: In a pilot study of pollutants in the milk of women living in 4 urban-industrial areas in the US, dichlorodifluoromethane was found in 2 of 8 samples [23].

Other Environmental Concentrations:

Probable Routes of Human Exposure: General population exposure occurs by inhalation of dichlorodifluoromethane in ambient air. Occupational exposure may occur via inhalation or dermal contact.

Average Daily Intake: AIR INTAKE (assume air concentration 350 ppt): 35 ug; WATER INTAKE - Insufficient data; FOOD INTAKE - Insufficient data.

Occupational Exposure: NIOSH (NOHS 1972-1974) has statistically estimated that 2,414,759 workers are exposed to dichlorodifluoromethane in the US [21]. NIOSH (NOES 1981-1983) has statistically estimated that 435,096 workers are exposed to dichlorodifluoromethane in the US [22]. Eighty-six percent of the exposures are with trade name mixtures containing dichlorodifluoromethane.

Body Burdens: In a pilot study of pollutants in the milk of women living in 4 urban-industrial areas in the US, dichlorodifluoromethane was found in 2 of 8 samples [23].

REFERENCES

1. Bopp RF et al; Org Geochem 3: 9-14 (1981)
2. Brodzinsky R, Singh HB; Volatile Organic Chemicals in the Atmosphere: An Assessment of Available Data Menlo Park CA: SRI Inter Contract 68-02-3452 (1982)
3. Bullister JL, Weiss RF; Science 221: 265-8 (1983)

4. Chemical Marketing Reporter; Chemical Profiles Aug 24 (1981)
5. Chou CC et al; J Phys Chem 82: 1-7 (1978)
6. Cole RH et al; J Water Pollut Control Fed 56: 898-908 (1984)
7. Cronn DR et al; Environ Sci Tech 17: 383-8 (1983)
8. Crutzen PJ, Gidel LT; J Geophys Res 88: 6641-61 (1983)
9. Dewalle FB et al; Determination of Toxic Chemicals in Effluent from Household Septic Tanks USEPA-600/S2-85-050. Cincinnati OH: p. 4 (1985)
10. DuPont de Nemours; Freon Product Information B-2 Wilmington, DE: EI Dupont de Nemours Co (1980)
11. Fabian P, Goemer D; Fresenius' Z Anal Chem 319: 890-7 (1984)
12. Gaeb S et al; Angew Chem 90: 398-9 (1978)
13. Gamlen PH et al; Atmos Environ 20: 1077-85 (1986)
14. Graedel TE; Chemical Compounds in the Atmosphere. NY, NY: Academic Press pp. 440 (1978)
15. Great Lakes Water Quality Board; Report to the Great Lake Water Quality Board Windsor Ontario Canada pp. 195 (1983)
16. Guicherit R, Schulting FL; Sci Tot Environ 43: 193-219 (1985)
17. Hampson RF; Chemical kinetic and photochemical data sheets for atmospheric reactions. FAA-EE-80-17. Washington DC: US Dept Transport (1980)
18. Hansch C, Leo AJ; MEDCHEM Project Claremont CA: Pomona College (1985)
19. Lyman WJ et al; Handbook of Chem Property Estimation Methods NY: McGraw-Hill pp. 4-1 to 4-33, 5-1 to 5-30, 15-1 to 15-34 (1982)
20. NASA; Chemical and Photochemical Data for Use in Stratospheric Modeling. Evaluation No. 4. NASA-CR-163973. JPL-PUB-81-3. pp.131 (1981)
21. NIOSH; National Occupational Health Survey (1975)
22. NIOSH; National Occupational Exposure Survey (1989)
23. Pellizzari ED et al; Bull Environ Contam Toxicol 28: 322-8 (1982)
24. Rasmussen RA et al; Environ Sci Tech 16: 124-6 (1982)
25. Riddick JA et al; Organic Solvents NY, NY: Wiley Interscience (1986)
26. Roberts PV, Dandliker PG; Environ Sci Technol 17: 484-9 (1983)
27. Roberts PV; Environ Sci Technol 18: 890 (1984)
28. Sabel GV, Clark TP; Waste Manag Res 2: 119-30 (1984)
29. Singh HB et al; J Geophys Res 88: 3675-83 (1983)
30. Smart BE; Kirk Othmer's Encycl Chem Tech 3rd NY, NY: Wiley Interscience 10: 829-70 (1980)
31. Staples CA et al; Environ Toxicol Chem 4: 131-42 (1985)
32. Su C, Goldberg ED; pp. 353-74 in Mar Pollut Transfer Windom HL et al eds Lexington, MA: DC Heath Co (1976)
33. Tyson BJ et al; Geophys Res Lett 5: 535-8 (1978)
34. Vogt WG, Walsh JJ; in Proc - APCA Annu Meet 78th p. 17 (1985)
35. Wakeham SG et al; Environ Sci Technol 20: 574-80 (1986)
36. Wayne RP; Chemistry of Atmospheres Oxford: Clarendon pp. 153-8 (1985)

1,1-Dichloro-1-fluoroethane

Synonyms: HCFC-141b

Structure:

CAS Registry Number: 1717-00-6

Molecular Formula: $C_2H_3Cl_2F$

SMILES Notation: C(Cl)(Cl)C(F)

CHEMICAL AND PHYSICAL PROPERTIES

Boiling Point: 32.9 °C

Melting Point:

Molecular Weight: 116.95

Dissociation Constants:

Log Octanol/Water Partition Coefficient: 2.04 (estimated) [9]

Water Solubility: 2632 mg/L at 25 °C (calculated from the estimated vapor pressure and Henry's Law constants)

Vapor Pressure: 412 mm Hg at 25 °C (estimated) [8]

Henry's Law Constant: 0.0241 atm-m^3/mol [5]

ENVIRONMENTAL FATE/EXPOSURE POTENTIAL

Summary: 1,1-Dichloro-1-fluoroethane is an anthropogenic compound that may be released to the environment as a fugitive emission during its production or use. If released to soil, 1,1-dichloro-1-fluoroethane will rapidly volatilize from both moist and dry soil to the atmosphere. It will display high mobility in soil. If released to water, 1,1-dichloro-1-fluoroethane will rapidly volatilize to the atmosphere. The estimated half-life for volatilization from a model river is 3.2 hr. 1,1-Dichloro-1-fluoroethane will not bioconcentrate in fish and aquatic organisms nor will it adsorb to sediment and suspended organic matter. If released to the atmosphere, 1,1-dichloro-1-fluoroethane will undergo a slow gas-phase reaction with photochemically produced hydroxyl radicals. The atmospheric half-life has been estimated to be 4.3 and 7.8 years. 1,1-Dichloro-1-fluoroethane may also undergo atmospheric removal by wet deposition processes; however, it should rapidly revolatilize to the atmosphere. Occupational exposure to 1,1-dichloro-1-fluoroethane may occur by inhalation or dermal contact during its production or use.

Natural Sources: 1,1-Dichloro-1-fluoroethane is of anthropogenic origin, and it is not known to be produced by natural sources.

Artificial Sources: 1,1-Dichloro-1-fluoroethane is a potential CFC replacement; if so, it may be released to the environment as a fugitive emission during its production and use.

Terrestrial Fate: The high Henry's Law constant and vapor pressure of 1,1-dichloro-1-fluoroethane indicates that it will rapidly volatilize from moist and dry soil. An estimated soil adsorption coefficient of 57 indicates that it will display high mobility in soil.

Aquatic Fate: If released to water, 1,1-dichloro-1-fluoroethane should rapidly volatilize to the atmosphere because of its high Henry's Law constant. Its estimated volatilization half-life in a model river 1 m deep flowing at 1 m/sec with a wind speed of 3 m/sec is 3.2 hr [2]. Bioconcentration factors ranging from 7 to 21 are estimated for 1,1-dichloro-1-fluoroethane, indicating that it should not bioconcentrate in fish and aquatic organisms. A soil adsorption coefficient of 57 can be estimated for 1,1-dichloro-1-fluoroethane. This indicates that 1,1-dichloro-1-fluoroethane should not adsorb to sediment and suspended organic matter.

Atmospheric Fate: If released to the atmosphere, 1,1-dichloro-1-fluoroethane will slowly react with photochemically produced hydroxyl radicals. The half-life of 1,1-dichloro-1-fluoroethane in the atmosphere has been estimated to be 4.3 years [2] and 7.8 years [1]. The high water solubility of 1,1-dichloro-1-fluoroethane indicates that it may undergo atmospheric removal by wet deposition. However, 1,1-dichloro-1-fluoroethane should rapidly revolatilize to the atmosphere.

Biodegradation:

Abiotic Degradation: The rate constant for the reaction of 1,1-dichloro-1-fluoroethane with photochemically produced hydroxyl radicals is 7×10^{-14} cm^3/molecule-sec [3,7]. This translates into an atmospheric half-life of 230 days using an average atmospheric hydroxyl radical concn of 5×10^5 radicals/cm^3 [7]. Other estimates of the atmospheric lifetime that use the hydroxyl rate constants and half-lifes of methyl chloroform to include other factors affecting the atmospheric lifetime other than reaction with hydroxyl radicals are 4.3 years [1] and 7.8 years [2].

Bioconcentration: Estimated bioconcentration factors ranging from 7 to 21 can be estimated for 1,1-dichloro-1-fluoroethane using regression equations with its log octanol/water partition coefficient or water solubility [4]. These values indicate that 1,1-dichloro-1-fluoroethane will not bioconcentrate in fish and aquatic organisms.

Soil Adsorption/Mobility: A soil adsorption coefficient of 57 can be estimated for 1,1-dichloro-1-fluoroethane using a regression equation with its water solubility [1]. According to a suggested classification scheme, these Koc indicate that 1,1-dichloro-1-fluoroethane will display high mobility in soil [6].

Volatilization from Water/Soil: Dichloro-1-fluoroethane will rapidly volatilize from water and moist soil to the atmosphere due to its high estimated Henry's Law constant. Its estimated half-life for volatilization from a model river 1 m deep flowing at 1 m/sec with a wind speed of 3 m/sec is 3.2 hr [4]. 1,1-Dichloro-1-fluoroethane's high estimated vapor pressure indicates that it will also rapidly volatilize from dry soil to the atmosphere.

Water Concentrations:

1,1-Dichloro-1-fluoroethane

Effluent Concentrations:

Sediment/Soil Concentrations:

Atmospheric Concentrations:

Food Survey Values:

Plant Concentrations:

Fish/Seafood Concentrations:

Animal Concentrations:

Milk Concentrations:

Other Environmental Concentrations:

Probable Routes of Human Exposure: Occupational exposure to 1,1-dichloro-1-fluoroethane may occur by inhalation or dermal contact during its production or use.

Average Daily Intake:

Occupational Exposure:

Body Burdens:

REFERENCES

1. Brown AC et al; Atmos Environ 24A: 2499-511 (1990)
2. Fisher DA et al; Nature 244: 508-12 (1990)
3. Liu R et al; J Phys Chem 94: 3247-9 (1990)
4. Lyman WJ et al; Handbook of Chemical Property Estimation Methods NY: McGraw-Hill Chapt 4, 5, 15 (1982)
5. Meylan WM, Howard PH; Environ Toxicol Chem 10: 1283-93 (1991)
6. Swann RL et al; Res Rev 85: 17-28 (1983)
7. Talukdar R et al; J Phys Chem 95: 5815-21 (1991)
8. USEPA; PCGEMS (Graphical Estimation Modeling System); PCCHEM (1987)
9. USEPA; PCGEMS (Graphical Estimation Modeling System); CLOGP (1986)

2,4-Dichlorotoluene

SUBSTANCE IDENTIFICATION

Synonyms:

Structure:

CAS Registry Number: 95-73-8

Molecular Formula: $C_7H_6Cl_2$

Smiles Notation: c(ccc(c1Cl)C)(c1)Cl

CHEMICAL AND PHYSICAL PROPERTIES

Boiling Point: 196-197 °C at 760 mm Hg

Melting Point: -13.5 °C

Molecular Weight: 161.03

Dissociation Constants:

Log Octanol/Water Partition Coefficient: 4.24 [7]

Water Solubility: Insoluble in water [10]

Vapor Pressure: 0.458 mm Hg at 25 °C [3]

Henry's Law Constant: 0.346 x 10^{-2} atm-m^3/mol at 25 °C [9]

ENVIRONMENTAL FATE/EXPOSURE POTENTIAL

Summary: 2,4-Dichlorotoluene may be released to the environment in emissions or in wastewater during its production and use as a chemical intermediate. If released to soil, 2,4-dichlorotoluene will be immobile.

Volatilization of 2,4-dichlorotoluene may be important from moist and dry soil surfaces. Insufficient data are available to determine the rate or importance of biodegradation of 2,4-dichlorotoluene in soil. If released to water, 2,4-dichlorotoluene will adsorb to suspended solids and sediment. Volatilization from water should be rapid with resistance in the liquid phase controlling its rate. The estimated volatilization half-life of 2,4-dichlorotoluene in a model river 1 m deep with a 1 m/sec current and a 3 m/sec wind is 4.0 hr. The half-life of 2,4-dichlorotoluene in the Rhine River in the Netherlands is 1 day, as determined by measuring concentration differences over a reach of the river. Bioconcentration may be an important fate process. Insufficient data are available to determine the rate or importance of biodegradation of 2,4-dichlorotoluene in water. In the atmosphere, it should partially exist adsorbed to particulate matter and be subject to gravitational settling. In the vapor phase, it will degrade by reaction with photochemically produced hydroxyl radicals (half-life 11.6 days). Removal of particulate-phase 2,4-dichlorotoluene can occur through wet and dry deposition. Exposure to 2,4-dichlorotoluene would be primarily occupational.

Natural Sources:

Artificial Sources: 2,4-Dichlorotoluene may be released as emissions or in wastewater during its production and use as a chemical intermediate [5].

Terrestrial Fate: Based on a recommended classification scheme [14], an estimated Koc value of 4,800, determined from an experimental log Kow [7] and a recommended regression-derived equation [11], suggests that 2,4-dichlorotoluene will be immobile in soil. Volatilization of 2,4-dichlorotoluene may be important from moist soil surfaces given an experimental Henry's Law constant of 0.346×10^{-2} atm-m^3/mol [9], and from dry soil surfaces based on an experimental vapor pressure of 0.458 mm Hg at 25 °C [3]. Insufficient data are available to determine the rate or importance of biodegradation of 2,4-dichlorotoluene in soil.

Aquatic Fate: Based on a recommended classification scheme [14], an estimated Koc value of 4,800, determined from an experimental log Kow [7] and a recommended regression-derived equation [11], suggests that 2,4-dichlorotoluene will adsorb to suspended solids and sediment. The Henry's Law constant for 2,4-dichlorotoluene is 0.346×10^{-2} atm-m^3/mol [9].

Volatilization from water should therefore be rapid with resistance in the liquid phase controlling its rate [11]. Using the Henry's Law constant, the estimated volatilization half-life of 2,4-dichlorotoluene in a model river 1 m deep with a 1 m/sec current and a 3 m/sec wind is 4.0 hr [11]. The half-life of 2,4-dichlorotoluene in the Rhine River in the Netherlands is 1 day, as determined by measuring concentration differences over a reach of the river [16]. According to a recommended classification scheme [4], an estimated BCF value of 983 calculated using a regression-derived equation [11] suggests that 2,4-dichlorotoluene would be expected to bioconcentrate moderately in aquatic organisms. Insufficient data are available to determine the rate or importance of biodegradation of 2,4-dichlorotoluene in water.

Atmospheric Fate: According to a recommended classification scheme [2], a vapor pressure of 0.458 mm Hg at 25 °C [3] indicates that 2,4-dichlorotoluene will partially exist in the atmosphere adsorbed to particulate matter and be subject to gravitational settling. It will degrade by reaction with photochemically produced hydroxyl radicals (half-life 11.6 days) [1] and possibly by photolysis [12,15]. Due to its relatively long half-life, considerable dispersion might be expected. Particulate-phase 2,4-dichlorotoluene may be physically removed from the air by wet and dry deposition.

Biodegradation:

Abiotic Degradation: 2,4-Dichlorotoluene reacts with photochemically produced hydroxyl radicals in the atmosphere by addition to the aromatic ring and H-atom abstraction. The estimated half-life for these reactions is 11.6 days [1]. 2,4-dichlorotoluene has an adsorption band at 279.2 nm in the vapor phase and 280.5 in methanol that extends to about 300 nm [12,15] and it is therefore a candidate for direct photolysis by sunlight. However, no information concerning photolysis rates could be found.

Bioconcentration: The log octanol/water partition coefficient for 2,4-dichlorotoluene is 4.24 [7], from which one can estimate a BCF of 983 using a recommended regression equation [11]. According to a recommended classification scheme [4], this estimated BCF value suggests that 2,4-dichlorotoluene would be expected to bioconcentrate moderately in aquatic organisms.

2,4-Dichlorotoluene

Soil Adsorption/Mobility: The log octanol/water partition coefficient for 2,4-dichlorotoluene is 4.24 [7], from which one can estimate a Koc of 4,800 using a recommended regression equation [11]. According to a recommended classification scheme [14], this estimated Koc value suggests that 2,4-dichlorotoluene would be expected to adsorb strongly to soil and sediment.

Volatilization from Water/Soil: The Henry's Law constant for 2,4-dichlorotoluene is 0.346×10^{-2} atm-m^3/mol [9]. Volatilization from water should therefore be rapid with resistance in the liquid phase controlling its rate [11]. Using the Henry's Law constant, the estimated volatilization half-life of 2,4-dichlorotoluene in a model river 1 m deep with a 1 m/sec current and a 3 m/sec wind is 4.0 hr [11]. Volatilization of 2,4-dichlorotoluene may be important from moist soil surfaces given its experimental Henry's Law constant [9], and from dry soil surfaces based on an experimental vapor pressure of 0.458 mm Hg at 25 °C [3].

Water Concentrations: SURFACE WATER: 2,4-Dichlorotoluene was detected, but not quantified, in the Niagara River of the Lake Ontario Basin [6].

Effluent Concentrations: In a comprehensive survey of wastewater from 4000 industrial and publicly owned treatment works (POTWs) sponsored by the Effluent Guidelines Division of the US EPA, 2,4-dichlorotoluene was identified in discharges of the following industrial category (frequency of occurrence, median concn in ppb): rubber processing (1; 314.9), organic chemicals (11; 20.9), publicly owned treatment works (5; 40.4) [13]. The highest effluent concentration was 1424 ppb in the paint and ink industry [13]. 2,4-Dichlorotoluene has been detected, but not quantified, in water samples at the Love Canal [8].

Sediment/Soil Concentrations: 2,4-Dichlorotoluene has been detected, but not quantified, in soil/sediment samples at the Love Canal [8].

Atmospheric Concentrations: SOURCE AREAS: 2,4-Dichlorotoluene has been detected but not quantified in air samples at the Love Canal [8].

Food Survey Values:

2,4-Dichlorotoluene

Plant Concentrations:

Fish/Seafood Concentrations:

Animal Concentrations:

Milk Concentrations:

Other Environmental Concentrations:

Probable Routes of Human Exposure: Exposure to 2,4-dichlorotoluene would be primarily occupational via inhalation and dermal contact.

Average Daily Intake:

Occupational Exposure:

Body Burdens:

REFERENCES

1. Atkinson R; Int J Chem Kinet 19: 799-828 (1987)
2. Bidleman TF; Environ Sci Technol 22:361-7 (1988)
3. Daubert TE, Danner RP; Physical & Thermodynamic Properties of Pure Chemicals Vol 4 NY, NY: Hemisphere Pub Corp (1989)
4. Franke C et al; Chemosphere 29:1501-14 (1994)
5. Gelfand S; Kirk-Othmer Encyclopedia of Chemical Technology 3rd ed 5: 825 (1979)
6. Great Lakes Water Quality Board; An Inventory of Chemical Substances Identified in the Great Lakes Ecosystem. Windsor Ontario: Great Lakes Water Quality Board (1983)
7. Hansch C et al; Exploring QSAR; Hydrophobic, Electronic and Steric Constants ACS Professional Reference Book Heller SR (consult ed), Washington, DC: Amer Chem Soc pg. 28 (1995)
8. Hauser TR, Bromberg SM; Environ Monit Assess 2: 249-72 (1982)
9. Hine J, Mookerjee PK; J Org Chem 40: 292-8 (1975)
10. Lide DR (ed); CRC Handbook of Chemistry and Physics 75th ed. Boca Raton, FL: CRC Press, pg. 3-40 (1994)
11. Lyman WJ et al; in Handbook of Chemical Property Estimation Methods Washington, DC: Amer Chem Soc (1990)
12. Sadtler; Sadtler Standard Spectra 211UV Philadelphia PA: Sadtler Research Lab (NA)
13. Shackelford WM et al; Analyt Chim Acta 146: 15-27 (1983)

2,4-Dichlorotoluene

14. Swann RL et al; Res Rev 85:23 (1985)
15. Thakur G, Upadhya KN; Curr Sci 37: 134-5 (1968)
16. Zoeteman BCJ et al; Chemosphere 9: 231-49 (1980)

2,6-Dichlorotoluene

SUBSTANCE IDENTIFICATION

Synonyms:

Structure:

CAS Registry Number: 118-69-4

Molecular Formula: $C_7H_6Cl_2$

Smiles Notation: c(c(c(cc1)Cl)C)(c1)Cl

CHEMICAL AND PHYSICAL PROPERTIES

Boiling Point: 198 °C at 760 mm Hg

Melting Point:

Molecular Weight: 161.03

Dissociation Constants:

Log Octanol/Water Partition Coefficient: 4.29 [6]

Water Solubility:

Vapor Pressure: 0.399 mm Hg at 25 °C [8] (estimated)

Henry's Law Constant: 0.346 x 10^{-2} atm-m³/mol at 25 °C [7]

ENVIRONMENTAL FATE/EXPOSURE POTENTIAL

Summary: 2,6-Dichlorotoluene may be released to the environment in emissions or in wastewater during its production and use as a chemical intermediate. If released to soil, 2,6-dichlorotoluene will be immobile.

Volatilization from moist and dry soils may occur. Insufficient data are available to determine the rate or importance of biodegradation of 2,6-dichlorotoluene in soil. If released to water, 2,6-dichlorotoluene may adsorb to suspended solids and sediment. 2,6-Dichlorotoluene may volatilize from water surfaces with an estimated volatilization half-life in a model river of 4.0 hr. 2,6-Dichlorotoluene is expected to bioconcentrate moderately in aquatic organisms. Insufficient data are available to determine the rate or importance of biodegradation of 2,6-dichlorotoluene in water. In the atmosphere, 2,6-dichlorotoluene will exist in the vapor phase. Vapor-phase 2,6-dichlorotoluene will degrade by reaction with photochemically produced hydroxyl radicals (half-life 11.6 days) and possibly photolysis. Particulate-phase 2,6-dichlorotoluene may be physically removed from the air by wet and dry deposition. Exposure to 2,6-dichlorotoluene would be primarily occupational.

Natural Sources:

Artificial Sources: 2,6-Dichlorotoluene may be released as emissions or in wastewater during its production and use as a chemical intermediate in the manufacture of dyestuffs and herbicides [4].

Terrestrial Fate: Based on a recommended classification scheme [12], an estimated Koc value of 5100, determined from a log Kow value of 4.29 [6] and a recommended regression equation [8], indicates that 2,6-dichlorotoluene will be immobile in soil, 2,6-Dichlorotoluene's high vapor pressure [8] and high Henry's Law constant [7] indicate that volatilization from moist and dry soils may occur. Insufficient data are available to determine the rate or importance of biodegradation of 2,6-dichlorotoluene in soil.

Aquatic Fate: Based on a recommended classification scheme [12], an estimated Koc value of 5100, determined from an experimental log Kow value [6] and a recommended regression-derived equation [8], indicates that 2,6-dichlorotoluene may adsorb to suspended solids and sediment in the water. 2,6-Dichlorotoluene may volatilize from water surfaces based on an experimental Henry's Law constant of 0.346×10^{-2} atm-m^3/mol [7]. Using the Henry's Law constant [7], the estimated volatilization half-life of 2,6-dichlorotoluene in a model river 1 m deep with a 1 m/sec current and a 3 m/sec wind is 4.0 hr [8]. An estimated BCF value of 1070 [8], from an

146

experimental log Kow value [6], suggests that 2,6-dichlorotoluene is expected to bioconcentrate moderately in aquatic organisms. Insufficient data are available to determine the rate or importance of biodegradation of 2,6-dichlorotoluene in water.

Atmospheric Fate: According to a suggested classification scheme [2], an estimated vapor pressure of 0.399 mm Hg at 25 °C [8] indicates that 2,6-dichlorotoluene will exist in the vapor phase in the ambient atmosphere. Vapor-phase 2,6-dichlorotoluene will degrade by reaction with photochemically produced hydroxyl radicals (half-life 11.6 days [1]) and possibly photolysis [10]. Particulate-phase 2,6-dichlorotoluene may be physically removed from the air by wet and dry deposition.

Biodegradation: Bacterial strains isolated from a landfill site previously used for disposal of chlorinated organic wastes were able to utilize 2,6-dichlorotoluene [13].

Abiotic Degradation: 2,6-Dichlorotoluene reacts with photochemically produced hydroxyl radicals in the atmosphere by addition to the aromatic ring and H-atom abstraction. The estimated half-life for these reactions is 11.6 days [1]. 2,6-Dichlorotoluene has an absorption band at 274 nm that extends to about 300 nm [10], and it is therefore a candidate for direct photolysis by sunlight. No information concerning photolysis rates could be found; however, on irradiation with shorter UV light in dry, nitrogen-purged methanol, rapid dehalogenation occurs at the 2 position [9].

Bioconcentration: The log octanol/water partition coefficient for 2,6-dichlorotoluene is 4.29 [6], from which one can estimate a BCF of 1070 using a recommended regression equation [8]. According to a recommended classification scheme [3], 2,6-dichlorotoluene is expected to bioconcentrate moderately in aquatic organisms.

Soil Adsorption/Mobility: The log octanol/water partition coefficient for 2,6-dichlorotoluene is 4.29, from which one can estimate a Koc of 5,100 using a recommended regression equation [8]. According to a recommended classification scheme [12], 2,6-dichlorotoluene will adsorb strongly to soil and sediment.

2,6-Dichlorotoluene

Volatilization from Water/Soil: The Henry's Law constant for 2,6-dichlorotoluene is 0.346×10^{-2} atm-m^3/mol [7]. Volatilization from water should therefore be rapid with resistance in the liquid phase controlling its rate [8]. Using the Henry's Law constant, the estimated volatilization half-life of 2,6-dichlorotoluene in a model river 1 m deep with a 1 m/sec current and a 3 m/sec wind is 4.0 hr [8]. 2,6-Dichlorotoluene's high vapor pressure [8] and high Henry's Law constant [7] indicate that volatilization from moist and dry soils may occur.

Water Concentrations: 2,6-Dichlorotoluene was detected, but not quantified, in the Niagara River of the Lake Ontario Basin [5].

Effluent Concentrations: In a comprehensive survey of wastewater from 4000 industrial and publicly owned treatment works (POTWs) sponsored by the Effluent Guidelines Division of the USEPA, 2,6-dichlorotoluene was identified in discharges of the following industrial category (frequency of occurrence, median concentration in ppb): organic chemicals (14; 26.7), publicly owned treatment works (5; 40.4) [11]. The highest effluent concentration was 1424 ppb in the paint and ink industry [11].

Sediment/Soil Concentrations:

Atmospheric Concentrations:

Food Survey Values:

Plant Concentrations:

Fish/Seafood Concentrations:

Animal Concentrations:

Milk Concentrations:

Other Environmental Concentrations:

Probable Routes of Human Exposure: Exposure to 2,6-dichlorotoluene would be primarily occupational via inhalation and dermal contact.

Average Daily Intake:

Occupational Exposure:

Body Burdens:

REFERENCES

1. Atkinson R; Int J Chem Kinet 19: 799-828 (1987)
2. Bidleman TF; Environ Sci Technol 22:361-7 (1988)
3. Franke C et al; Chemosphere 29:1501-14 (1994)
4. Gelfand S; Kirk-Othmer Encyclopedia of Chemical Technology 3rd ed 5: 25 (1979)
5. Great Lakes Water Quality Board; An Inventory of Chemical Substances Identified in the Great Lakes Ecosystem. Windsor Ontario: Great Lakes Water Quality Board (1983)
6. Hansch C et al; Exploring QSAR; Hydrophobic, Electronic, and Steric Constants ACS Professional Reference Book, Heller SR (consult ed), Washington, DC: Amer Chem Soc pg. 28 (1995)
7. Hine J, Mookerjee PK; J Org Chem 40: 292-8 (1975)
8. Lyman WJ et al; Handbook of Chemical Property Estimation Methods Washington, DC: Amer Chem Soc (1990)
9. Mansour M et al; Chemosphere 9: 59-60 (1980)
10. Sadtler; Sadtler Standard Spectra 5404UV Philadelphia PA: Sadtler Research Lab (NA)
11. Shackelford WM et al; Analyt Chim Acta 146: 15-27 (1983)
12. Swann RL et al; Res Rev 85:23 (1983)
13. Vandenbergh PA et al; Appl Environ Microbiol 42: 737-9 (1981)

2,2-Dichloro-1,1,1-trifluoroethane

SUBSTANCE IDENTIFICATION

Synonyms: HFC-123

Structure:

CAS Registry Number: 306-83-2

Molecular Formula: $C_2HCl_2F_3$

SMILES Notation: FC(F)(F)C(Cl)Cl

CHEMICAL AND PHYSICAL PROPERTIES

Boiling Point: 44.88 °C [13]

Melting Point: -111.83 °C [13]

Molecular Weight: 152.93

Dissociation Constants:

Log Octanol/Water Partition Coefficient: 2.17 (estimated) [10]

Water Solubility: 185.7 mg/L at 25 °C (estimated) [12]

Vapor Pressure: 575 mm Hg at 25.5 °C [16]

Henry's Law Constant: 0.0955 atm-m^3/mol at 25 °C [9]

ENVIRONMENTAL FATE/EXPOSURE POTENTIAL

Summary: 2,2-Dichloro-1,1,1-trifluoroethane is an anthropogenic compound that is used in the production of trifluoroacetic acid. It may be released to the environment as a fugitive emission during its production or use. If released to soil, 2,2-dichloro-1,1,1-trifluoroethane will rapidly volatilize from either moist or dry soil to the atmosphere. It will display moderate mobility in soil. If released to water, 2,2-dichloro-1,1,1-trifluoroethane will rapidly volatilize to the atmosphere. The estimated half-life for volatilization from a model river is 3.6 hr. 2,2-Dichloro-1,1,1-trifluoroethane will not bioconcentrate in fish and aquatic organisms nor will it adsorb to sediment or suspended organic matter. If released to the atmosphere, 2,2-dichloro-1,1,1-trifluoroethane will undergo a slow gas-phase reaction with photochemically produced hydroxyl radicals with an estimated half-life of 479 days. The atmospheric life-time of 2,2-dichloro-1,1,1-trifluoroethane has been estimated to range from 1.2 to 6.3 years. 2,2-Dichloro-1,1,1-trifluoroethane may undergo atmospheric removal by wet deposition processes; however, any removed by this process is expected to rapidly revolatilize to the atmosphere. Occupational exposure to 2,2-dichloro-1,1,1-trifluoroethane may occur by inhalation or dermal contact during its production or use.

Natural Sources: 2,2-Dichloro-1,1,1-trifluoroethane is of anthropogenic origin, and it is not known to be produced by natural sources.

Artificial Sources: 2,2-Dichloro-1,1,1-trifluoroethane is an anthropogenic compound that is used in the production of trifluoroacetic acid [1]. It may be released to the environment as a fugitive emission during its production and use.

Terrestrial Fate: If released to soil, the estimated vapor pressure for 2,2-dichloro-1,1,1-trifluoroethane indicates that it will rapidly volatilize from dry soil to the atmosphere. Estimated soil adsorption coefficients of 150-360 [7,8] indicate that it will display moderate to high mobility in soil [14]. The estimated Henry's Law constant indicates that it will also rapidly volatilize from moist soil to the atmosphere.

Aquatic Fate: If released to water, the estimated Henry's Law constant for 2,2-dichloro-1,1,1-trifluoroethane indicates that it will rapidly volatilize to the atmosphere. The estimated half-life for volatilization from a model river is 3.6 hr [7]. Estimated bioconcentration factors ranging from 26 to 32 [7] indicate that 2,2-dichloro-1,1,1-trifluoroethane will not bioconcentrate in fish and aquatic organisms. Estimated soil adsorption coefficients of 150 to 360 [7,8] indicate that it will not adsorb to sediment or suspended organic matter.

Atmospheric Fate: If released to the atmosphere, 2,2-dichloro-1,1,1-trifluoroethane will undergo a slow gas-phase reaction with photochemically produced hydroxyl radicals. The recommended rate constant for this process of 3.35×10^{-14} cm^3/molecule-sec [2] translates to an atmospheric half-life of 479 days using an average atmospheric hydroxyl radical concn of $5 \times 10^{+5}$ molec/cm^3 [2]. The estimated water solubility of 2,2-dichloro-1,1,1-trifluoroethane indicates that it may undergo atmospheric removal by wet deposition processes; however, any removed is expected to rapidly revolatilize to the atmosphere.

Biodegradation:

Abiotic Degradation: Experimental rate constants for the gas-phase reaction of 2,2-dichloro-1,1,1-trifluoroethane with photochemically produced hydroxyl radicals of 1.4×10^{-14} cm^3/molecule-sec [4,5] and 5.9×10^{-14} cm^3/molecule-sec at 303 K [3] have been reported. The recommended value of 3.35×10^{-14} cm^3/molecule-sec [2] translates to an atmospheric half-life of 479 days using an average atmospheric hydroxyl radical concn of $5 \times 10^{+5}$ molec/cm^3 [2]. An atmospheric lifetime of 6.3 yrs has been estimated as a result of this reaction [11]. A calculated atmospheric lifetime for 2,2,-dichloro-1,1,1-trifluoroethane, obtained using both 1-dimensional and 2-dimensional models, ranges from 1.2-2.4 yrs [6]. The major degradation product is suggested to be trifluoroacetyl chloride [15]

Bioconcentration: Estimated bioconcentration factors of 26 to 32 can be calculated for 2,2-dichloro-1,1,1-trifluoroethane based on its estimated log octanol/water partition coefficient and estimated water solubility. These values indicate that 2,2-dichloro-1,1,1-trifluoroethane will not bioconcentrate in fish and aquatic organisms.

Soil Adsorption/Mobility: Estimated soil adsorption coefficients of 150 to 360 can be calculated for 2,2-dichloro-1,1,1-trifluoroethane based on a structure estimation method [8] and its estimated log octanol/water partition coefficient [7]. This value indicates that 2,2-dichloro-1,1,1-trifluoroethane will display moderate to high mobility in soil [14].

2,2-Dichloro-1,1,1-trifluoroethane

Volatilization from Water/Soil: The estimated Henry's Law constant indicates that 2,2-dichloro-1,1,1-trifluoroethane will rapidly volatilize from water and moist soil to the atmosphere [7]. The estimated half-life for volatilization from a model river 1 m deep flowing at 1 m/sec with a wind speed of 3 m/sec is 3.6 hr [7]. The estimated vapor pressure of 2,2-dichloro-1,1,1-trifluoroethane indicates that 2,2-dichloro-1,1,1-trifluoroethane will rapidly volatilize from dry soil to the atmosphere.

Water Concentrations:

Effluent Concentrations:

Sediment/Soil Concentrations:

Atmospheric Concentrations:

Food Survey Values:

Plant Concentrations:

Fish/Seafood Concentrations:

Animal Concentrations:

Milk Concentrations:

Other Environmental Concentrations:

Probable Routes of Human Exposure: Occupational exposure to 2,2-dichloro-1,1,1-trifluoroethane may occur by inhalation or dermal contact during its production or use.

Average Daily Intake:

Occupational Exposure:

Body Burdens:

2,2-Dichloro-1,1,1-trifluoroethane

REFERENCES

1. Astrologes G; p. 894 in Ullman's Encycl Indust Chem A10 NY: VCH Publ (1987)
2. Atkinson R; J Chem Phys Ref Data Monograph 1 (1989)
3. Brown AC et al; Atmos Environ 24A: 2499-511 (1990)
4. Cohen N, Benson SW; J Phys Chem 91: 171-5 (1987)
5. Cohen N, Benson SW; J Phys Chem 91: 162-70 (1987)
6. Fisher DA et al; Nature 344: 508-12 (1990)
7. Lyman WJ et al; Handbook of Chemical Property Estimation Methods NY: McGraw-Hill Chapt 15 (1982)
8. Meylan WM et al; Environ Sci Technol 26:1560-1567 (1992)
9. Meylan WM, Howard PH; Environ Toxicol Chem 10: 1283-93 (1991)
10. Meylan W, Howard PH; J Pharm Sci 84: 83-92 (1995)
11. Montague DC, Perrine RL; Atmos Environ 24: 1331-9 (1990)
12. SRC; WS/KOW Program, Syracuse Research Corp. (1995)
13. SRC; MPBPVP Program, Syracuse Research Corp. (1995)
14. Swann RL et al; Res Rev 85: 17-28 (1983)
15. Tuazon ER, Atkinson, RR; J Atmos Chem 17: 179-199 (1993)
16. Weber LA; Fluid Phase Equilbia 80: 141-48 (1992)

1,2-Diethylbenzene

SUBSTANCE IDENTIFICATION

Synonyms:

Structure:

CAS Registry Number: 135-01-3

Molecular Formula: $C_{10}H_{14}$

SMILES Notation: c1cc(CC)c(CC)cc1

CHEMICAL AND PHYSICAL PROPERTIES

Boiling Point: 183.4 °C

Melting Point: -31.2 °C

Molecular Weight: 134.22

Dissociation Constants:

Log Octanol/Water Partition Coefficient: 4.50 (estimated) [5]

Water Solubility: 71.1 mg/L at 25 °C [19]

Vapor Pressure: 1.05 mm Hg at 25 °C [3]

Henry's Law Constant: 2.61 x 10^{-3} atm-m^3/mol (estimated from water solubility and vapor pressure)

ENVIRONMENTAL FATE/EXPOSURE POTENTIAL

Summary: 1,2-Diethylbenzene may be released to the environment in effluents at sites where it is produced or used as a chemical intermediate or solvent. Diethylbenzenes are components of gasoline, kerosine and No. 2 Fuel oil, and 1,2-diethylbenzene is released to the environment as a product of combustion engines. It may also be released to the environment as volatile emissions from landfills and offshore oil drilling platforms. 1,2-Diethylbenzene is not expected to hydrolyze, but has the potential to undergo direct photolysis in sunlit environmental media because it absorbs UV light at wavelengths in the environmentally significant range, >290 nm. Limited aqueous grab sample data for gas oil mixtures containing 1,2-diethylbenzene suggest it should biodegrade in soil and water. An estimated Koc indicates 1,2-diethylbenzene should have a medium mobility in soil and it may partition from the water column to organic matter in sediments and suspended solids. The potential for bioconcentration of 1,2-diethylbenzene in aquatic organisms is low. 1,2-Diethylbenzene's Henry's Law constant suggests that it should rapidly volatilize from natural waters. The volatilization half-lives from a model river and a model pond, the latter including the effect of adsorption, have been estimated to be about 3.8 hr and 4.25 days, respectively. Based on its vapor pressure, 1,2-diethylbenzene should evaporate from dry surfaces, especially when present in high concn such as in spill situations. 1,2-Diethylbenzene is expected to exist entirely in the vapor phase in ambient air. Vapor phase reactions with photochemically produced hydroxyl radicals in the atmosphere may be important (estimated half-life of 1.9 days). Physical removal from air by rainfall and dissolution in clouds, etc. may occur; however, the short atmospheric residence time of 1,2-diethylbenzene suggests that wet deposition will be of limited importance. The most probable exposure to 1,2-diethylbenzene would be occupational exposure, which may occur through dermal contact or inhalation at places where it is produced or used. Workplace exposures have been documented. Common nonoccupational exposure would include inhalation; 1,2-diethylbenzene is a widely occurring atmospheric pollutant.

Natural Sources:

Artificial Sources: 1,2-Diethylbenzene may be released to the environment via effluents at sites where it is produced or used as a chemical intermediate or a solvent [6]. Diethylbenzenes are components of gasoline, kerosine and No. 2 Fuel oil [2], and 1,2-diethylbenzene has been identified as a product of combustion engines [14]. It may also be released to the environment as volatile emissions from landfills [17]. Offshore oil drilling platforms may release 1,2-diethylbenzene to surrounding waters [10].

Terrestrial Fate: 1,2-Diethylbenzene is not expected to hydrolyse in soils [9], but has the potential to undergo photolysis on sunlit soil surfaces (it absorbs UV light at wavelengths in the environmentally significant range, >290 nm [12]). Limited aqueous grab sample data for gas oil mixtures containing 1,2-diethylbenzene suggest it should biodegrade in terrestrial environments. An estimated Koc of 418 [9] indicates 1,2-diethylbenzene should have a medium mobility in soil [13]. Based upon an estimated Henry's Law constant, 1,2-diethylbenzene should rapidly volatilize from moist soils [9]. Based on its vapor pressure, 1,2-diethylbenzene should evaporate from dry surfaces, especially when present in high concn such as in spill situations.

Aquatic Fate: 1,2-Diethylbenzene is not expected to hydrolyze [9], but has the potential to undergo direct photolysis in aquatic systems (it absorbs UV light at wavelengths in the environmentally significant range, >290 nm [12]). Ground water [7] and marine water grab sample tests [16], in which 1,2-diethylbenzene was a constituent of gas oil mixtures, suggest it will biodegrade in natural waters. The estimated log bioconcentration factor (log BCF) of 1.75 indicates the potential for 1,2-diethylbenzene to bioconcentrate in aquatic organisms is low. An estimated Koc of 418 [9] indicates 1,2-diethylbenzene can partition from the water column to organic matter contained in sediments and suspended solids. An estimated Henry's Law constant suggests 1,2-diethylbenzene should rapidly volatilize from natural waters [9]. Based on this Henry's Law constant, the volatilization half-life from a model river has been estimated to be 3.8 hr [9]. The volatilization half-life from a model pond, which considers the effect of adsorption, has been estimated to be about 4.25 days [15].

Atmospheric Fate: Based on its vapor pressure, 1,2-diethylbenzene is expected to exist entirely in the vapor phase in ambient air [4]. Vapor phase reactions with photochemically produced hydroxyl radicals in the atmosphere may be important. The rate constant for 1,2-diethylbenzene was estimated to be 8.28×10^{-12} cm^3/molecule-sec at 25 °C, which corresponds to an atmospheric half-life of about 1.9 days at an atmospheric concn of $5 \times 10^{+5}$ hydroxyl radicals per cm^3[1]. The water solubility of 1,2-diethylbenzene indicates that physical removal from air by rainfall and dissolution in clouds, etc. may occur; however, the short atmospheric residence time of 1,2-diethylbenzene suggests that wet deposition is of limited importance. 1,2-Diethylbenzene has the potential to undergo direct photolysis in air (it absorbs UV light at wavelengths in the environmentally significant range, >290 nm [12]).

Biodegradation: 1,2-Diethylbenzene at a concn of 22.7 ug/L was completely removed within 5 days from a gas oil mixture at a concn of about 2 mg/L added to acclimated, fresh, well water grab samples from Tuffenwies and Zurich, Switzerland, with a pH of 8.0, at 10 and 25 °C and microbial populations of 300-400 cells/mL [7]. A gas oil with a concn of 0.5 ppm contained 1,2-diethylbenzene, which degraded at a moderate rate in North Sea coastal water maintained at 20 °C for 14 days [16].

Abiotic Degradation: Alkyl benzenes are generally resistant to hydrolysis [9]. Alkyl benzenes absorb UV light at wavelengths in the environmentally significant range, >290 nm [12]; therefore, 1,2-diethylbenzene will not hydrolyze, but has the potential to undergo direct photolysis in the environment. In the atmosphere, vapor phase reactions with hydroxyl radicals may be important. The rate constant for the vapor-phase reaction of 1,2-diethylbenzene with photochemically produced hydroxyl radicals in air has been estimated to be 8.28×10^{-12} cm^3/molecule-sec at 25 °C, which corresponds to an atmospheric half-life of about 1.9 days at an atmospheric concn of $5 \times 10^{+5}$ hydroxyl radicals per cm^3 [1].

Bioconcentration: Based upon its water solubility, the bioconcentration factor (log BCF) for 1,2-diethylbenzene has been calculated to be 1.75, from a recommended regression-derived equation [9]. This BCF value indicates the potential for 1,2-diethylbenzene to bioconcentrate in aquatic organisms is low.

Soil Adsorption/Mobility: Based on its water solubility, a Koc value of 418 for 1,2-diethylbenzene has been calculated from a regression-derived equation [9]. This Koc value indicates 1,2-diethylbenzene will have a medium mobility in soil [13].

Volatilization from Water/Soil: 1,2-Diethylbenzene is reported to be a product of combustion engines [14]. 1,2-Diethylbenzene was also identified as a vapor emitted from landfills [17]. Formation water from offshore oil production platforms were found to contain 1,2-diethylbenzene at a concn of 140 ug/L [10].

Sediment/Soil Concentrations:

Atmospheric Concentrations: SOURCE DOMINATED: According to the National Ambient Volatile Organic Compounds (VOCs) Database, the median concn of 1,2-diethylbenzene for source-dominated atmospheres is below the detection limit for 7 samples [11]. 1,2-Diethylbenzene was detected in the air at the Gatwick Airport, UK, in 1979; airplane engines were the source [14]. URBAN/SUBURBAN: According to the National Ambient Volatile Organic Compounds (VOCs) Database, the median urban atmospheric concn of 1,2-diethylbenzene is 0.232 ppbV for 342 samples [11]. Diethylbenzenes were detected in 4 of 4 outdoor air samples and 8 of 12 indoor air samples from 10 public access buildings (offices, schools, elderly homes and a hospital) [18]. The atmospheres of Pretoria, Johannesburg, and Durban, South Africa, were shown to contain 1,2-diethylbenzene [18]. According to the National Ambient Volatile Organic Compounds (VOCs) Database, the median suburban atmospheric concn of 1,2-diethylbenzene is 0.172 ppbV for 202 samples [11]. RURAL/REMOTE: According to the National Ambient Volatile Organic Compounds (VOCs) Database, the median remote atmospheric concn of 1,2-diethylbenzene is below the detection limit for 1 sample [11].

Food Survey Values:

Plant Concentrations:

Fish/Seafood Concentrations:

Animal Concentrations:

Milk Concentrations:

Other Environmental Concentrations:

Probable Routes of Human Exposure: The most probable route of human exposure to 1,2-diethylbenzene is by inhalation or dermal contact. Atmospheric workplace exposures have been documented [14,18].

1,2-Diethylbenzene

Monitoring data indicate that 1,2-diethylbenzene can be an atmospheric pollutant.

Average Daily Intake:

Occupational Exposure: The most probable human exposure to 1,2-diethylbenzene would be occupational exposure. Diethylbenzenes were detected in 8 of 12 indoor air samples from 10 public access buildings (offices, schools, elderly homes and a hospital) [18]. 1,2-Diethylbenzene was detected in the air at the Gatwick Airport, UK, in 1979; airplane engines were the source [14]. Nonoccupational exposures may include the inhalation of 1,2-diethylbenzene as an atmospheric pollutant.

Body Burdens:

REFERENCES

1. Atkinson R; Intern J Chem Kin 19: 799-828 (1987)
2. Coleman WE et al; Arch Environ Contam Toxicol 13: 171-8 (1984)
3. Daubert TE, Danner RP; Data Compilation, Tables of Properties of Pure Cmpds, Design Inst for Phys Prop Data, Am Inst for Phys Prop Data, NY,NY (1989)
4. Eisenreich SJ et al; Environ Sci Technol 15: 30-8 (1981)
5. GEMS; Graphical Exposure Modeling System. CLOGP. USEPA (1987)
6. Hawley GG; Condensed Chemical Dictionary 10th ed Van Nostrand Reinhold NY p. 375 (1981)
7. Kappeler T, Wuhrmann K; Water Res 12: 327-33 (1978)
8. Lucas SV; GC/MS Anal of Org in Drinking Water Concentrates and Advanced Treatment Concentrates Vol 1 USEPA-600/1-84-020A (NTIS PB85-128239) p. 397 (1984)
9. Lyman WJ et al; Handbook of Chemical Property Estimation Methods NY: McGraw-Hill pp 4-9, 5-4, 5 -10, 7-4, 15-15 to 15-29 (1982)
10. Sauer TC Jr; Environ Sci Technol 15: 917-23 (1981)
11. Shah JJ, Heyerdahl EK; National Ambient VOC Database Update USEPA-600/3-88/010 (1988)
12. Silverstein RM, Bassler GC; Spectrometric Ident of Org Cmpd, J Wiley & Sons Inc p. 148-169 (1963)
13. Swann RL et al; Res Rev 85: 16-28 (1983)
14. Tsani-Bazaca E et al; Chemosphere 11: 11-23 (1982)
15. USEPA; EXAMS II Computer Simulation (1987)
16. Van der Linden AC; Dev Biodegrad Hydrocarbons 1: 165-200 (1978)

1,2-Diethylbenzene

17. Vogt WG, Walsh JJ Volatile Organic Compounds in Gases from Landfill Simulators Proc APCA Annu Meet 78th 6: 2-17 (1985)
18. Wallace L; Volatile Organic Chemicals in 10 Public-Access Buildings USEPA-600/D-87/152 (1987)
19. Yalkowsky SH et al; Arizona Data Base of Water Solubility (1989)

1,3-Diethylbenzene

SUBSTANCE IDENTIFICATION

Synonyms:

Structure:

CAS Registry Number: 141-93-5

Molecular Formula: $C_{10}H_{14}$

SMILES Notation: c1cc(CC)cc(CC)c1

CHEMICAL AND PHYSICAL PROPERTIES

Boiling Point: 181 °C at 760 mm Hg

Melting Point: -83.89 °C

Molecular Weight: 134.22

Dissociation Constants:

Log Octanol/Water Partition Coefficient: 4.50 (estimated) [4]

Water Solubility: 170 mg/L (estimated) [17]

Vapor Pressure: 1.13 mm Hg at 25 °C [6]

Henry's Law Constant: 1.17×10^{-3} atm-m^3/mol at 25 °C (estimated from vapor pressure and water solubility)

ENVIRONMENTAL FATE/EXPOSURE POTENTIAL

Summary: 1,3-Diethylbenzene may be released to the environment via effluents at sites where it is produced or used as a chemical intermediate or a solvent. Diethylbenzenes are components of gasoline, kerosine and No. 2 Fuel oil, and 1,3-diethylbenzene is released to the environment as a product of combustion engines. 1,3-Diethylbenzene may also be released to the environment via wastewater effluents from oil refineries, paint and ink industries, textile mills, automobile and other laundries, plastic and organic chemicals. 1,3-Diethylbenzene is not expected to hydrolyze, but has the potential to undergo direct photolysis in sunlit environmental media (it absorbs UV light at wavelengths in the environmentally significant range, >290 nm). Limited aqueous grab sample data for gas oil mixtures containing 1,3-diethylbenzene suggest it should biodegrade in soil and water. An estimated Koc indicates 1,3-diethylbenzene should have a medium mobility in soil and it can partition from the water column to organic matter in sediments and suspended solids. 1,3-Diethylbenzene has the potential to bioconcentrate in aquatic organisms. An estimated Henry's Law constant of 1.17×10^{-3} atm-m^3/mol at 25 °C suggests that 1,3-diethylbenzene should rapidly volatilize from natural waters. The volatilization half-lives from a model river and a model pond, the latter considers the effect of adsorption, have been estimated to be about 4.3 hr and 3.7 days, respectively. Based on its vapor pressure, 1,3-diethylbenzene should evaporate from dry surfaces, especially when present in high concentrations such as in spill situations. 1,3-Diethylbenzene is expected to exist entirely in the vapor phase in ambient air. Vapor phase reactions with photochemically produced hydroxyl radicals in the atmosphere may be important (estimated half-life of 1.1 days). The most probable human exposure to 1,3-diethylbenzene would be occupational exposure, which may occur through dermal contact or inhalation at places where it is produced or used. Workplace exposures have been documented. Common nonoccupational exposure would include inhalation; 1,3-diethylbenzene is a widely occurring atmospheric pollutant.

Natural Sources:

Artificial Sources: 1,3-Diethylbenzene may be released to the environment via effluents at sites where it is produced or used as a chemical intermediate or a solvent [9]. Diethylbenzenes are components of gasoline, kerosine and No. 2 Fuel oil [5], and 1,3-diethylbenzene has been identified as a product of combustion engines [24]. 1,3-Diethylbenzene may also be released to the environment via wastewater effluents from oil refineries [22], paint and ink industries, textile mills, automobile and other laundries, plastic and organic chemicals [3].

Terrestrial Fate: 1,3-Diethylbenzene is not expected to hydrolyze in soils [16], but has the potential to undergo photolysis on sunlit soil surfaces (it absorbs UV light at wavelengths in the environmentally significant range, >290 nm [21]). Limited aqueous grab sample data for gas oil mixtures containing 1,3-diethylbenzene suggest it should biodegrade in terrestrial environments. An estimated Koc of 260 [1] indicates 1,3-diethylbenzene should have a medium mobility in soil [23]. Based upon an estimated Henry's Law constant, 1,3-diethylbenzene should rapidly volatilize from moist soils [16]. Based on its vapor pressure, 1,3-diethylbenzene should evaporate from dry surfaces, especially when present in high concentrations such as in spill situations.

Aquatic Fate: 1,3-Diethylbenzene is not expected to hydrolyze [1], but has the potential to undergo direct photolysis in aquatic systems (it absorbs UV light at wavelengths in the environmentally significant range, >290 nm [21]). Ground water [13] and marine water grab sample tests [26], in which 1,3-diethylbenzene was a constituent of gas oil mixtures, suggest it will biodegrade in natural waters. The estimated log bioconcentration factor (log BCF) of 3.19 indicates 1,3-diethylbenzene has the potential to bioconcentrate in aquatic organisms. An estimated Koc of 260 [16] indicates 1,3-diethylbenzene can partition from the water column to organic matter contained in sediments and suspended solids. An estimated Henry's Law constant suggests 1,3-diethylbenzene should rapidly volatilize from natural waters [16]. Based on the Henry's Law constant, the volatilization half-life from a model river has been estimated to be 4.3 hr [16]. The volatilization half-life from a model pond, which considers the effect of adsorption, has been estimated to be about 3.7 days [25].

Atmospheric Fate: Based on its vapor pressure, 1,3-diethylbenzene is expected to exist entirely in the vapor phase in ambient air [7]. Vapor phase reactions with photochemically produced hydroxyl radicals in the atmosphere may be important. The rate constant for 1,3-diethylbenzene was estimated to be 1.48×10^{-11} cm^3/molecule-sec at 25 °C, which corresponds to an atmospheric half-life of about 1.1 days at an atmospheric concentration of $5 \times 10^{+5}$ hydroxyl radicals per cm^3 [1]. 1,3-Diethylbenzene has the potential to undergo direct photolysis in air (it absorbs UV light at wavelengths in the environmentally significant range, >290 nm [21]).

Biodegradation: 1,3-Diethylbenzene at a concentration of 0.5 ppm C was completely removed within 5 days from a gas oil mixture added to acclimated fresh-wellwater grab samples from Tuffenwies and Zurich, Switzerland, with a pH of 8.0, at 10 and 25 °C and microbial populations of 300-400 cells/mL [13]. A gas oil sample with an overall concentration of 0.5 ppm contained 1,3-diethylbenzene, which degraded at a moderate rate in North Sea coastal water maintained at 20 °C for 14 days [26].

Abiotic Degradation: Alkyl benzenes are generally resistant to hydrolysis [16]. Alkyl benzenes absorb UV light at wavelengths in the environmentally significant range, >290 nm [21]; therefore, 1,3-diethylbenzene will not hydrolyze, but has the potential to undergo direct photolysis in the environment. In the atmosphere, vapor-phase reactions with hydroxyl radicals may be important. The rate constant for the vapor-phase reaction of 1,3-diethylbenzene with photochemically produced hydroxyl radicals in air has been estimated to be 1.48×10^{-11} cm^3/molecule-sec at 25 °C, which corresponds to an atmospheric half-life of about 1.1 days at an atmospheric concentration of $5 \times 10^{+5}$ hydroxyl radicals per cm^3 [1].

Bioconcentration: Based upon an estimated log Kow, the bioconcentration factor (log BCF) for 1,3-diethylbenzene has been calculated to be 3.19, from a recommended regression-derived equation [16]. This BCF value indicates 1,3-diethylbenzene has the potential to bioconcentrate in aquatic organisms.

Soil Adsorption/Mobility: Based on an estimated water solubility, a Koc value of 260 for 1,3-diethylbenzene has been calculated from a regression-derived equation [16]. This Koc value indicates 1,3-diethylbenzene should have a medium mobility in soil [16].

Volatilization from Water/Soil: The estimated Henry's Law constant for 1,3-diethylbenzene indicates that it should rapidly volatilize from natural waters [16]. The volatilization half-life from a model river (1 m deep flowing 1 m/sec with a wind speed of 3 m/sec) has been estimated to be 4.3 hr [16]. The volatilization half-life from a model pond, which considers the effect of adsorption, has been estimated to be about 3.7 days [25]. Based on the vapor pressure [2], 1,3-diethylbenzene should evaporate from dry surfaces, especially when present in high concentrations such as in spill situations.

Water Concentrations: DRINKING WATER: 1.3-Diethylbenzene was listed as a contaminant found in drinking water for a survey of US cities including Pomona, Escondido, Lake Tahoe and Orange County, CA, and Dallas, Washington, DC, Cincinnati, Philadelphia, Miami, New Orleans, Ottumwa, IA, and Seattle [15]. In February 1980, 1,3-diethylbenzene was detected in drinking water from Cincinnati, OH at a concentration of 14 ng/L [5].

Effluent Concentrations: 1,3-Diethylbenzene is reported to be a product of combustion engines [24]. Motorboats emitted 1,3-diethylbenzene to canal water with resultant concentrations ranging from 8 to 60 ng/L with an average of 41 ng/L for 8 samples [12]. 1,3-Diethylbenzene was detected at a concentration of 2.0 ug/m^3 in a plume at a distance of 1 mile from its source, a General Motors facility located in Janesville, WI [19]. 1,3-Diethylbenzene was contained in the Dissolved Air Flotation treatment effluent of a Class B oil refinery at a concentration of 13 ng/g [22]. 1,3-Diethylbenzene was detected in 5 of 21 industrial categories of wastewater effluents [3]. Extract from the wastewater of paint and ink industry contained 1,3-diethylbenzene at an average concentration of 15 mg/L; the extract from wastewater of a textile mill contained 1,3-diethylbenzene at an average concentration of 238 mg/L; the extract from wastewater of automatic and other laundries contained 1,3-diethylbenzene at an average concentration of 149 mg/L; the extract from wastewater of plastic manufacturing contained 1,3-diethylbenzene at an average concentration of 46 mg/L; and the extract from wastewater of organic chemicals manufacturing contained 1,3-diethylbenzene at an average concentration of 24 mg/L [3].

Sediment/Soil Concentrations:

Atmospheric Concentrations: SOURCE DOMINATED: At a distance of 1 mile from its source, 1,3-diethylbenzene was detected at a concentration of 2.0 ug/m^3 in the plume emitted from a General Motors plant located in Janesville, WI [19]. 1,3-Diethylbenzene was detected in the air at the Gatwick Airport, UK, in 1979; airplane engines were the source [24]. Air at a distance of 1 km from Volvo and Saab automobile manufacturing plants in SW Sweden contained 1,3-diethylbenzene at average concentrations of 0.9 and 3.0 ug/m^3; compared to regional air which contained 1,3-diethylbenzene at an average concentration of <0.01 ug/m^3 [18]. URBAN/SUBURBAN: The median urban atmospheric concentration of 1,3-diethylbenzene is 0.051 ppbV for 379 samples and the median suburban atmospheric concentration of 1,3-diethylbenzene is 0.067 ppbV for 165 samples [20]. The 1,3-diethylbenzene concentration ranged from 0 to 3 ppbV at a downtown Los Angeles location where it was detected in 16 of 17 samples in the Fall of 1981 [8]. Diethylbenzenes were detected in 4 of 4 outdoor air samples and 8 of 12 indoor air samples from 10 public access buildings (offices, schools, elderly homes and a hospital) [27]. 1,3-Diethylbenzene was detected in 2 of 21 air samples from Houston, TX, at concentrations of 108 to 97 ppm [14]. Between April and August 1975, ambient air at Delft, the Netherlands, contained 1,3-diethylbenzene at an average concentration of 0.8 ppb with a maximum concentration of 3.0 ppb [2]. 1,3-Diethylbenzene was qualitatively detected in the suburban air of Tubingen, Germany [12]. RURAL/REMOTE: 1,3-Diethylbenzene was qualitatively detected in ambient air of the Black Forest, Germany [12]. Rural air in Sweden contained 1,3-diethylbenzene at concentrations < 0.01 ug/ m^3 [18].

Food Survey Values:

Plant Concentrations:

Fish/Seafood Concentrations: A 10 g tissue sample of carp from Las Vegas, NV, contained 1,3-diethylbenzene at a concentration of 40 ppb [10].

Animal Concentrations:

Milk Concentrations:

Other Environmental Concentrations:

Probable Routes of Human Exposure: The most probable route of human exposure to 1,3-diethylbenzene is by inhalation or dermal contact. Atmospheric workplace exposures have been documented [24,27].

1,3-Diethylbenzene

Monitoring data indicate that 1,3-diethylbenzene can be an atmospheric pollutant.

Average Daily Intake:

Occupational Exposure: The most probable human exposure to 1,3-diethylbenzene would be occupational exposure. Diethylbenzenes were detected in 8 of 12 indoor air samples from 10 public access buildings (offices, schools, elderly homes and a hospital) [27]. 1,3-Diethylbenzene was detected in the air at the Gatwick Airport, UK, in 1979; airplane engines were the source [2]. Nonoccupational exposures may include the inhalation of 1,3-diethylbenzene as an atmospheric pollutant.

Body Burdens:

REFERENCES

1. Atkinson R; Intern J Chem Kin 19: 799-828 (1987)
2. Bos R et al; Sci Total Environ 7: 269-81 (1977)
3. Bursey JT, Pellizzari ED; Analysis of Industrial Wastewater for Organic Pollutants in Consent Decree Survey. Contract No 68-03-2867. Athens, GA: USEPA Environ Res Lab (1982)
4. CLOGP; PCGEMS Graphical Exposure Modeling System USEPA (1986)
5. Coleman WE et al; Arch Environ Contam Toxicol 13: 174-8 (1984)
6. Daubert TE, Danner RP; Data Compilation, Tables of Properties of Pure Cmpds, Design Inst for Phys Prop Data, Am Inst for Phys Prop Data, NY, NY (1989)
7. Eisenreich SJ et al; Environ Sci Technol 15: 30-8 (1981)
8. Grosjean D, Fung K; J Air Pollut Control Assoc 34: 537-43 (1984)
9. Hawley GG; Condensed Chemical Dictionary 10th ed Van Nostrand Reinhold NY p. 375 (1981)
10. Hiatt MH, Anal Chem 55: 506-16 (1983)
11. Hine J, Mookerjee PK; J Org Chem 40: 292-8 (1975)
12. Juttner F; Chemosphere 15: 985-92 (1986)
13. Kappeler T, Wuhrmann K; Water Res 12: 327-33 (1978)
14. Lonneman WA et al; Hydrocarbons in Houston Air USEPA-600/3-79/018 p. 44 (1979)
15. Lucas SV; GC/MS Anal of Org in Drinking Water Concentrates and Advanced Treatment Concentrates Vol 1 USEPA-600/1-84--020A (NTIS PB85-128239) p. 397 (1984)
16. Lyman WJ et al; Handbook of Chemical Property Estimation Methods NY: McGraw-Hill p. 4-9 (1982)

1,3-Diethylbenzene

17. Neely WB, Blau GE; Environmental Exposure from Chemical Vol 1, Boca Baton, FL CRC Press p. 35 (1985)
18. Petersson G; Environ Pollut Series B: 207-17 (1982)
19. Sexton K, Westberg H; Environ Sci Technol 14: 329-32 (1980)
20. Shah JJ, Heyerdahl EK; National Ambient VOC Database Update USEPA-600/3-88/010 (1988)
21. Silverstein RM, Bassler GC; Spectrometric Ident Org Cmpd, J Wiley & Sons Inc p. 148-69 (1963)
22. Snider EH, Manning FS; Environ Int 7: 237-58 (1982)
23. Swann RL et al; Res Rev 85: 16-28 (1983)
24. Tsani- Bazaca E et al; Chemosphere 11: 11-23 (1982)
25. USEPA; EXAMS II Computer Simulation (1987)
26. Van der Linden AC; Dev Biodegrad Hydrocarbons 1: 165-200 (1978)
27. Wallace L; Volatile Organic Chemicals in 10 Public-Access Buildings USEPA-600/D-87/152 (1987)

1,4-Diethylbenzene

SUBSTANCE IDENTIFICATION

Synonyms:

Structure:

CAS Registry Number: 105-05-5

Molecular Formula: $C_{10}H_{14}$

SMILES Notation: c1(CC)ccc(CC)cc1

CHEMICAL AND PHYSICAL PROPERTIES

Boiling Point: 183.8 °C at 760 mm Hg

Melting Point: -42.8

Molecular Weight: 134.22

Dissociation Constants:

Log Octanol/Water Partition Coefficient: 4.50 (estimated) [6]

Water Solubility: 24.8 mg/L at 25 °C [22]

Vapor Pressure: 1.1 mm Hg at 25 °C

Henry's Law Constant: 7.54 x 10^{-3} atm-m³/mol at 25 °C (estimated from vapor pressure and water solubility)

ENVIRONMENTAL FATE/EXPOSURE POTENTIAL

Summary: 1,4-Diethylbenzene may be released to the environment via effluents at sites where it is produced or used as a chemical intermediate or a solvent. Diethylbenzenes are components of gasoline, kerosine and No. 2 Fuel oil, and 1,4-diethylbenzene is released to the environment as a product of combustion engines. 1,4-Diethylbenzene is not expected to hydrolyze, but has the potential to undergo direct photolysis in sunlit environmental media (it absorbs UV light at wavelengths in the environmentally significant range, >290 nm). Limited aqueous grab sample data for gas oil mixtures containing 1,4-diethylbenzene suggest it should biodegrade in soil and water. A high estimated Koc indicates 1,4-diethylbenzene should have a low mobility in soil and it should partition from the water column to organic matter in sediments and suspended solids. The potential for bioconcentration of 1,4-diethylbenzene in aquatic organisms is low. A Henry's Law constant of 7.54×10^{-3} atm-m^3/mol at 25 °C suggests that 1,4-diethylbenzene should rapidly volatilize from natural waters. The volatilization half-lives from a model river and a model pond, the latter considers the effect of adsorption, have been estimated to be about 3.5 hr and 6 days, respectively. Based on its vapor pressure, 1,4-diethylbenzene should evaporate from dry surfaces, especially when present in high concentration such as in spill situations. 1,4-Diethylbenzene is expected to exist entirely in the vapor phase in ambient air. Vapor phase reactions with photochemically produced hydroxyl radicals in the atmosphere may be important (estimated half-life of 1.9 days). The short atmospheric residence time of 1,4-diethylbenzene suggests that wet deposition is of limited importance. The most probable human exposure to 1,4-diethylbenzene would be occupational exposure, which may occur through dermal contact or inhalation at places where it is produced or used. Workplace exposure has been documented. Common nonoccupational exposure would include inhalation; 1,4-diethylbenzene is a widely occurring atmospheric pollutant.

Natural Sources:

Artificial Sources: 1,4-Dicthylbenzene may be released to the environment via effluents at sites where it is produced or used as a chemical intermediate or a solvent [8]. Diethylbenzenes are components of gasoline, kerosine and No. 2 Fuel oil [3], and 1,4-diethylbenzene has been identified as a product of combustion engines [18].

Terrestrial Fate: 1,4-Diethylbenzene is not expected to hydrolyze in soils [12]. It has the potential to undergo direct photolysis on sunlit soil surfaces because it absorbs UV light at wavelengths in the environmentally significant range (>290 nm) [16]. Limited aqueous grab sample data for gas-oil mixtures containing 1,4-diethylbenzene suggest it should biodegrade in terrestrial environments. An estimated Koc of 746 [1] indicates 1,4-diethylbenzene should have a low mobility in soil [17]. Based upon its estimated Henry's Law constant, 1,4-diethylbenzene should rapidly volatilize from moist soils [16]. Based on the vapor pressure, 1,4-diethylbenzene should evaporate from dry surfaces, especially when present in high concentration such as in spill situations.

Aquatic Fate: 1,4-Diethylbenzene is not expected to hydrolyze [12]. It has the potential to undergo direct photolysis in aquatic systems because it absorbs UV light at wavelengths in the environmentally significant range (>290 nm) [16]. Ground water [10] and marine water grab sample tests [20] with gas-oil mixtures, in which 1,4-diethylbenzene was a constituent, suggest it will biodegrade in natural waters. The estimated log bioconcentration factor (log BCF) of 2.00 indicates the potential for 1,4-diethylbenzene to bioconcentrate in aquatic organisms is low. An estimated Koc of 746 [1] indicates 1,4-diethylbenzene should partition from the water column to organic matter contained in sediments and suspended solids. Its estimated Henry's Law constant suggests that 1,4-diethylbenzene should rapidly volatilize from natural waters [1]. Based on this Henry's Law constant, the volatilization half-life from a model river has been estimated to be 3.5 hr [1]. The volatilization half-life from a model pond, which considers the effect of adsorption, has been estimated to be about 6 days [19].

Atmospheric Fate: Based on its vapor pressure, 1,4-diethylbenzene is expected to exist entirely in the vapor phase in ambient air [5]. Vapor phase reactions with photochemically produced hydroxyl radicals in the atmosphere may be important. The rate constant for 1,4-diethylbenzene for this reaction was estimated to be 8.28×10^{-12} cm^3/molecule-sec at 25 °C, which corresponds to an atmospheric half-life of about 1.9 days using a hydroxyl radical concentration of $5 \times 10^{+5}$ radicals per cm^3 [1]. 1,4-Diethylbenzene's water solubility indicates that physical removal from air by rainfall and dissolution in clouds may occur; however, the short atmospheric residence time of 1,4-diethylbenzene suggests that wet deposition is of limited importance. 1,4-Diethylbenzene has the potential to undergo direct photolysis in air because it absorbs UV light at wavelengths in the environmentally significant range, >290 nm [16].

Biodegradation: 1,4-Diethylbenzene at a concentration of 0.5 ppmC was completely removed within 5 days from a gas oil mixture added to acclimated fresh-wellwater grab samples from Tuffenwies and Zurich, Switzerland, with a pH of 8.0, at 10 and 25 °C and microbial populations of 300-400 cells/mL [10]. A gas oil sample with an overall concentration of 0.5 ppm contained 1,4-diethylbenzene, which degraded at a moderate rate in North Sea coastal water maintained at 20 °C for 14 days [20].

Abiotic Degradation: Alkyl benzenes are generally resistant to hydrolysis [12]. Alkyl benzenes absorb UV light at wavelengths in the environmentally significant range, >290 nm [16]; therefore, 1,4-diethylbenzene will not hydrolyze, but has the potential to undergo direct photolysis in the environment. In the atmosphere, vapor-phase reactions with hydroxyl radicals may be important. The rate constant for the vapor-phase reaction of 1,4-diethylbenzene with photochemically-produced hydroxyl radicals in air has been estimated to be 8.28 x 10^{-12} cm^3/molecule-sec at 25 °C. This corresponds to an atmospheric half-life of about 1.9 days at an atmospheric concentration of 5 x 10^{+5} hydroxyl radicals per cm^3 [1].

Bioconcentration: Based upon its water solubility, the bioconcentration factor (log BCF) for 1,4-diethylbenzene has been calculated to be 2.00, from a recommended regression-derived equation [12]. This BCF value indicates the potential for 1,4-diethylbenzene to bioconcentrate in aquatic organisms is low.

Soil Adsorption/Mobility: Based on its water solubility, a Koc value of 746 for 1,4-diethylbenzene has been calculated from a regression-derived equation [12]. This Koc value indicates that 1,4-diethylbenzene will have a low mobility in soil [17].

Volatilization from Water/Soil: Using its Henry's Law constant, the volatilization half-life for 1,4-diethylbenzene from a model river (1 m deep flowing 1 m/sec with a wind speed of 3 m/sec) is estimated to be 3.5 hr [12]. The volatilization half-life from a model pond, which considers the effect of adsorption, has been estimated to be about 6 days [19]. Based on its vapor pressure, 1,4-diethylbenzene should evaporate from dry surfaces, especially when present in high concentration such as in spill situations.

Water Concentrations: DRINKING WATER: 1,4-Diethylbenzene was listed as a contaminant found in drinking water for a survey of US cities including Pomona, Escondido, Lake Tahoe and Orange County, CA, and Dallas, Washington, DC, Cincinnati, Philadelphia, Miami, New Orleans, Ottumwa, IA, and Seattle [11]. In February 1980, 1,4-diethylbenzene was detected in drinking water from Cincinnati, OH, at a concentration of 12 ng/L [3].

Effluent Concentrations: Vulcanization and extrusion operations during rubber and synthetic production with electrical insulation emits 1,4-diethylbenzene to the air [2]. 1,4-Diethylbenzene is reported to be a product of combustion engines [18]. Motorboats emitted 1,4-diethylbenzene to canal water with resultant concentration ranging from 9 to 29 ng/L with an average of 17 ng/L for 7 samples [9]. 1,4-Diethylbenzene was detected at a concentration of 2.0 ug/m^3 in a plume at a distance of 1 mile from its source, a General Motors facility located in Janesville, WI [14].

Sediment/Soil Concentrations:

Atmospheric Concentrations: SOURCE DOMINATED: At a distance of 1 mile from its source, 1,4-diethylbenzene was detected at a concentration of 2.0 ug/m^3 in the plume emitted from a General Motors plant located in Janesville, WI [14]. 1,4-Diethylbenzene was detected in the air at the Gatwick Airport, UK, in 1979; airplane engines were the source [18]. URBAN/SUBURBAN: 1,4-Diethylbenzene was listed as one of 64 most abundant air pollutants in US cities [13]. For 821 air samples, collected from 1984-1986, in 39 US cities, the median 1,4-diethylbenzene concentration was 2.4 ppbC, with a minimum and maximum concentration of 4.0 and 33 ppbC [13]. According to the National Ambient Volatile Organic Compounds (VOCs) Database, the median urban atmospheric concentration of 1,4-diethylbenzene below the detection limit for 250 samples and the median suburban atmospheric concentration of 1,4-diethylbenzene is 0.250 ppbV for 71 samples [15]. The 1,4-diethylbenzene concentration ranged from 0 to 3 ppbV at a downtown Los Angeles location where it was detected in 16 of 17 samples in the Fall of 1981 [7].

1,4-Diethylbenzene

Diethylbenzenes were detected in 4 of 4 outdoor air samples and 8 of 12 indoor air samples from 10 public access buildings (offices, schools, homes for the elderly and a hospital) [21].

Food Survey Values:

Plant Concentrations:

Fish/Seafood Concentrations:

Animal Concentrations:

Milk Concentrations:

Other Environmental Concentrations:

Probable Routes of Human Exposure: The most probable route of human exposure to 1,4-diethylbenzene is by inhalation or dermal contact. Atmospheric workplace exposures have been documented [2,21]. Monitoring data indicate that 1,4-diethylbenzene can be an atmospheric pollutant [15].

Average Daily Intake:

Occupational Exposure: The most probable human exposure to 1,4-diethylbenzene would be occupational exposure. The atmospheric concentration of 1,4-diethylbenzene ranged from 0 to 2 ug/m^3 for the extrusion area of an electrical insulation manufacturing plant [2]. The source of 1,4-diethylbenzene was reported to be aromatic oil used as a plasticizer [2]. Diethylbenzenes were detected in 8 of 12 indoor air samples from 10 public access buildings (offices, schools, homes for the elderly and a hospital) [21]. 1,4-Diethylbenzene was detected in the air at the Gatwick Airport, UK, in 1979: airplane engines were the source [18]. Nonoccupational exposures may include the inhalation of 1,4-diethylbenzene as an atmospheric pollutant.

Body Burdens:

1,4-Diethylbenzene

REFERENCES

1. Atkinson R; Intern J Chem Kin 19: 799-828 (1987)
2. Cocheo V et al; Am Ind Hyg Assoc J 44: 521-7 (1983)
3. Coleman WE et al; Arch Environ Contam Toxicol 13: 171-8 (1984)
4. Daubert TE, Danner RP; Data Compilation, Tables of Properties of Pure Cmpds, Design Inst for Phys Prop Data, Am Inst for Phys Prop Data, NY, NY (1989)
5. Eisenreich SJ et al; Environ Sci Technol 15: 30-8 (1981)
6. GEMS; Graphical Exposure Modeling System. CLOGP. USEPA (1987)
7. Grosjean D, Fung K; J Air Pollut Control Assoc 34: 537-43 (1984)
8. Hawley GG; Condensed Chemical Dictionary 10th ed Van Nostrand Reinhold NY p. 375 (1981)
9. Juttner F; Z Wasser-Abwasser-Forrsch 21: 36-9 (1988)
10. Kappeler T, Wuhrmann K; Water Res 12: 327-33 (1978)
11. Lucas SV; GC/MS Anal of Org in Drinking Water Concentrates and Advanced Treatment Concentrates Vol 1 USEPA-600/1-84-020A (NTIS PB85-128239) p. 397 (1984)
12. Lyman WJ et al; Handbook of Chemical Property Estimation Methods NY: McGraw-Hill pp. 4-9, 5-4, 5-10, 7-4, 15-15 to 15-19 (1982)
13. Seila RL et al; Determination of C2 to C12 Ambient Air Hydrocarbons in 39 US Cities, from 1984 through 1986. USEPA-600/S3-89/058 (1989)
14. Sexton K, Westberg H; Environ Sci Technol 14: 329-32 (1980)
15. Shah JJ, Heyerdahl EK; National Ambient VOC Database Update USEPA-600/3-88/010 (1988)
16. Silverstein RM, Bassler GC; Spectrometric Ident Org Cmpd, J Wiley & Sons Inc p. 148-69 (1963)
17. Swann RL et al; Res Rev 85: 16-28 (1983)
18. Tsani-Bazaca E et al; Chemosphere 11: 11-23 (1982)
19. USEPA; EXAMS II Computer Simulation (1987)
20. Van der Linden AC; Dev Biodegrad Hydrocarbons 1: 165-200 (1978)
21. Wallace L; Volatile Organic Chemicals in 10 Public-Access Buildings USEPA-600/D-87/152 (1987)
22. Yalkowsky SH et al; Arizona Data Base of Water Solubility (1989)

Dipropylamine

SUBSTANCE IDENTIFICATION

Synonyms:

Structure:

CAS Registry Number: 142-84-7

Molecular Formula: $C_6H_{15}N$

SMILES Notation: N(CCC)CCC

CHEMICAL AND PHYSICAL PROPERTIES

Boiling Point: 109-110 °C

Melting Point: -39.6 °C

Molecular Weight: 101.19

Dissociation Constants: pKa = 11.00 [15]

Log Octanol/Water Partition Coefficient: 1.67 [5]

Water Solubility: 29,200 ppm at room temperature [11]

Vapor Pressure: 24.1 mm Hg at 25 °C [15]

Henry's Law Constant: 5.1 x 10^{-5} atm-m^3/mol at 25 °C [6]

ENVIRONMENTAL FATE/EXPOSURE POTENTIAL

Summary: Dipropylamine occurs naturally in various plants, such as tobacco. It is released to the environment by humans in effluents from its industrial production and use. If released to the atmosphere, dipropylamine is rapidly degraded (estimated half-life of 4.4 hr) by reaction with photochemically produced hydroxyl radicals. If released to water, dipropylamine is physically removed by volatilization. Volatilization half-lives of 0.83 and 9.5 days have been estimated for a shallow (1 m deep) model river and pond, respectively. If released to soil, dipropylamine is expected to be moderately to very mobile and easily leached based upon estimated Koc values of 15-393. Evaporation from dry soil is likely to occur. A single screening study has demonstrated that dipropylamine is readily biodegraded by activated sludge inocula. The general population is primarily exposed through consumption of food products in which dipropylamine apparently occurs as a natural product. Occupational exposure is possible through inhalation and dermal contact at sites of commercial production and use.

Natural Sources: Dipropylamine has been found to occur naturally in tobacco [7,16].

Artificial Sources: Dipropylamine is present in various waste effluents generated at commercial sites of propylamines manufacture [9].

Terrestrial Fate: When released to soil, dipropylamine is expected to be moderately to very mobile and easily leached based upon estimated Koc values of 15-393. Its relatively high vapor pressure suggests that significant evaporation from dry surfaces may occur. Little data is available on biodegradation, but one screening study [8] suggests that dipropylamine may be biodegradable in soil.

Aquatic Fate: Dipropylamine is physically removed from water by volatilization. Volatilization half-lives of 0.83 and 9.5 days have been estimated for a shallow (1 m deep) model river and an environmental pond, respectively. A single screening study [8] has demonstrated that dipropylamine is readily biodegraded by activated sludge inocula. Aquatic bioconcentration and adsorption to sediment are not expected to be important.

Atmospheric Fate: Based on the vapor pressure, dipropylamine is expected to exist almost entirely in the vapor phase in the ambient atmosphere [2].

Dipropylamine

Vapor-phase dipropylamine is degraded rapidly in the atmosphere by reaction with photochemically produced hydroxyl radicals (estimated half-life of 4.4 hr).

Biodegradation: Dipropylamine was determined to be biodegradable using the Japanese MITI test protocol [8].

Abiotic Degradation: The rate constant for the vapor-phase reaction of dipropylamine with photochemically produced hydroxyl radicals has been estimated to be 87×10^{-12} cm^3/molecule-sec at 25 °C [1,12]; assuming an average atmospheric hydroxyl radical concn of $5 \times 10^{+5}$ molecules/cm^3, the half-life for this reaction is estimated to be 4.4 hr [1,12].

Bioconcentration: Based on the log Kow, the log BCF for dipropylamine can be estimated to be 1.04 from a recommended regression-derived equation [10], and therefore, dipropylamine will not be expected to bioconcentrate.

Soil Adsorption/Mobility: Based on the water solubility and the log Kow, the Koc of dipropylamine can be estimated to range from 15-193 from a regression-derived equation [10]. These estimated Koc values are indicative of very high to medium soil mobility [18].

Volatilization from Water/Soil: The value of the Henry's Law constant is indicative of potentially significant, but not rapid, volatilization from environmental waters [10]. The volatilization half-life from a model river (1 m deep flowing 1 m/sec with a wind speed of 3 m/sec) has been estimated to be 20 hr [10]. The volatilization half-life from an environmental pond has been estimated to be 9.5 days [19].

Water Concentrations: SURFACE WATER: Dipropylamine was detected at levels ranging from 0.3-3 ppb in various river waters from W. Germany [13].

Effluent Concentrations:

Sediment/Soil Concentrations: Dipropylamine was qualitatively detected in a loam topsoil collected near Moscow [3].

179

Atmospheric Concentrations:

Food Survey Values: The following concns (in mg/kg) of dipropylamine were detected in various food products from W. Germany: preserved broken beans, 1.0; preserved shelled peas, 0.1; preserved red cabbage, 0.1; paprika, 0.3; paprika brine, 2.0; cucumber, 0.1-1.4; pickled onions, 1.1; celery, 0.9; cheese, 8.4; brown bread, 0.4 [13]. Trace levels (<0.1 ppm) detected in baked ham [17]. Dipropylamine was qualitative detection in boiled beef [4].

Plant Concentrations: Latakia tobacco leaf has been found to contain dipropylamine [7]. Dipropylamine was identified as a naturally occurring compound (concn <0.1 ppm) in samples of American tobacco [16].

Fish/Seafood Concentrations: Dipropylamine concns of 0.2-0.4 ppm have been detected in whole samples of spotted trout and small mouth bass [17].

Animal Concentrations:

Milk Concentrations:

Other Environmental Concentrations:

Probable Routes of Human Exposure: The general population is primarily exposed to dipropylamine through consumption of food products. Dipropylamine apparently occurs as a natural product in many food items. Occupational exposure is possible through inhalation and dermal contact at sites of commercial production and use.

Average Daily Intake:

Occupational Exposure: 18,018 workers are potentially exposed to dipropylamine, based on statistical estimates derived from the NIOSH survey conducted between 1972 and 1974 in the US [14].

Body Burdens:

REFERENCES

1. Atkinson R; Inter J Chem Kinet 19: 799-828 (1987)
2. Eisenreich SJ et al; Environ Sci Technol 15: 30-8 (1981)
3. Golovnya RV et al; p. 327-35 in USSR Acad Med Sci (1982)
4. Golovnya RV et al; Chem Senses Flavour 4: 97-105 (1979)
5. Hansch C, Leo AJ; Medchem Project Issue No. 26 Claremont, CA: Pomona College (1985)
6. Hine J, Mookerjee PK; J Org Chem 40: 292-98 (1975)
7. Irvinc WJ, Saxby MJ; Phytochemistry 8: 473-6 (1969)
8. Kawasaki M; Ecotoxic Environ Safety 4: 444-54 (1980)
9. Liepins R et al; Industrial Process Profiles for Environmental Use: USEPA-600/2-77-023f Chpt 6. p. 6-563, 6-566 (1977)
10. Lyman WJ et al; Handbook of Chemical Property Estimation Methods NY: McGraw-Hill p. 5-4 (1982)
11. Merck; An Encyclopedia of Chemicals, Drugs and Biologicals 10th ed pp. 1130 (1983)
12. Meylan WM, Howard PH; Chemosphere 26:2293-2299 (1993)
13. Neurath GB et al; Food Cosmet Toxicol 15: 275-82 (1977)
14. NIOSH; National Occupational Hazard Survey (NOHS) (1974)
15. Riddick JL et al; Organic Solvents: Physical Properties and Methods of Purification. 4th ed. NY: Wiley-Interscience (1986)
16. Singer GM, Lijinsky W; J Agric Food Chem 24: 553-5 (1976)
17. Singer GM, Lijinsky W; J Agric Food Chem 24: 550-3 (1976)
18. Swann RL et al; Res Rev 85: 16-28 (1983)
19. USEPA; EXAMS II Computer Simulation (1987)

Dodecane

Synonyms:

Structure:

CAS Registry Number: 112-40-3

Molecular Formula: $C_{12}H_{26}$

SMILES Notation: C(CCCCCCCCCC)C

CHEMICAL AND PHYSICAL PROPERTIES

Boiling Point: 216.3 °C

Melting Point: -9.6

Molecular Weight: 170.34

Dissociation Constants:

Log Octanol/Water Partition Coefficient: 6.10 [3]

Water Solubility: 0.0034 mg/L at 25 °C [21]

Vapor Pressure: 0.12 mm Hg at 25 °C [21]

Henry's Law Constant: 9.35 atm-m^3/mol at 25 °C [22]

ENVIRONMENTAL FATE/EXPOSURE POTENTIAL

Summary: n-Dodecane will enter the air primarily from fugitive emissions and exhaust associated with the use of gasoline and diesel fuel. In addition, n-dodecane is released on land and in waterways from spills and in wastewater. Releases into water will decrease in concentration (half-life 0.5-4 days) due to adsorption to sediment and particulate matter in the water column, biodegradation, and possibly volatilization, particularly from oil slicks. Dodecane released on land will be retained in the upper layers of soil and biodegrade within several months, especially if microbial populations are acclimated. In the atmosphere, n-dodecane is most likely associated with particulate matter and will be subject to gravitational settling. The vapor-phase compound will slowly photooxidize with an estimated half-life of 4 to 17 hr. n-Dodecane did not bioconcentrate in the one species of fish studied but does bioconcentrate in algae and mussels. The primary source of exposure is from the air, especially in areas of high traffic. Away from urban areas, air is free of dodecane.

Natural Sources: Paraffin fraction of petroleum [30].

Artificial Sources: Wastewater and spills from laboratory and general use of paraffins, petroleum oils, tars, etc; municipal waste; highway runoff; auto and motorboat exhaust [1] result in environmental release of n-dodecane. Diesel vehicle exhaust [11] and tobacco smoke [7] also contain n-dodecane.

Terrestrial Fate: n-Decane will be adsorbed by soil and also be biodegraded in a period of under 4 months. Its presence in percolation water under a landfill may not reflect dodecane's usual behavior in soil since high concentrations of chemicals may destroy soil microorganisms and the disposal of solvents in a landfill will increase leaching.

Aquatic Fate: When introduced to water, n-dodecane disappears due to biodegradation, adsorption to sediments and particulate matter in the water column, and evaporation [31]. The dominant mechanism will depend on the weather conditions and characteristics of the body of water. The half-life of dodecane was 0.5 days in a river in the Netherlands based on sampling between points in the river [35]. The half-life in a mesocosm simulating seasonal conditions in Narragansett Bay, RI, was 1.1, 0.7 and 3.6 days under spring, summer, and winter conditions, respectively [31]. Based on its accelerated degradation in the presence of microorganisms and the fact that its half-life was longest in winter, it was suggested that biodegradation and adsorption to particulate matter were key factors in dodecane's disappearance [31].

Dodecane

Atmospheric Fate: When released into the atmosphere, n-dodecane degrades by reaction with photochemically produced hydroxyl radicals (half-life: 17 hr in clean air, 4 hr in moderately polluted air). The experimental hydroxyl radical rate constant for n-dodecane is 14.2×10^{-12} cm^3/mol-sec [18] for an estimated half-life of 1.2 days assuming a hydroxyl radical concentration of $5 \times 10^{+5}$ radicals/cm^3. Part of the dodecane will be associated with particulate matter. The photodegradation of adsorbed dodecane has not been studied; however, it will be subject to gravitational settling. However, its reactivity when adsorbed to particulate matter has not been studied.

Biodegradation: Dodecane biodegrades in sewage, sediment, soil, and fresh and marine water, with the rate of degradation being strongly influenced by the acclimation of the degrading microorganisms [5,6,10,19,23,26,30,31,32]. Thirty-seven percent of the dodecane was mineralized in a 5 day biodegradability test using activated sludge, with most of the remaining radioactivity from the labeled substrate being bound to the sludge as unextractable residue [5]. In other studies, 74% of the theoretical BOD was achieved in 24 hr [6]; 22 and 67% of the theoretical BOD was attained in 2 and 10 days, respectively, in a soil suspension [10]; and 40 and 46% degradation occurred when n-dodecane was exposed to microorganisms from polluted estuarial water and oil-rich sediment in Chesapeake Bay for an unspecified time period. Less degradation occurred in less contaminated water [32]. In similar studies, 16 and 49% of dodecane in crude oil exposed to harbor water degraded in 5 and 15 days, respectively, whereas 21 and 87% exposed to harbor sediment degraded in the same time period [23], and 95.1% degradation occurred in 21 days in seawater inoculated with oil-oxidizing microorganisms [30]. When anaerobically digested sewage sludge was amended on soil, all dodecane had disappeared from 0-1 and 6-7 cm core sections 1 yr after the last treatment [19]. Anaerobic conditions tend to retard hydrocarbon degradation [19]. Dodecane in jet fuels applied to soil cores and subjected to simulated rain had disappeared when tested after 131 days [26]. In a mesocosm experiment that simulated seasonal conditions in Narragansett Bay, RI, the half-life for dodecane was 1.1, 0.7, and 3.6 days under spring, summer, and winter conditions, respectively. Under summer conditions the half-life was tripled when mercury-chloride was added to eliminate biodegradation [31]. The degradation proceeded without any lag and mineralization was rapid [31].

Dodecane

Abiotic Degradation: Alkanes degrade in the atmosphere by reacting with photochemically produced hydroxyl radicals to form water and the corresponding alkyl radical. The half-life for n-dodecane was calculated to be 17 hr in clean air and 4 hr in moderately polluted air [13]. The experimental hydroxyl radical rate constant for n-dodecane is 14.2×10^{-12} cm^3/mol-sec [18], for an estimated half-life of 1.2 days assuming a hydroxyl radical concentration of $5 \times 10^{+5}$ radicals/cm^3. However, its reactivity when adsorbed to particulate matter has not been studied. When dodecane adsorbed on silica gel is irradiated with UV light (290 nm) at 15 °C, 2.9% is mineralized in 17 hr [5]. Whether dodecane also photolyzes in the free state or sorbs on particulate matter in the atmosphere has not been established. Alkanes are generally resistant to hydrolysis [20].

Bioconcentration: The log of the bioconcentration factor in static tests was 1.72 for golden orfes after 3 days and 3.80 for green algae after 24 hr [5]. Only traces of dodecane were taken up by a marine diatom from crude oil [16].

Soil Adsorption/Mobility: There is little data on the adsorption of dodecane by soil. It is slowly intercalated into well dried montmorillonite clay [4]. When JP-4 and JP-5 jet fuels were applied to soil cores and subjected to simulated rain, the dodecane component of the fuels was transported only in the top 10 cm after 50-65 days [26]. Koc has been estimated to be 67,000, which would indicate that in a bay with a typical concentration of suspended particles of 6 mg/L, 2.4% of the dodecane would be in the sorbed state [31].

Volatilization from Water/Soil: No specific information could be found on the evaporation of dodecane from water or soil; however, it has been suggested that its rapid removal from river and seawater (half-life 0.5-4 days) was not partially due to evaporation [31,35]. Since undecane evaporated from an oil slick (half-life ca 24 hr in a 21 km/hr wind) [30], dodecane can be presumed to evaporate in a somewhat longer time. The estimated Henry's Law constant of 9.35 atm-m^3/mol at 25 °C [22] would suggest that n-dodecane will evaporate rapidly from water when it is not adsorbed to suspended solids [20]. Using a model river (1 m deep, 3 m wind speed, and 1 m velocity) and the Henry's Law constant, the half-life for evaporation would be 3.8 hr.

185

Dodecane

Water Concentrations: DRINKING WATER: 3 New Orleans water treatment plants 0.1-0.37 ppb; however, not cited as being identified in drinking water of 10 US cities [17]. GROUND WATER: percolation water 30-500 m from dump 35 ppb [30]. SURFACE WATER: Identified but not quantified in Lower Fox River, WI [25], and Delaware River, PA [29]. SEAWATER: Coastal water Vineland Sound, MA, 5 ppt max, usually undetectable [8]. Surface coastal waters of Gulf of Mexico 8 stations, 0-9.8 ppt [27].

Effluent Concentrations: Dodecane was detected in municipal waste incineration plant at 0.05 ug/m^3 [15].

Sediment/Soil Concentrations: Puget Sound (March Point) 2.5-4.8 ppb [2].

Atmospheric Concentrations: RURAL/REMOTE: US - 85 samples 0 ppt median and max [1]. URBAN/SUBURBAN: US - 936 samples 0-23,000 ppt, 250 ppt median [1]. INDOOR AIR: Identified in indoor air in a larger but unspecified fraction of Chicago homes, indoor air concn 2.4 times outdoor concn [14]. Concn of dodecane in a telephone switching center in Neenah, WI was 1.1 to 1.7 ug/m^3 from 3/87 to 5/87 (1.0-1.2 ug/m^3 outdoor) and 3.0 to 4.8 (0.8 to 1.1 ug/m^3 outdoor) from 5/87 to 7/88 [28]. Concn of dodecane was 2.5, 2.1, and 2.2 ug/m3 in personal air samples from Los Angeles (February 1984), Los Angeles (May 1984), and Contra Costa (June 1984), respectively [33]. The concn of dodecane in a building with health and comfort complaints was 4.4 to 10 ug/m3 from 6/87 to 3/88 [34]. SOURCE AREAS: US - 14 samples 310-4900 ppt, 1100 ppt avg [1]. Allegheny Mountain Tunnel, PA 520-700 ppt [11]. Concn of dodecane was 0.7, 0.7, and 0.2 ug/m^3 in outdoor air samples from Los Angeles (Feb 1984), Los Angeles (May 1984), and Contra Costa (June 1984), respectively [33].

Food Survey Values:

Plant Concentrations:

Fish/Seafood Concentrations: Mussels from Puget Sound (March Point) 79-360 ppb [2]. Three species of fish from the Canary Islands contained 325 to 1063 ng/g wet wt [9].

Dodecane

Animal Concentrations:

Milk Concentrations:

Other Environmental Concentrations:

Probable Routes of Human Exposure: Since n-dodecane is present in emissions from auto and diesel exhaust and evaporation of gasoline, humans are primarily exposed to n-dodecane in air, especially in areas with heavy traffic and near filling stations. Tobacco smoke is another source of exposure.

Average Daily Intake:

Occupational Exposure: Estimated concn at an oil shale wastewater facility; 10 ppb outdoors, 2.3 ppb indoors [12].

Body Burdens: Detected, not quantified in 7 of 8 samples of mothers' milk of women from 4 urban areas in US [24]. Concn of dodecane was 0.2, 0.4, and 0.4 ug/m^3 in breath samples from Los Angeles (February 1984), Los Angeles (May 1984), and Contra Costa (June 1984), respectively [33].

REFERENCES

1. Brodzinsky R, Singh HB; Volatile Organic Chemicals in the Atmosphere: an assessment of available data; pp. 131-2 SRI contract 68-02-3452 Menlo Park, CA (1982)
2. Brown DW et al; Investigation of Petroleum in the Marine Environs of the Strait of Juan de Fuca and Northern Puget Sound; p.107 USEPA-600/7-79-164 (1979)
3. Coates et al; Environ Sci Technol 19: 628-32 (1985)
4. Eltantawy IM, Arnold PW; Nature (London) Phys Sci 237: 123-5 (1972)
5. Freitag D et al; Ecotox Environ Safety 6: 60-81 (1982)
6. Gerhold RM, Malaney EW; J Water Pollut Control 38: 562-79 (1966)
7. Graedel TE; Chemical Compounds in the Atmosphere; Academic Press New York NY p. 64 (1978)
8. Gschwend PM et al; Environ Sci Technol 16: 31-8 (1982)
9. Guintero S, Diaz C; Mar Pollut Bull 28: 44-9 (1994)
10. Haines JR, Alexander M; Appl Microbiol 28: 1084-5 (1974)
11. Hampton CV et al; Environ Sci Technol 17: 699-708 (1983)
12. Hawthorne SB, Sievers RE; Environ Sci Technol 18: 483-90 (1984)
13. Hendry DG, Kenley,RA; Atmospheric Reactions of Organic Compounds; pp. 14-29

USEPA-560/12-79-001 (1979)

14. Jarke FH et al; ASHRAE Trans 87: 153-66 (1981)
15. Jay K et al; Chemosphere 30: 1249-1260 (1995)
16. Karydis M; Microb Ecol 5: 287-93 (1980)
17. Keith LH et al; pp. 329-73 in Identification and Analysis of Organic Chemicals in Water; Keith LH ed Ann Arbor Press Ann Arbor MI (1976)
18. Kwok ESC, Atkinson R; Atmos Environ (in press) Final Report CMA ARC-8-0-OR (1995)
19. Liu D; Bull Environ Contam Toxicol 25: 616-22 (1980)
20. Lyman WJ et al; Handbook of Chemical Property Estimation Methods. Environmental behavior of organic compounds McGraw-Hill New York NY (1982)
21. Mackay D, Shiu WY; J Phys Chem Ref Data 10: 1175-1198 (1981)
22. Meylan W, Howard PH; Environ Toxicol Chem 10: 1283-93 (1991)
23. Nagata S, Kondo G; 1977 Oil Spill Conf Amer Petrol Inst pp. 617-20 (1977)
24. Pellizzari ED et al; Bull Environ Contam Toxicol 28: 322-8 (1982)
25. Peterman DH et al; pp.145-60 in Hydrocarbons and Halogenated Hydrocarbons in the Aquatic Environment Afghan BK, Mackay D eds Plenum Press New York NY (1980)
26. Ross WD et al; Environmental Fate and Biological Consequences of Chemicals Related to Air Force Activities; p. 173 NTIS AD-A121 28815 (1982)
27. Sauer TC Jr et al; Mar Chem 7: 1-16 (1978)
28. Shields HC, Weschler CJ; J Air Waste Manage Assoc 42: 792-804 (1992)
29. Suffet IH et al; pp. 375-97 in Identification and Analysis of Organic Chemicals in Water; Keith LH ed Ann Arbor Press Ann Arbor MI (1976)
30. Verscheuren K; Handbook of Environmental Data on Organic Chemicals; 2nd ed Van Nostrand Reinhold Co New York NY pp. 595-6 (1983)
31. Wakeham SG et al; Environ Sci Technol 17: 611-7 (1983)
32. Walker JD, Colwell RR; Prog Water Technol 7: 783-91 (1975)
33. Wallace LA; Toxicol Environ Chem 12: 215-236 (1986)
34. Weschler CJ et al; Am Ind Hyg Assoc J 51: 261-268 (1990)
35. Zoeteman BCJ et al; Chemosphere 9: 231-49 (1980)

Ethyl n-Butyrate

SUBSTANCE IDENTIFICATION

Synonyms:

Structure:

CAS Registry Number: 105-54-4

Molecular Formula: $C_6H_{12}O_2$

SMILES Notation: O=C(OCC)CCC

CHEMICAL AND PHYSICAL PROPERTIES

Boiling Point: 120 °C

Melting Point: -97.8 °C

Molecular Weight: 116.16

Dissociation Constants:

Log Octanol/Water Partition Coefficient: 1.85 [15]

Water Solubility: 4,900 mg/L at 20 °C [3]

Vapor Pressure: 12.8 mm Hg at 20 °C [4]

Henry's Law Constant: 3.99×10^{-4} atm-m^3/mol at 20 °C (calculated from vapor pressure and water solubility)

ENVIRONMENTAL FATE/EXPOSURE POTENTIAL

Summary: Ethyl butyrate may be released to the environment in effluents and emissions from its manufacturing plants, in spills during transport of bulk quantities, and from the land disposal of unused products that contain this compound. It is also a natural product of certain plants. If released to soil, it will not be expected to strongly adsorb to soil but will be expected to exhibit very high mobility in the soil based upon an estimated Koc of 41. It may be susceptible to volatilization from near-surface soils based upon the predicted rapid volatilization rates from surface water and relatively high vapor pressure. Hydrolysis will not be expected to be an important removal process except in alkaline soils based upon hydrolysis rates measured in water. It is not known whether it will be subject to biodegradation in soil. If released to water, it will not be expected to strongly adsorb to sediment or suspended particulate matter based upon an estimated Koc of 41 or to bioconcentrate in aquatic organisms based upon an estimated BCF of 12. It is not known whether it will be subject to biodegradation in natural waters or soils, but based upon the chemical structure it should biodegrade rapidly. It will be subject to rapid volatilization from surface water based upon an estimated half-life of 5.5 hr for volatilization from a model river (1 m deep flowing 1 m/sec with a wind speed of 3 m/sec) calculated using an estimated Henry's Law constant. The volatilization half-life from a model pond has been estimated to be 65 hr. Indirect photooxidation with hydroxyl radicals in water is not expected to be an important removal process. It should not directly photolyze in water. Hydrolysis will not be expected to be an important removal process except in alkaline waters based upon hydrolysis half-lives of 101 years, 6.3 years, 229 days, and 23 days at pH 5, 7, 8, and 9, respectively, calculated using measured base and acid hydrolysis rate constants. If ethyl butyrate is released to the atmosphere, it will be expected to exist almost entirely in the vapor phase. It will be susceptible to rapid reaction with photochemically produced hydroxyl radicals in the atmosphere based upon a half-life of 6 days, which was calculated from a measured rate constant. It should not directly photolyze in the atmosphere. General population exposure to ethyl butyrate will occur mainly via the ingestion of certain foods that contain the compound. Occupational exposure may occur through inhalation of contaminated air and dermal contact with solutions containing the compound.

Natural Sources: Ethyl butyrate is a natural product of certain plants and has been detected in the volatile components from the following natural foods: US blue cheese [5]; Beaufort mountain cheese [6]; dalieb fruit

Ethyl n-Butyrate

(*Borassus aethiopum* L.) [9]; ripening bananas [14]; commercial and concentrated aqueous orange essences [17]; Concord grape essence [22]; tree-ripened nectarines [24]; and ripening kiwi fruit [2].

Artificial Sources: Ethyl butyrate may be released to the environment in effluents and emissions from its manufacturing plants, in spills during transport of bulk quantities, and from the land disposal of unused perfumes, lacquers, solvents, natural and synthetic resins, flavorings, artificial rum, and other products that contain this compound [1,2].

Terrestrial Fate: If ethyl butyrate is released to soil, it will not be expected to strongly adsorb to soil but will be expected to exhibit very high mobility in the soil based upon an estimated Koc of 41 derived from the water solubility [12]. Therefore, it would be expected to leach through soil. It may be susceptible to volatilization from near-surface soils based upon the Henry's Law constant. Hydrolysis will not be expected to be an important removal process except in alkaline soils based upon hydrolysis rates measured in water [13]. It is not known whether it will be subject to biodegradation in soil, but based upon the chemical structure it should biodegrade rapidly [3].

Aquatic Fate: If ethyl butyrate is released to water, it will not be expected to strongly adsorb to sediment or suspended particulate matter based upon an estimated Koc of 41 [12] from the water solubility or to bioconcentrate in aquatic organisms based upon an estimated BCF of 15 [12]. It is not known whether it will be subject to biodegradation in natural waters, but based upon the chemical structure it should biodegrade rapidly [3]. It will be subject to rapid volatilization from surface water based upon an estimated half-life of 5.5 hr [12] for volatilization from a model river (1 m deep flowing 1 m/sec with a wind speed of 3 m/sec), calculated [12] using the estimated Henry's Law constant. The volatilization half-life from a model pond, which considers the effect of adsorption, has been estimated to be 65 hr [25]. Indirect photooxidation with hydroxyl radicals in water is not expected to be an important removal process based upon a half-life of 1.14 years calculated using a measured rate constant [1] and assuming a concentration of 2×10^{-17} moles hydroxyl radicals/L. It should not directly photolyze in water because simple alkyl esters do not absorb light at wavelengths >290 nm [21]. Hydrolysis will not be expected to be an important removal process except in alkaline waters based upon hydrolysis half-lives of 101 years, 6.3 years, 229 days, and 23 days at pH 5, 7, 8, and 9, respectively, calculated using measured base and acid hydrolysis rate constants [13].

Atmospheric Fate: If ethyl butyrate is released to the atmosphere, it will be expected to exist almost entirely in the vapor phase [7] based upon the experimental vapor pressure. It will be susceptible to rapid reaction with photochemically produced hydroxyl radicals in the atmosphere based upon a half-life of 6 days, which was calculated from a measured rate constant [26]. It should not directly photolyze in the atmosphere because simple alkyl esters do not absorb light at wavelengths >290 nm [21].

Biodegradation: No data regarding the biodegradation of ethyl butyrate in natural media or in laboratory screening studies were located, but based upon the chemical structure it should biodegrade rapidly [3].

Abiotic Degradation: The rate constant for the vapor-phase reaction of ethyl butyrate with photochemically produced hydroxyl radicals has been measured to be 2.7×10^{-12} cm^3/molecule-sec at 25 °C [26], which corresponds to an atmospheric half-life of 6 days at an atmospheric concentration of $5 \times 10^{+5}$ hydroxyl radicals per cm^3. The rate constant for reaction of ethyl butyrate with photochemically produced hydroxyl radicals in water has been measured to be $9.6 \times 10^{+8}$ 1/mol-sec at 25 °C [1], which corresponds to a half-life of 1.14 years at a concentration of 2×10^{-17} moles hydroxyl radicals/L. Hydrolysis of ethyl butyrate is expected to be a relatively slow process in natural waters except in alkaline waters based upon hydrolysis half-lives of 101 years, 6.3 years, 229 days, and 23 days at pH 5, 7, 8, and 9, respectively, calculated using measured base and acid hydrolysis rate constants of 3.5×10^{-2} L/mol-sec and 1.8×10^{-5} L/mol-sec, respectively [13]. Direct photolysis will not be expected to be an important degradation process in the environment because simple alkyl esters do not absorb light at wavelengths >290 nm [21].

Bioconcentration: No data regarding the bioconcentration of ethyl butyrate were located. Based upon an estimated log Kow, a BCF of 15 has been estimated using a recommended regression equation [12]. Based upon the experimental water solubility, a BCF of 5 has been estimated using a recommended regression equation [12]. Based upon these estimated BCF, ethyl butyrate will not be expected to bioconcentrate in aquatic organisms.

Ethyl n-Butyrate

Soil Adsorption/Mobility: No data regarding the adsorption of ethyl butyrate to soils, sediments or suspended particulate matter were located. Based upon the experimental water solubility, a Koc of 41 has been estimated using a recommended regression equation [12]. Based upon this estimated Koc, ethyl butyrate will be expected to exhibit very high mobility in soil [23].

Volatilization from Water/Soil: No experimental data were located regarding the volatilization rates of ethyl butyrate from water or soil. Based upon the Henry's Law constant for ethyl butyrate, the volatilization half-life from a model river (1 m deep flowing 1 m/sec with a wind speed of 3 m/sec) has been estimated to be 5.5 hr [12]. The volatilization half-life from a model pond, which considers the effect of adsorption, has been estimated to be 65 hr [25].

Water Concentrations: Ethyl butyrate was detected, not quantitated, in one river of eight weakly polluted small rivers and brooks in SW Germany [11].

Effluent Concentrations: Ethyl butyrate has been qualitatively detected in leachate from a municipal waste disposal site in the Netherlands [8].

Sediment/Soil Concentrations:

Atmospheric Concentrations:

Food Survey Values: Ethyl butyrate was detected, not quantified (detection limits listed if specified in source), in the following foods: US blue cheese aroma fraction [5]; Beaufort mountain cheese volatiles [6]; volatile flavor components of dalieb fruit (*Borassus aethiopum* L.) [9]; volatiles of ripening bananas (qualitatively detected 120 hr after unripened bananas were placed in glass test chamber; relative concn increased after initially detected through end of experiment - 10 days) [14]; commercial and concentrated aqueous orange essences [17]; Concord grape essence [22]; tree-ripened nectarines [24]. It was detected in the volatile components of ripening kiwi fruit at levels of 0.6% of the volatiles in mature fruit and 14.2% of the volatiles in ripe fruit [2]. Ethyl butyrate was detected at 0.43 to 1.53 ppm in fresh-squeezed unpasteurized orange juice [16].

Plant Concentrations:

Fish/Seafood Concentrations: Ethyl butyrate was detected at a concn of 0.29 ppm in a sample of mussel (*Mytilus edulis*) collected on July 31, 1985, at the Oarai Coast in Ibaraki, Japan [28]. It was not detected (detection limit not specified) in a sample of mussel collected at the same location on July 31, 1986 [28].

Animal Concentrations:

Milk Concentrations:

Other Environmental Concentrations:

Probable Routes of Human Exposure: General population exposure to ethyl butyrate will occur mainly via the ingestion of certain foods that contain the compound [2,5,6,9,14,17,22,24]. Occupational exposure may occur through inhalation of contaminated air and dermal contact with solutions containing the compound.

Average Daily Intake:

Occupational Exposure: NIOSH (NOES 1981-1983) has statistically estimated that 18,266 workers are potentially exposed to ethyl butyrate in the US [18]. NIOSH (NOHS 1972-1974) has statistically estimated that 5,049 workers are potentially exposed to ethyl butyrate in the US [18].

Body Burdens:

REFERENCES

1. Anbar M, Neta P; Int J Appl Radiation Isotopes 18: 493-523 (1967)
2. Bartley JP, Schwede AM; J Agric Food Chem 37: 1023-5 (1989)
3. Boethling RS et al; Environ Sci Technol 28: 459-465 (1994)
4. Daubert TE, Danner RP; Physical and Thermodynamic Properties of Pure Compounds Am Inst Chem Eng (1989)
5. Day EA, Anderson DF; J Agric Food Chem 13: 2-4 (1965)
6. Dumont JP, Adda J; J Agric Food Chem 26: 364-7 (1978)
7. Eisenreich SJ et al; Environ Sci Technol 15: 30-8 (1981)

8. Harmsen J; Water Res 17: 699-705 (1983)
9. Harper DB et al; J Sci Food Agric 37: 685-88 (1986)
10. Hawley GG; Condensed Chemical Dictionary 10th ed NY: Van Nostrand Reinhold p. 426-7 (1981)
11. Juttner F; Wat Sci Tech 25: 155-164 (1992)
12. Lyman WJ et al; Handbook of Chem Property Estimation Methods NY: McGraw-Hill p. 5-5 (1982)
13. Mabey W, Mill T; J Phys Chem Ref Data 7: 383-415 (1978)
14. Macku C, Jennings WG; J Agric Food Chem 35: 845-8 (1987)
15. Meylan W, Howard PH; J Pharm Sci 84: 83-92 (1995)
16. Moshohas MG, Shaw PE; J Agric Food Chem 42: 1525-28 (1994)
17. Moshonas MG, Shaw PE; J Agric Food Chem 38: 2181-4 (1990)
18. NIOSH; The National Occupational Hazard Survey (NOHS) (1974)
19. NIOSH; The National Occupational Exposure Survey (NOES) (1983)
20. Riddick JA et al; Organic Solvents NY: John Wiley & Sons Inc (1984)
21. Silverstein RM et al; Spectrometric Id of Org Cmpd NY: J Wiley & Sons Inc 3rd ed p. 246 (1974)
22. Stevens KL et al; J Food Sci 30: 1006-7 (1965)
23. Swann RL et al; Res Rev 85: 17-28 (1983)
24. Takeoka GR et al; J Agric Food Chem 36: 553-60 (1988)
25. USEPA; EXAMS II Computer Simulation (1987)
26. Wallington, TJ et al; J Phys Chem 92: 5024-8 (1988)
27. Windholz M ed; The Merck Index 10th ed Rahway, NJ: Merck and Co p. 547 (1983)
28. Yasuhara A, Morita M; Chemosphere 16: 2559-65 (1987)

Ethyl Chloroacetate

SUBSTANCE IDENTIFICATION

Synonyms:

Structure:

CAS Registry Number: 105-39-5

Molecular Formula: $C_4H_7ClO_2$

SMILES Notation: O=C(OCC)CCl

CHEMICAL AND PHYSICAL PROPERTIES

Boiling Point: 144-146 °C

Melting Point:

Molecular Weight: 122.56

Dissociation Constants:

Log Octanol/Water Partition Coefficient: 1.12 [11] (estimated)

Water Solubility: 19,400 mg/L at 25 °C [12]

Vapor Pressure: 5 mm Hg at 25 °C [1]

Henry's Law Constant: 4.77E-005 atm-m³/mol [10]

ENVIRONMENTAL FATE/EXPOSURE POTENTIAL

Summary: Ethyl chloroacetate use as a solvent and in the synthesis of vat dyestuffs may result in releases to the environment. If released to the soil, ethyl chloroacetate will not readily evaporate but will leach if complete hydrolysis does not occur first. Ethyl chloroacetate released to water will result in volatilization (half-life in model river - 24 hr), but is not likely to adsorb to suspended solids or sediments. The ester will hydrolyze in water, especially at more alkaline pHs. Release of ethyl chloroacetate to the atmosphere would result in reaction with photochemically generated hydroxyl radicals (estimated half-life of 12 days). No information on the biodegradation of ethyl chloroacetate in the soil or water was available. It is unknown whether ethyl chloroacetate will photolyze at environmental wavelengths.

Natural Sources:

Artificial Sources: Ethyl chloroacetate use as a solvent, in organic synthesis, as a military poison, and in the synthesis of vat dyestuffs [5] may result in release to the environment from various waste streams.

Terrestrial Fate: Ethyl chloroacetate is likely to hydrolyze in moist soils. The half-life for the soil matrix cannot be estimated from the available data, but since esters generally hydrolyze more rapidly at acidic and basic pHs and due to the nonneutral state of moist soils, the hydrolysis should be at least as fast as in water. The ester is expected to leach extensively and may reach ground water where it should hydrolyze completely.

Aquatic Fate: Ethyl chloroacetate is expected to hydrolyze completely when released to water. At 25 °C and neutral pH, ethyl chloroacetate hydrolyzes with a half-life of 74 days [4]. An alkaline hydrolysis rate constant of 1.56/M-sec at 25 °C [2] has been reported, which corresponds to half-lives of 5.1 and 0.5 days at pH 8 and 9, respectively. Ethyl chloroactetate is not expected to adsorb to sediments or bioconcentrate in fish and no significant volatilization is expected to occur.

Atmospheric Fate: The atmospheric chemistry of esters is expected to resemble that of the alkanes except that esters will react more rapidly with hydroxyl radicals than will the corresponding alkane [3]. Ethyl chloroacetate is expected to degrade ultimately to small, oxygenated organic molecules. The half-life for the reaction of ethyl chloroacetate with atmospheric hydroxyl radicals was estimated to be 12 days, assuming an average concentration of $5 \times 10^{+5}$ hydroxyl radicals/cm^3 [9]. Although ethyl chloroacetate does photolyze with 254 nm irradiation [7], the importance of sunlight (>295 nm) induced photolysis is unknown.

Biodegradation:

Abiotic Degradation: Ethyl chloroacetate will hydrolyze in water with an ionic strength of 1.0 at pH 7 and 25 °C with a half-life of 74 days [4]. An alkaline hydrolysis rate constant of 1.56/M-sec at 25 °C [2] has been reported which corresponds to half-lives of 5.1 and 0.5 days at pH 8 and 9, respectively. Ethyl chloroacetate was found to photolyze in cyclohexane at a wavelength of 254 nm and at temperatures of 15-35 °C [8]. Photolysis of ethyl chloroacetate in the gas phase was found to occur at 143 °C upon irradiation at 254 nm [7] but no insight into the importance of sunlight (>290 nm) induced photolysis is provided by these data. No data was available on the photolysis of ethyl chloroacetate in water or its reaction with hydroxyl radicals. The estimated hydroxyl radical rate constant for ethyl chloroacetate is 1.4×10^{-12} cm^3/molecule-sec [9] resulting in a half-life of 12 days, assuming an average concentration of $5 \times 10^{+5}$ hydroxyl radicals/cm^3 [9].

Bioconcentration: The estimated BCF value of 4 using a recommended regression equation [6] and the estimated octanol/water partition coefficient suggest that ethyl chloroacetate will not bioconcentrate.

Soil Adsorption/Mobility: An estimated Koc value of approximately 12 [6] from the octanol/water partition coefficient suggests that ethyl chloroacetate will not adsorb to soils or sediments. It is expected, therefore, that ethyl chloroacetate will leach if complete hydrolysis does not occur first.

Volatilization from Water/Soil: The vapor pressure and the Henry's Law constant of ethyl chloroacetate suggest that the ester may volatilize from water or soils. Based on the estimated Henry's Law constant, the half-life for the volatilization of ethyl chloroacetate from a model river of depth 1 m, flowing at 1 m/sec with an overhead wind speed of 5 m/sec, has been estimated to be 24 hr [6]; the volatilization rate from a model lake of depth 1 m, flowing at 0.05 m/sec with an overhead wind speed of 0.5 m/sec, is 10.6 days [6].

Ethyl Chloroacetate

Water Concentrations:

Effluent Concentrations:

Sediment/Soil Concentrations:

Atmospheric Concentrations:

Food Survey Values:

Plant Concentrations:

Fish/Seafood Concentrations:

Animal Concentrations:

Milk Concentrations:

Other Environmental Concentrations:

Probable Routes of Human Exposure:

Average Daily Intake:

Occupational Exposure:

Body Burdens:

REFERENCES

1. Boublik T et al; The Vapor Pressure of Pure Substances. Vol 17 Elsevier Sci Publ Amsterdam, Netherlands (1984)
2. Collette TW; Environ Sci Technol 24: 1671-1676 (1990)
3. Graedel TE; Chemical Compounds in the Atmosphere Academic Press NY p.224 (1978)
4. Jencks WP, Carriulo J; J Am Chem Soc 83: 1743-50 (1960)
5. Lewis RJ; Hawley's Condensed Chemical Dictionary 12th ed. NY: Van Nostrand Reinhold Co pg. 483 (1993)
6. Lyman WJ et al; Handbook of Chemical Property Estimation Methods. Environment Behavior of Organic Compounds. McGraw-Hill NY (1982)

Ethyl Chloroacetate

7. Matuszewski B; Tow Przyj Nauk Pr Kom Mat Przyr, Pr Chem 13: 109-116 (1972)
8. Matuszewski B; Rpcz Chem 45: 2141-48 (1971)
9. Meylan WM, Howard PH; Chemosphere 26:2293-2299 (1993)
10. Meylan W, Howard PH; Environ Toxicol Chem 10: 1283-93 (1991)
11. Meylan W, Howard PH; J Pharm Sci 84: 83-92 (1995)
12. Nanda AK, Sharma MM; Chem Engr Sci 22: 769-775 (1967)

Ethylene Dibromide

SUBSTANCE IDENTIFICATION

Synonyms:

Structure:

CAS Registry Number: 106-93-4

Molecular Formula: $C_2H_4Br_2$

SMILES Notation: BrCCBr

CHEMICAL AND PHYSICAL PROPERTIES

Boiling Point: 131-132 °C

Melting Point: 9.8 °C

Molecular Weight: 187.88

Dissociation Constants:

Log Octanol/Water Partition Coefficient: 1.96 [23]

Water Solubility: 4,150 mg/L at 25 °C [27]

Vapor Pressure: 11.2 mm Hg at 25 °C [10]

Henry's Law Constant: 6.67×10^{-4} atm-m^3/mol at 25 °C (calculated from vapor pressure and water solubility)

ENVIRONMENTAL FATE/EXPOSURE POTENTIAL

Summary: Ethylene dibromide (EDB) will enter the atmosphere primarily from fugitive emissions and exhaust associated with its use as a scavenger in leaded gasoline. Another former important but localized source is emissions from fumigation centers for citrus, grain, etc. and soil fumigation operations. In the atmosphere, ethylene dibromide will degrade by reaction with photochemically produced hydroxyl radicals (half-life 67 days). The water solubility for ethylene dibromide indicates that physical removal from air by precipitation and dissolution in clouds may occur. When spilled in water, EDB will be removed by evaporation (half-life 1-5 days). When spilled on land or applied to land during soil fumigation, ethylene dibromide will exhibit low to moderate adsorption and leach. EDB has been reported to biodegrade in aquifer samples with half-lives of 75 and 50 days under aerobic and anaerobic conditions, respectively. A selected field half-life of 100 days has been suggested. Little bioconcentration into the food chain is expected. Humans were exposed to EDB in the past from the air, especially in areas of high traffic where it was used as a lead scavenger. Another source of exposure is from ingesting fumigated food (former use), which can contain ppm levels of EDB. With current restrictions on the use of leaded gasoline and EDB fumigation, exposure to this substance should decrease.

Natural Sources:

Artificial Sources: Evaporative losses of EDB can occur from the use, storage, and transport of leaded gasoline, in which it is used as a lead scavenger [19,55]. Spills and leaking storage tanks for leaded gasoline [19,55] and exhaust from vehicles using leaded gasoline [19,55] may also result in environmental releases. Emissions have occurred in the past from its former use as a fumigant for soil, grain, fruits, vegetables, tobacco, and seed uses; these uses have been restricted or discontinued [19,55]. Wastewater and emissions from its use as a solvent for resins, gums, and waxes and as a chemical intermediate in the synthesis of dyes and pharmaceuticals may result in releases [19].

Terrestrial Fate: When spilled on land, ethylene dibromide will partially evaporate. EDB has been detected in agricultural top soils up to 19 years after the last known application [50] and concns were stable over the study period (1.5-2 yr) [43] This was suggested to be due to entrapment of EDB in micropores of soil; freshly added EDB was rapidly degraded in soil suspensions (half-lives of 4 to 8 days in two soils), while native EDB was not degraded [50]. Although some information suggests that ethylene dibromide will slowly degrade in soil, its low sorption to soil and monitoring data demonstrate that it will leach into the soil and get into ground water. EDB has been reported to biodegrade in aquifer samples with half-lives of 75 and 50 days under aerobic and anaerobic conditions, respectively [44]. Degradation does occur in anaerobic aquifer material (concn at 0, 3, 7, and 16 weeks was 194, 78, 35, and <1 ug/L) [57]. A selected field half-life of 100 days has been suggested [2]. EDB residues that persisted in soil for many years were not taken up by plants [20].

Aquatic Fate: The primary removal process for ethylene dibromide in water is evaporation. Under natural conditions it would have a half-life of slightly over a day in a typical river and about 5 days in a lake. Adsorption to sediment should be relatively low. In anoxic sediment-water suspensions, ethylene dibromide has a half-life of 55 hr [30]. Limited experimental evidence suggests that EDB does not degrade under anaerobic conditions such as may occur in aquifers. In addition, ground water monitoring shows that EDB is persistent in aquifers.

Atmospheric Fate: Ethylene dibromide degrades in the atmosphere (half-life 67 days) by reaction with photochemically produced hydroxyl radicals. The water solubility for ethylene dibromide indicates that physical removal from air by precipitation and dissolution in clouds may occur.

Biodegradation: 97% degradation in soil to ethylene in 8 weeks has been reported in laboratory experiments [11]. Under anaerobic conditions in the presence of denitrifying bacteria, no degradation was observed in 8 weeks [6]. However, under methanogenic batch culture conditions, ethylene dibromide was undetectable after 2 weeks (initial concn 89 ug/L) [6]. EDB has been reported to biodegrade in aquifer samples with half-lives of 75 and 50 days under aerobic and anaerobic conditions, respectively [44]; degradation was faster within the contaminant plume suggesting acclimation was important [44]. An aqueous biodegradation screening test resulted in zero % theoretical BOD [12].

Abiotic Degradation: Ethylene dibromide is exceedingly stable towards hydrolysis having reported half-lives of 13.2 yrs at pH 7 and 20 °C [17], 8 yrs at pH 7 and 20 °C [31], and 2.5 yrs at pH 7.5 and 25 °C [56]. Degradation of 100 ug/L of ethylene dibromide was reported, forming 30 ug/L of ethylene glycol and 6 ug/L of vinyl bromide [48]. Under conditions typical of hypoxic aqueous environments (both pristine and contaminated ground water - presence of bisulfide ion, HS⁻), ethylene dibromide had a half-life of 37 to 70 days, with the major product being 1,2-dithioethane [4]. It reacts with photochemically produced hydroxyl radicals with a half-life of 32 days or a 2.2% loss per sunlit day [28,49]. The experimental rate constant for hydroxyl radicals with ethylene dibromide is 2.5×10^{-13} cm^3/molecule-sec at 25 °C [2], which corresponds to an atmospheric half-life of about 67 days at an atmospheric concentration of $5 \times 10^{+5}$ hydroxyl radicals per cm^3 [2].

Bioconcentration: The measured log BCF in fish is < 1 [32]. BCFs of 1.6-3.2 and <3.5-14.9 were reported for EDB for 150 ug/L and 15 ug/L, respectively [12]. These BCF values indicate little potential for EDB to bioconcentrate in fish

Soil Adsorption/Mobility: Ethylene dibromide exhibits low to moderate adsorption to soil with measured Koc values ranging from 14 to 160 [13,40,47]. Gaseous ethylene dibromide has been used as a soil fumigant. For typical field soils, 99% of the ethylene dibromide used in fumigation is in the sorbed state [10]. The selected Koc value for ethylene dibromide is 34 mL/g [2]; values ranged from 11 to 161 [2].

Volatilization from Water/Soil: The half-life for evaporation of ethylene dibromide from a model water 1 m deep with a 3 m/sec wind and a current of 1 m/sec is 6.1 hr [36]. Experiments performed in a wind-wave tank yielded a half-life for evaporation of 4.26 hr from water 1 m deep with a wind speed of 8.6 m/sec [37]. To estimate the half-life for evaporation from water in the environment, one can take the rate of evaporation of ethylene dibromide from water relative to the reaeration rate, 0.53 [38], and combine that with the reaeration rate for typical bodies of water [36] to obtain a half-life of 32 hr in a river and 130 hr in a lake. Due to its relatively high vapor pressure, ethylene dibromide will volatilize readily from near surface soil.

Water Concentrations: DRINKING WATER: 3 drinking water wells in California and Hawaii 35-300 ppb [9]. The Netherlands 0.1 ppb max [34]. EDB was positively detected in 328 of 5571 public drinking water samples in California with a mean concn of 0.08 ug/L [51]. GROUND WATER: EDB was identified, but not quantified in ground water in New Jersey in a survey including 408 samples of well water [22]. EDB was detected in 5 to

8% of ground water samples from Polk County and Jackson County, FL [14]. SURFACE WATER: EDB was identified in a stream near oil refining and manufacturing 1.05-1.13 ppb [21]. US - 14 heavily industrialized river basins 2 of 204 sites > 1 ppb [18]. RAIN: Samples near fumigation center - 1 ppb [1].

Effluent Concentrations: Oil refinery effluent < 0.2 ppb [1]; runoff water from area with several gasoline stations < 0.2 ppb [1]; runoff from fumigation center 2 ppb [1].

Sediment/Soil Concentrations: No detectable residues of EDB were found in soil or dustfall at bulk gasoline handling facilities in New Jersey and Oklahoma. Minimum detectable quantity was 10-15 ng/sample [1]. Nanogram per gram levels were found in the soil at two citrus fumigation centers in Florida where the dustfall ranged from 6 to 363 pg/cm^2/hr [1]. EDB was detected in vadose cores, particularly topsoils, at concns up to 32 ug/kg [43]. EDB was detected in soils of one of two locations with histories of EDB applications at concns of 0.2-0.6 ppb [16].

Atmospheric Concentrations: RURAL REMOTE: US (12 sites) 0 ppt median, 25% of samples > 3.8 ppt; 9 ppt max [8]. Concn of 1.0-1.9 pptV were detected during 1983 at Barrow, Alaska [46]. EDB concn of ND-24 pptV were detected in air samples from the Arctic during March and April 1983 [5]. URBAN/SUBURBAN: US (676 sites) 0-130 ppt, 26 ppt median [8]. 0.001-0.17 ug/m^3 of ethylene dibromide detected in London, England and as high as 1.2-1.8 ug/m^3 on a garage forecourt [35]. 2201 Air samples from 65 locations had no detectable concn of EDB [33]. SOURCE AREAS: US (242 sites) 0-31,000 ppt, 190 ppt median [8]; Lipari and BFI landfills in New Jersey 0-770 ppt, 350 ppt avg [7]; manufacturing site in US 300-15,000 ppt; urban location near gas stations and highway 9-14 ppt, oil refinery 29-210 ppt [21]. Generally present in air in US, according to the Environmental Protection Agency. Levels near groups of petroleum stations and along well traveled highways were around 11 ug/m^3. It has been found in concns of up to 96 ug/m^3 up to a mile away from a US Dept of Agriculture fumigation center [29]. No EDB was detected in personal, indoor, outdoor, or breath air samples taken in Los Angeles, CA [24].

Food Survey Values: Residue levels in edible portions of fruit is the following: grapefruit/Not detected (ND) - 0.326 ppm; limes/ND - 0.01 ppm;

papaya/ND - 0.102 ppm; mangos/0.001 - 0.27 ppm; oranges/ND - 0.24 ppm [39]. Apples: after fumigation with 12 or 24 mg/L EDB at 13 °C for 4 hr and stored at this temperature: 36 and 75 ppm after 1 day, 1.2 and 1.6 ppm after 6 days [29]. Fumigated oats used as chicken feed 10-15 ppm several weeks after fumigation [1]. Citrus fruits 4 days postfumigation peel 1-43 ppm, pulp 0.4-2.4 ppm with the residue dependent on the rate and duration of fumigation and temperature and length of postfumigation aeration [1]. After a 10-day fumigation of wheat followed by a 2-4 or 10-12 week aeration period, residues found were 5-30 ppm in the whole wheat, which resulted in 2-4 ppm in the flour milled from it, 18-23 ppm in the shorts and bran, 0.002-0.04 ppm in white bread, whole meal bread 0.006-0.026 ppm, with the lower values corresponding to the longer aeration period [1]. In commercially fumigated wheat 3.26 ppm was found 1 week postfumigation and 1.36 ppm after 7 weeks with residues in flour of 0.01 to 0.29 ppm, bran from 0-0.40 ppm, middlings 0-0.30 ppm [1]. No EDB residues were found in bread baked with the flour [1]. EDB was not detected in any table-ready foods analyzed [26]; another study reported that 2 out of 549 food items had detectable concns of EDB (avg concn of 7 ng/g) [15]. A total of 24 whole grains, milled grain products, intermediate grain-based foods, and animal feeds analyzed contained EDB levels up to 540 ppb (wheat) [25].

Plant Concentrations: No detectable residues of EDB are found in plants grown in EDB fumigated soils [1].

Fish/Seafood Concentrations:

Animal Concentrations:

Milk Concentrations:

Other Environmental Concentrations: Motor fuel antiknock mixes contain about 18% or 2.8 g/L ethylene dibromide and aviation fuel antiknock mixes contain 36% ethylene dibromide [29].

Probable Routes of Human Exposure: Humans have been exposed to EDB in the past during application as a soil or commodity fumigant, from residues in and on raw agricultural commodities and in processed grain commodities following fumigation [1]. Exposure to EDB in ambient air occurs from the use of leaded gasoline and therefore exposure is highest in areas with heavy traffic, filling stations, etc.

Average Daily Intake: Estimated daily intake (ug/kg/day) of EDB by geographic region in the US: Northeast, citrus: 0.00052, grain: 0.0051; North Central, citrus: 0.005, grain: 0.0052; South, citrus: 0.003, grain: 0.0059; West, citrus: 0.00074, grain: 0.0052 [53].

Occupational Exposure: Citrus fumigation centers in Florida 48-403 ppb inside, 0.09-3.8 ppb outside [1,19]. Following broadcast fumigation of soil by injection after which an air inversion developed, the breathing zone of applicator was 414 ppb avg 1 ft above, treated field 432 ppb avg, and adjacent untreated field 49 ppb avg [1]. In two manufacturing plants and two plants using EDB, short-term exposure was as high as 23.4 ppm while median 8 hr time weighted average (TWA) concentration ranged from < 0.02 to 160 ppb [52]. Yearly exposure of workers connected with soil fumigation ranged from 1.8 to 27 mg/kg while that for citrus fumigation ranged from 0.44 to 287 mg/kg [52]. Gas station operators - 7 ppb while near gas pump for 8 hr TWA; exposure of 5 ppb, garage mechanic 3 ppb TWA [41]. Two ethylene dibromide manufacturing plants 0-18.25 ppm [41]. Workers in the following occupations have exposure levels to ethylene dibromide: private farmer/applicator - 0.01-1.8 mg/kg/yr; fumigation station/corridor operators - 2.3-7.5 mg/kg/yr; fumigation station/outdoor operators - 5.1-6.5 mg/kg/yr; truckers/station personnel - 0.42-2.1 mg/kg/yr; warehouse/indoor laborers - 287 mg/kg/yr; warehouse/outdoor laborers - 75.3 mg/kg/yr; warehouse/stickmen - 151.8 mg/kg/yr; spot treatment/applicator - 4.3-59.3 mg/kg/yr; spot treatment/mill worker - 9.4-10.9 mg/kg/yr [54]. In the past, workers transporting and distributing fumigated citrus may have been routinely exposed to airborne ethylene dibromide at greater than the allowable limit of 130 ppb [45]. NIOSH (NOES 1981-83) has estimated that 6912 workers are potentially exposed to ethylene dibromide in the US [42].

Body Burdens:

REFERENCES

1.   42 FR 63134; 12/14 (1977)

2. Atkinson R; Kinetics and mechanisms of the gas-phase reactions of the hydroxyl radical with organic compounds. J Phys Chem Ref Data Monograph No. 1. (1989)
3. Augustijn-Beckers PWM, et al; Rev Environ Contam Toxicol 137:1-82 (1994)
4. Barbash JE, Reinhard M; Environ Sci Technol 23: 1349-1355 (1989)
5. Berg WW et al; Geophys Res Lett 11: 429-432 (1984)
6. Bouwer EJ, McCarty PL; Appl Environ Microbiol 45: 1295-9 (1983)
7. Bozzelli JW et al; Analysis of Selected Toxic and Carcinogenic Substances in Ambient Air in New Jersey; State of New Jersey Dept Environ Prob (1980)
8. Brodzinsky R, Singh HB; Volatile Organic Chemical in the Atmosphere: an assessment of available data; pp.15.1-15.34 SRI contract 68-02-3452 (1982)
9. Burmaster DE; Environ 24: 6-13, 33-36 (1982)
10. Call F; J Sci Food Agricult 8: 81-5 (1957)
11. Castro CE, Belser NO; Environ Sci Technol 2: 779-83 (1968)
12. Chem Inspect Testing Inst; Biodegradation and bioaccumulation data of existing chemicals based on the CSCL Japan. Japan Chem Indust Ecol - Toxicol Inform Center. ISBN 4-89074-101-1 (1992)
13. Chiou CT et al; Science 206: 831-2 (1979)
14. Choquette AF, Katz BG; Grid-based groundwater sampling: lessons from an extensive regional network for 1,2-dibromoethane (EDB) in Florida. IAHS Publ 1989, 182(Reg Charact Water Qual), 79-86 (1989)
15. Daft JL; Sci Total Environ 100: 501-518 (1991)
16. Duncan DW, Oshima RJ; 1,2-Dibromoethane (EDB) in two soil profiles In: ACS Symp Ser, Garner WY et al (ed) 315(Eval Pestic Ground Water):282-93 (1986)
17. Ehrenberg L et al; Rad Botany 15: 185-94 (1974)
18. Ewing BB et al; Monitoring to Detect Previously Unrecognized Pollutants in Surface Waters; pp.75 USEPA-560/6-77-015, Appendix USEPA-560/6-77015a (1977)
19. Fishbein L; Sci Total Environ 11: 223-57 (1979)
20. Frink CR, Bugbee GJ; Soil Sci 148: 303-307 (1989)
21. Going J, Long S; Sampling and Analysis of Selected Toxic Substances Task II - Ethylene Dibromide; final report; pp.39 USEPA-560/6-75-001 (1975)
22. Greenberg M et al; Environ Sci Technol 16: 14-9 (1982)
23. Hansch C et al; Exploring QSAR Hydrophobic, Electronic, and Steric Constants, ACS, Washington, DC (1995)
24. Hartwell TC et al; Atm Environ 26A: 1519-1527 (1992)
25. Heikes DL; J Assoc Off Anal Chem 68: 1108-11 (1985)
26. Heikes DL; J Assoc Off Anal Chem 70: 215-226 (1987)
27. Horvath AL; Halogenated Hydrocarbons: Solubility-Miscibility with Water, Marcel Dekker, NY (1982)
28. Howard CJ, Evenson KM; J Chem Phys 64: 4303-6 (1976)
29. IARC; Some Fumigants, the Herbicides 2,4-D and 2,4,5-T, Chlorinated Dibenzodioxins and Miscellaneous Industrial Chemicals; 15: 195-209 (1977)
30. Jafvert CT, Wolfe NL; Environ Toxicol Chem 6: 827-837 (1987)
31. Jungclaus GA, Cohen SZ; Hydrolysis of ethylene dibromide. In: Am Chem Soc Div Environ Chem 191st Natl Meet 26:12-6 (1986)
32. Kawasaki M; Ecotox Environ Safety 4: 444-54 (1980)

33. Kelly TJ et al; Environ Sci Technol 28: 378A-387A (1994)
34. Kraybill HF; Annals NY Acad Sci 298: 80-9 (1977)
35. Leinster P et al; Atmos Environ 12 (12): 2383-7 (1978)
36. Lyman WJ et al; Handbook of Chemical Property Estimation Methods. Environmental behavior of organic chemicals; McGraw-Hill New York NY pp. 15.1-15.34 (1982)
37. Mackay D, Yeun ATK; Environ Sci Technol 17: 211-7 (1983)
38. Mackay D et al; Volatilization of Organic Pollutants from Water; USEPA-600/53-82-019 (1982)
39. Maddy KT et al; Recent Studies and Eval by the Calif Dept of Food and Agric on EDB with Emphasis of Use in Fumigation of Fruit, Calif Dept Food Agric Report HS-956 (1982)
40. Mingelgrin U, Gerstl Z; J Environ Qual 12: 1-11 (1983)
41. NIOSH; Criteria for a Recommended Standard Occupational Exposure to Ethylene Dibromide; (1977)
42. NIOSH; National Occupational Exposure Survey (NOES) (1989)
43. Pignatello, JJ et al; J Contam Hydrol 5:195-214 (1990)
44. Pignatello, JJ; J Environ Qual 16: 307-312 (1987)
45. Rappaport SM et al; J Agric Food Chem 32: 1112-6 (1984)
46. Rasmussen RA, Khalil MAK; Geophys Res Lett 11: 433-436 (1984)
47. Reinbold KA et al; Absorption of Energy - Related Organic Pollutants: a Literature Review; p.95 USEPA-600/3-79-080 (1979)
48. Reinhard M, Vogel TM; Environ Sci Technol 22: 231 (1988)
49. Singh HB et al; Atmos Environ 15: 601-12 (1981)
50. Steinberg SM et al; Environ Sci Technol 21: 1201-1208 (1987)
51. Storm DL; Chemical monitoring of California's public drinking water sources: Public exposure and health impacts. In: Water Contam Health, Wang RGM (ed), Marcel Dekker: NY pp 67-124 (1994)
52. Syracuse Research Corp; pp.6-10 SRC TR 80-586 Contract 210-78-009 (1980)
53. USEPA, Office of Drinking Water; Criteria Document (Draft): Ethylene dibromide p. IV-7 (1985)
54. USEPA; Ethylene Dibromide (EDB) Position Doc #4 p.7 (1982)
55. Verschueren K; Handbook of Environmental Data on Organic Chemicals; 2nd ed Van Nostrand Reinhold Co New York NY pp.635-7 (1983)
56. Vogel TM, Reinhard M; Environ Sci Technol 20:992-997 (1986)
57. Wilson BH et al; Environ Sci Technol 20: 997-1002 (1986)

Formic Acid

SUBSTANCE IDENTIFICATION

Synonyms:

Structure:

CAS Registry Number: 64-18-6

Molecular Formula: CH_2O_2

SMILES Notation: O=CO

CHEMICAL AND PHYSICAL PROPERTIES

Boiling Point: 100.5 °C

Melting Point: 8.4 °C

Molecular Weight: 46.02

Dissociation Constants: pKa = 3.75 [31]

Log Octanol/Water Partition Coefficient: -0.54 [16]

Water Solubility: infinite [31]

Vapor Pressure: 42.6 mm Hg at 25 °C [9]

Henry's Law Constant: 1.67×10^{-7} atm-m^3/mol [11]

ENVIRONMENTAL FATE/EXPOSURE POTENTIAL

Summary: Formic acid occurs naturally in plants and insects. It is also a product of microbial metabolism of organic matter and is produced in the photooxidation of biogenic and anthropogenic compounds. Formic acid will also be released to the environment as emissions and in wastewater in its production and various industrial uses. If released on land, formic acid should leach into some soils where it would probably biodegrade. In natural water it has been shown to adsorb to sediment and would probably also biodegrade. Bioconcentration in aquatic organisms is not important. In the atmosphere, formic acid would be scavenged by rain and dissolve in cloud water where it reacts with dissolved hydroxyl radicals. It also reacts in the vapor phase with hydroxyl radicals (half-life 34 days). Humans are exposed to formic acid in ambient air and food, as well as occupationally via inhalation and dermal contact.

Natural Sources: Formic acid occurs in fruits, vegetables, and leaves and roots of plants, and also in the secretions of numerous insects [35]. The secretion of the bombardier beetle is 75% formic acid [35]. Formic acid is also an intermediate product in the decomposition of organic matter in lake sediment [20] and a photooxidation product of alkanes, alkenes, and biogenic terpenes by hydroxyl radical [2,13] and ozone-olefin reactions [4]. Formic acid is also produced in clouds by the oxidation of formaldehyde by hydroxyl radicals, oxygen, or hydrogen peroxide [1]. Formic acid is an intermediary human metabolite that is immediately transformed to formate [22]. Formate can be excreted in human and animal urine [34].

Artificial Sources: Formic acid is produced in large quantities (48 million lbs in 1984 [6]) and may be released to the environment in emissions and in wastewater during its production and uses (% of production) in: textile dyeing and finishing (21%), pharmaceuticals (20%), rubber intermediate (16%), leather and tanning treatment (15%), and catalysts (12%) [6]. Formic acid is also a component added to certain paint strippers [14]. It is also released in photo processing effluents [8].

Terrestrial Fate: If released on land, formic acid should leach into some soils where it would probably biodegrade based upon the results of screening studies. A field study was conducted in which an industrial acid waste that was disposed of by deep well injection was followed as it traveled a distance of 427-823 meters over a 2-4 yr period [18]. The concn of formic acid was not detectable in two observation wells while in a third well it was 0.4% of dissolved organic carbon [18]. Before injection the formic acid content was 11.4% of DOC [18]. The disappearance of the acid may have been a result of anaerobic degradation, or reaction with the mineral material in the ground water [18].

Aquatic Fate: Based on the results of screening studies, formic acid should biodegrade in water. It should not adsorb significantly to sediment.

Atmospheric Fate: In the atmosphere, formic acid will be rapidly scavenged by rain and dissolve in cloud water and aerosols. It participates in reactions in clouds and aerosols involving dissolved hydroxyl radicals. The vapor phase acid also reacts with photochemically produced hydroxyl radicals (half-life 34 days) and possibly with alkenes that may be present in polluted urban air.

Biodegradation: Formic acid biodegrades readily in screening tests. Specific results include: 4.3 and 38.8% of theoretical BOD after 5 and 10 days using a sewage seed [10]; 43.7-77.6% of theoretical BOD after 5 days with a sewage inoculum [17]; 70% of theoretical BOD in 24 hr using activated sludge [21]; 66% of theoretical BOD in 12 hr using an activated sludge inoculum [24]; 39.9% of theoretical BOD in 24 hr with activated sludge [28]; 48 and 51% of theoretical BOD after 5 days with unacclimated and acclimated sewage seed, respectively [29]; and 40.5 and 51.7% of theoretical BOD after 5 days with sewage seed in fresh water and synthetic seawater, respectively [32]. Formic acid is also amenable to anaerobic biodegradation [7,18]. In one study 89% degradation was obtained after a 4 day lag by methane cultures [7].

Abiotic Degradation: Formic acid is the strongest unsubstituted carboxylic acid and will exist almost entirely as the anion at environmental pHs. The anhydrous material decomposes to carbon monoxide and water [30]. Traces of water, including that formed during the decomposition, inhibit the reaction [30]. Decomposition rates are measurable at room temperature, and storage for more than 6 months is not recommended when the temperature exceeds 30 °C [30]. The anhydrous acid catalyzes its own esterification with alcohols and polyols but often also promotes dehydration to the ether or olefin [30]. Dry formic acid readily adds to olefins to form formate esters [30]. In the atmosphere, formic acid reacts with photochemically produced hydroxyl radicals with a half-life of 34 days [36]. Hydrogen atoms are a major free radical product of this reaction [36]. Reactions between hydroxyl radicals and formic acid also occur in cloud water [3]. During daylight hours, aqueous-phase OH radical reactions can both produce and destroy formic acid in cloud drops and may control the formic acid levels in rain [3].

Bioconcentration: A BCF of 0.22 is estimated for formic acid using a recommended regression equation [19]. Therefore formic acid should not bioconcentrate in aquatic organisms.

Soil Adsorption/Mobility: Formic acid is miscible in water. Chemicals that are soluble in water do not usually adsorb significantly to soil or sediment due to hydrophobic forces [19]. However, at ambient pHs, formic acid exists as the ion and may therefore have the potential to bind to clay or humic material by coulombic forces. However, no evidence of such binding has been reported.

Volatilization from Water/Soil: Formic acid's low Henry's Law constant indicates that volatilization from water would not be a significant transport process [19]. Its high vapor pressure indicates that volatilization would be rapid from dry soil and surfaces.

Water Concentrations: GROUND WATER: Ground water in a sand aquifer at Pensacola, FL, the site of a wood treatment facility, contained 0.13 ppb of formic acid at 6 m depth [12]. No formic acid was found at 18 m depth [12]. SURFACE WATER: Concentrations of formic acid in the Ohio River, Little Miami River and Tannes Creek were 12-39 ppb, 18.4-25.2 ppb, and 22.3 ppb, respectively [25]. In Lake Kizaki in Japan, surface concn of formic acid was 115 ppb [25]. Although the concentration varied with depth (0-28 m) between 0 and 115 ppb, the variation was not a smoothly decreasing one [15]. RAIN/SNOW: Precipitation samples collected at two Wisconsin lakes on the Wisconsin Acid Deposition Monitoring Network contained formic acid ranging from the detection limit (20 ppb) to 2,576 ppb, median 382 ppb [5]. In remote areas it may be the major organic acidic component in rain [4]. Other results for formic acid in rain include: 18.4 ppb in Ithaca, NY; 0-5.5 ppb in Los Angeles; 20-880 ppb in the South Indian Ocean; 330-980 ppb in remote Venezuela; and 0-1630 ppb in remote Australia [23].

Effluent Concentrations: Secondary effluents of four sewage treatment plants ranged from 51-144 ppb whereas primary effluents from three of the plants ranged from 46-587 ppb [25]. With increased treatment long chain fatty acids are broken down to shorter chain acids, thereby increasing the concn of lower acids [25].

Sediment/Soil Concentrations: The formic acid concns in the 0-1 cm layer of sediment at 2 stations in Lake Biwa in Japan were 437 and 64 ppm (wet weight), while the respective concn, in the 9-10 cm layer were 23 and 110 ppm [20]. No formic acid was in the interstitial water, indicating that the chemical was adsorbed on the sediment particles [20].

Atmospheric Concentrations: RURAL/REMOTE: Concns of formic acid at various rural and remote sites in Arizona range from a few tenths of a ppb to 3.5 ppb [2]. URBAN/SUBURBAN: Maximum daily formic acid concentrations observed in Claremont, CA, during 5 days of an air pollution episode in October 1979 ranged from 5-19 ppb [33].

Food Survey Values:

Plant Concentrations:

Fish/Seafood Concentrations:

Animal Concentrations:

Milk Concentrations:

Other Environmental Concentrations: Several commercial paint strippers contain up to 15% formic acid [14]. Formic acid was identified in infusion liquids packaged in PVC bags, autoclaved at 120 °C for 20 min and analyzed immediately and after storage for 3, 6, 12 and 24 mo at 25 °C [3]. Release occurred during autoclaving and the first 3 mo of storage.

Probable Routes of Human Exposure: The general public is exposed to formic acid from ambient air and in some fruits and vegetables. Workers will be occupationally exposed to formic acid via dermal contact and inhalation.

Average Daily Intake: AIR INTAKE (assume air concn of 0.5-3.5 ppb):

19-134 ug: WATER INTAKE - Insufficient data: FOOD INTAKE - Insufficient data.

Occupational Exposure: NIOSH (NOHS 1972-1974) has statistically estimated that 533,799 workers are exposed to formic acid in the US [26]. NIOSH (NOES 1981-1983) has statistically estimated that 106,875 workers are exposed to formic acid in the US [27].

Body Burdens:

REFERENCES

1. Adewuyi YG, et al; Atmos Environ 18: 2413 (1984)
2. Altshuller AP; Atmos Environ 17: 2131-65 (1983)
3. Arbin A; Sven Farm Tidskr 87 (9-10): 17-18 (1983)
4. Chameides WL, Davis DD; Nature 304: 427-9 (1983)
5. Chapman EG et al; Atmos Environ 20: 1717-27 (1986)
6. Chemical Marketing Reporter; Chemical Profile: Formic Acid. December 31 (1984)
7. Chou WL, et al; Bioeng Symp 8: 391-414 (1979)
8. Dagon TJ; J Water Pollut Contr Fed 45: 2123-35 (1973)
9. Daubert TE, Danner RP; Data Compilation Tables of Properties of Pure Compounds Amer Institute of Chem Eng (1985)
10. Gaffney PE, Ingols RS; Water Sew Works 108: 91 (1961)
11. Gaffney JS et al; Environ Sci Technol 21: 519-23 (1987)
12. Goerlitz DF et al; Environ Sci Tech 19: 955-61 (1985)
13. Graedel TE; Chemical Compounds In The Atmosphere. New York, NY: Academic Press pp. 50-90 (1978)
14. Hahn WJ, Werschulz PO; Evaluation Of Alternatives To Toxic Organic Paint Strippers USEPA-600/52-85/118 (1986)
15. Hama T, Handa N; Rikusiugaku Zasshi 42: 8-19 (1981)
16. Hansch C, Leo AJ; MEDCHEM Project Claremont CA: Pomona College (1985)
17. Heukelekian H, Rand MC; J Water Pollut Contr Assoc 29: 1040-53 (1955)
18. Leenheer JA, et al; Environ Sci Technol 10: 445-51 (1976)
19. Lyman WJ et al; Handbook of Chemical Property Estimation Methods New York: McGraw-Hill pp. 4-1 to 4-33, 5-1 to 5-30, 15-1 to 15-34 (1982)
20. Maeda H, Kawai A; Nippon Suisan Gakkaishi 52: 1205-8 (1986)
21. Malaney GW, Gerhold RM; J Water Pollut Control Fed 41: R18-R33 (1969)
22. Malorny G; Z Ernaehrungswiss 9: 340-48 (1969)
23. Mazurek MA, Simoneit BRT; CRC Critical Review Environ Control 16: 140 (1986)
24. McKinney RE, et al; Sew Indust Wastes 28: 547-57 (1956)
25. Murtaugh JJ, Bunch RL; J Water Pollut Control Fed 37: 410-5 (1965)
26. NIOSH; National Occupational Health Survey NOHS (1975)

27. NIOSH; National Occupational Exposure Survey NOES (1988)
28. Placak OR, Ruchhoft CC; Sewage Works J 19: 423-40 (1947)
29. Price KS, et al; J Water Pollut Contr Fed 46: 63-77 (1974)
30. Pryde EH; Kirk-Othmer Encycl Chem Tech 3rd ed 4: 814-34 (1978)
31. Riddick JA et al; Organic Solvents 4th ed, pp. 360-1. NY, NY: Wiley (1986)
32. Takemoto S, et al; Suishitsu Odaku Kenkyu 4: 80-90 (1981)
33. Tuazon EC et al; Atmospheric Measurement Of Trace Pollutants: Long Path Fourier Transform Infrared Spectroscopy USEPA-600/S3-81-026 (1981)
34. Van Oettingen WF; Am. Med Assoc Arch Ind Health 20: 517-531 (1959)
35. Wagner FS; Kirk-Othmer Encycl Chem Technol 3rd ed. 11: 251-8 (1980)
36. Wine PH, et al; Phys Chem 89: 2620-4 (1985)

Fumaric Acid

Synonyms:

Structure:

CAS Registry Number: 110-17-8

Molecular Formula: $C_4H_4O_4$

SMILES Notation: O=C(O)C=CC(=O)O

CHEMICAL AND PHYSICAL PROPERTIES

Boiling Point: Sublimes at 165 °C at 1.7 mm Hg

Melting Point: 287 °C (closed capillary, rapid heating)

Molecular Weight: 116.07

Dissociation Constants: $pKa_1 = 3.03$; $pKa_2 = 4.44$ at 18 °C [26]

Log Octanol/Water Partition Coefficient: 0.46 [6]

Water Solubility: 7000 mg/L at 25 °C [19]

Vapor Pressure: 1.7 mm Hg at 165 °C [10]

Henry's Law Constant: 8.5×10^{-14} atm-m^3/mol [16]

ENVIRONMENTAL FATE/EXPOSURE POTENTIAL

Summary: Fumaric acid is commonly produced by organisms. Its presence in soils, dusts and plants can result from biogenic sources. Anthropogenic sources of environmental release include gasoline and diesel engine exhaust and aqueous effluents from pulp mills. If released to soil or water, biodegradation is expected to be the major fate process. River die-away studies have yielded half-lives ranging from 1-15 days, with degradation rates increasing as the pollution in the water source increases. If released to the atmosphere, fumaric acid will exist primarily in the particulate phase where it can be physically removed via wet and dry deposition. Vapor-phase fumaric acid is readily degraded by sunlight-formed hydroxyl radicals (estimated half-life of 1.8 days). The major route of exposure to the general population is through consumption of food and beverages containing fumaric acid acidulants. The general population is also exposed through inhalation of air containing fumaric acid from both biogenic and human sources.

Natural Sources: Fumaric acid is commonly produced by organisms [8]; its presence in soils, dusts and plants can result from biogenic sources [8]. Fumaric acid is found naturally in many plants and is named after the genus Fumaria [19].

Artificial Sources: Fumaric acid has been detected in gasoline and diesel engine exhausts [8] and in aqueous effluents from pulp mills [23].

Terrestrial Fate: A variety of biodegradation studies have demonstrated that fumaric acid is readily biodegradable [5,7,12,14,21,25]. Therefore, biodegradation is expected to be the major degradation process in soil. Based upon measured pKa's of 3.03 (step 1) and 4.44 (step 2) [26], fumaric acid will exist predominantly in the ionized state in moist soils. The ability of ionized fumaric acid to leach in soil cannot be predicted adequately without experimental data.

Aquatic Fate: River die-away studies with fumaric acid have indicated that biodegradation is the major degradation process in natural water [21]; the half-life of fumaric acid in various natural waters ranged from 1-15 days, with faster degradation occurring in more polluted waters [21]; degradation half-life in distilled water controls was 55 days [21]. Abiotic degradation in natural waters can occur via reaction with sunlight-formed oxidants such as hydroxyl radicals and singlet oxygen [3,17]. Aquatic hydrolysis, volatilization, direct photolysis, bioconcentration, and adsorption to sediment are not expected to be important fate processes.

Atmospheric Fate: Atmospheric monitoring studies have indicated that fumaric acid is associated primarily with the particulate phase in the ambient atmosphere [8]. Particulates are physically removed from air via wet and dry deposition mechanisms. Fumaric acid has been detected in rain and snow water samples [13,22], indicating that wet deposition does occur. The major degradation process for atmospheric vapor-phase fumaric acid is expected to be reaction with sunlight-formed hydroxyl radicals; the half-life for this process in typical air has been estimated to be 1.8 days [2].

Biodegradation: In river die-away studies using various natural waters, the degradation half-life of fumaric acid ranged from 1-15 days, with faster degradation occurring in more polluted waters [21]; degradation half-life in distilled water controls was 55 days [21]. Using a microbe inocula taken from three polluted surface waters, a 5 day theoretical BOD of 34% was measured [5]. Using a Warburg respirometer and a sewage inocula, 5 day theoretical BODs of 69-85% were measured at concns of 3.75-75 ppm [7]. At 500 ppm, a theoretical BOD of 2.7% was measured after a 24-hr inoculation period in a Warburg respirometer using an activated sludge inocula [12]. A 5 day theoretical BOD of 78.6% was reported for a sewage inocula [25]. Using an activated sludge adapted to phenol, a theoretical BOD of 41% was measured after a 23 hr inoculation period in a Warburg respirometer [14].

Abiotic Degradation: Fumaric acid does not absorb UV light above 290 nm in methanol, acidic methanol, or basic methanol solution [20]; therefore, direct photolysis in the environment is unlikely to occur. Hydrolysis of fumaric acid in the environment is not expected to be important since carboxylic acid and alkene functional groups are resistant to environmental hydrolysis [11]. The rate constant for the vapor-phase reaction of fumaric acid with photochemically produced hydroxyl radicals has been estimated to be 8.9×10^{-12} cm^3/molecule-sec at 25 °C which corresponds to an atmospheric half-life of about 1.8 days at an atmospheric concn of $5 \times 10^{+5}$ hydroxyl radicals per cm^3 [2]. The rate constant for the vapor-phase reaction of fumaric acid with ozone has been estimated to be 1.75×10^{-18} cm^3/molecule-sec at 25 °C, which corresponds to an atmospheric half-life of about 6.5 days at an atmospheric concn of $7 \times 10^{+11}$ molecules per cm^3 [1]. The rate constant for the aqueous reaction of fumaric acid with photochemically produced hydroxyl radicals (pH 4.5-10) is $6.0 \times 10^{+9}$ 1/M-sec [3]; assuming a hydroxyl radical concn of 3×10^{-17} M in brightly sunlit natural water [17], the half-life would be about 45 days. Since fumaric acid is a substituted olefin, reaction with sunlight-formed singlet oxygen in water may be just as fast or faster than reaction with OH radicals [17].

Bioconcentration: Based upon the water solubility, the BCF of fumaric acid can be estimated to be approximately 4.2 [11]. This BCF value suggests that bioconcentration in aquatic organisms is not environmentally important.

Soil Adsorption/Mobility: Based upon the water solubility, the Koc of fumaric acid can be estimated to be 33.5 [11]; this estimated Koc value indicates very high soil mobility [24]. However, with step 1 and step 2 pKa's of 3.03 and 4.44 at 18 °C [26], fumaric acid will exist predominantly in the ionized state at environmental pHs. The ability of ionized fumaric acid to leach in soil cannot be predicted adequately without experimental data.

Volatilization from Water/Soil: The Henry's Law constant for fumaric acid indicates that the compound will not volatilize from water [11].

Water Concentrations: RAIN/SNOW: Rain and snow water collected from rural areas near Hubbard Brook, NH, and semi-rural areas near Ithaca, NY, between June 1976 and May 1977 was found to contain fumaric acid [13]. In snow/sleet and rain samples from Toyko, fumaric acid was detected at concns ranging from 0.22 to 4.64 ug/L [22].

Effluent Concentrations: A fumaric acid concn of 0.94 ug/m^3 was detected in the motor exhaust from a 1982 Toyota Corolla [8]; a concn of 3.2 ug/m^3 was detected in the exhaust from a diesel engine 1971 Mercedes Benz [8].

Sediment/Soil Concentrations: Bog sediments collected in the foothills of the Sierra Nevada Mountains contained fumaric acid levels of 4.76 mg/kg [8]. Soil samples collected on the campus of UCLA in Los Angeles, CA, contained fumaric acid levels of 0.2-0.6 mg/kg [8].

Atmospheric Concentrations: Air samples collected in west and downtown Los Angeles, CA, during June and October 1984 contained fumaric acid concns of 3.5-147.4 ng/m³ [8]; the fumaric acid detected was associated primarily with atmospheric particles rather than the vapor-phase [8]. Aerosol samples from Toyko contained 10 to 44 ng/m³ of fumaric acid [22] and 0.7 to 15 ng/m³ (21 samples collected during 1988-1989) [9].

Food Survey Values:

Plant Concentrations:

Fish/Seafood Concentrations:

Animal Concentrations:

Milk Concentrations:

Other Environmental Concentrations: Dust collected from the outside window ledge and balcony of two buildings in Los Angeles, CA, contained fumaric acid levels that ranged from 2.65 to 6.67 mg/kg [8].

Probable Routes of Human Exposure: Large volumes of fumaric acid are used as a food acidulant in applications such as beverages, baking powders, and fruit drinks [4,15]; therefore, the major route of exposure to the general population is through consumption of food and beverages. The general population is also exposed through inhalation of air containing fumaric acid from both biogenic and anthropogenic sources [8].

Average Daily Intake:

Occupational Exposure: NIOSH (NOES 1981-1983) has statistically estimated that 73,849 workers are potentially exposed to fumaric acid in the US [18].

Body Burdens:

REFERENCES

1. Atkinson R, Carter WPL; Chem Rev 84: 437-80 (1984)
2. Atkinson R; J Inter Chem Kinet 19: 799-828 (1987)
3. Buxton GC et al; J Phys Chem Ref Data 17: 722 (1988)

Fumaric Acid

4. Chem Mkt Rep; Chemical Profile-Fumaric Acid, July 3 p. 42 (1989)
5. Dore M et al; Trib Cebedeau 28: 3-11 (1975)
6. Hansch C et al; Exploring QSAR Hydrophobic, Electronic, and Steric Constants, ACS, Washington, DC (1995)
7. Heukelekian H, Rand MD; J Water Pollut Control Assoc 27: 1040-53 (1955)
8. Kawamura K, Kaplan IR; Environ Sci Technol 21: 105-10 (1987)
9. Kawamure K, Ikushima K; Environ Sci Technol 27: 2227-35 (1993)
10. Leung H-W, Paustenbach DJ; Am J Ind Med 18: 717-35 (1990)
11. Lyman WJ et al; Handbook of Chemical Property Estimation Methods NY: McGraw-Hill (1982)
12. Malaney GW, Gerhold RM; J Water Pollut Control Fed 41: R18-R33 (1969)
13. Mazurek MA, Simoneit BR; CRC Critical Reviews in Environmental Control 16: 74 (1986)
14. McKinney RE et al; Sewage Ind Wastes 28: 547-57 (1956)
15. Merck; The Merck Index, 10th ed. p. 612 (1983)
16. Meylan W, Howard PH; Environ Toxicol Chem 10: 1283-93 (1991)
17. Mill T, Mabey W; pp. 207-11 in Environmental Exposure From Chemicals Vol 1 Boca Raton, FL: CRC Press (1985)
18. NIOSH; National Occupational Exposure Survey (NOES) (1983)
19. Robinson WD, Mount RA; Kirk-Othmer Encycl of Chem Technol, 3rd ed. 14: 770-3 (1981)
20. Sadtler Research Lab; UV 165 (1966)
21. Saito N, Nagao M; Okayama-Ken Kankyo Hoken Senta Nempo 2: 274-6 (1978)
22. Sempere R, Kawamura K; Atmos Environ 28:449-459 (1994)
23. Suntio LR et al; Chemosphere 17: 1249-90 (1988)
24. Swann RL et al; Res Rev 85: 16-28 (1983)
25. Swope HG, Kenna M; Sewage Ind Waste 21: 467-8 (1950)
26. Weast RC; CRC Handbook of Chemistry and Physics, 66th ed. p. D-162 Boca Raton, FL: CRC Press (1985)

2-Heptanone

SUBSTANCE IDENTIFICATION

Synonyms:

Structure:

CAS Registry Number: 110-43-0

Molecular Formula: $C_7H_{14}O$

SMILES Notation: O=C(CCCCC)C

CHEMICAL AND PHYSICAL PROPERTIES

Boiling Point: 151.5 °C at 760 mm Hg

Melting Point: -35.5 °C

Molecular Weight: 114.18

Dissociation Constants:

Log Octanol/Water Partition Coefficient: 1.98 [17]

Water Solubility: 4300 mg/L at 25 °C [37]

Vapor Pressure: 3.86 mm Hg at 25 °C [37]

Henry's Law Constant: 1.77 x 10^{-4} atm-m^3/mol at 25 °C [26]

ENVIRONMENTAL FATE/EXPOSURE POTENTIAL

Summary: 2-Heptanone is a naturally occurring compound present in foods and essential oils. It is also used commercially as a solvent in a wide number of industrial applications. 2-Heptanone may be released to the environment as a fugitive emission during its production, formulation, use or transport, and in the effluent of industrial process and landfills. If released to soil, 2-heptanone is expected to display moderate to high mobility. Volatilization from both moist and dry soils may be a significant fate process. 2-Heptanone may biodegrade in soil under aerobic conditions. If released to water, 2-heptanone is expected to volatilize to the atmosphere. The estimated half-life for volatilization from a model river is 8.4 hr. 2-Heptanone is not expected to significantly bioconcentrate in fish and aquatic organisms nor is it expected to adsorb to sediment and suspended organic matter. 2-Heptanone is expected to biodegrade under aerobic conditions in aquatic systems. If released to the atmosphere, 2-heptanone is expected to undergo a gas-phase reaction with photochemically produced hydroxyl radicals; the estimated half-life for this process is 1.9 days. It may undergo atmospheric removal by wet deposition but it is not expected to undergo significant atmospheric removal by direct photolytic processes. Occupational exposure to 2-heptanone may occur by inhalation or dermal contact during its production, formulation or transport. Exposure to the general population may occur by ingestion of food in which it occurs naturally, by inhalation during the use of commercial products in which it is used as a solvent, or by the ingestion of contaminated drinking water.

Natural Sources: 2-Heptanone is a natural product and is a component of foods [23], oil of cloves and cinnamon-bark oil [45]. It is listed as a volatile compound emitted by vegetation [16,46] and has been detected in air samples from a forest [18].

Artificial Sources: 2-Heptanone is used as an industrial solvent for a wide variety of applications [45] and as a component of artificial flavors, oils, and perfumes [39,45]. It may be released to the atmosphere as a fugitive emission during its production, formulation, or use as a solvent. 2-Heptanone may also be released to the environment as a volatile emission during loading and transport [23], in industrial effluent [4], and in the effluent from landfills [3].

Terrestrial Fate: If released to soil, calculated soil adsorption coefficients ranging from 44-285 [25] indicate that 2-heptanone may display moderate to high mobility [40]. 2-Heptanone has the potential to biodegrade in soil

[34]. The vapor pressure of 2-heptanone and the calculated Henry's Law constant indicate that it is expected to volatilize from both dry and moist soil, respectively, to the atmosphere.

Aquatic Fate: If released to water, 2-heptanone is expected to rapidly volatilize to the atmosphere. The half-life for volatilization from a model river 1 m deep, flowing at 1 m/sec with a wind speed of 3 m/sec is 8.4 hr [25]. Calculated bioconcentration factors ranging from 5.5 to 19 [25] indicate that 2-heptanone is not expected to significantly bioconcentrate in fish and aquatic organisms. Calculated soil adsorption coefficients ranging from 44-285 [25] indicate that 2-heptanone will not significantly adsorb to sediment and suspended organic matter. Screening studies indicate that 2-heptanone is likely to biodegrade in aquatic systems under aerobic conditions [9,14,19,27].

Atmospheric Fate: If released to the atmosphere, 2-heptanone is expected to undergo a gas-phase reaction with photochemically produced hydroxyl radicals; the estimated half-life for this process is 1.9 days [1,43]. 2-Heptanone's relatively high water solubility, indicates that it may undergo atmospheric removal by wet deposition processes. Although 2-heptanone has the potential of being removed from the atmosphere by direct photochemical degradation, the rate of this process is not expected to be able to compete with atmospheric removal by the reaction with hydroxyl radicals [2].

Biodegradation: 2-Heptanone was found to undergo aerobic oxidation after a short lag period when inoculated with an unactivated sewage sludge seed [27]. 2-Heptanone had a theoretical biochemical oxygen demand (BOD) of 1.4%, 2.4%, and 4.8% after 6, 12, and 24 hrs, respectively, when incubated with an activated sludge seed at an initial concentration of 500 ppm [14]. 2-Heptanone underwent a 5-day theoretical BOD of 44% [9]. In a screening study using a sewage seed, 2-heptanone had a 10-day BOD of 0.50 g/g [19]. 2-Heptanone was qualitatively described as degrading in the presence of a sewage sludge seed [28]. Organisms isolated from soil and raised on C_1-C_8 straight chain paraffins were found to oxidize 2-heptanone [34].

Abiotic Degradation: An experimental rate constant for the gas-phase reaction of 2-heptanone with photochemically produced hydroxyl radicals of 8.67 x 10^{-11} cm^3/molecule-sec at 23 °C [43] translates to a half-life of 4.5 hrs using an average atmospheric hydroxyl radical concentration of 5 x 10^5 molec/cm^3 [1]. Although 2-heptanone is capable of direct photochemical degradation because of the reaction of its carbonyl functional group, the rate of atmospheric removal by this process is not expected to compete with degradation by the reaction with photochemically produced hydroxyl radicals [2].

Bioconcentration: Based on the log octanol/water partition coefficient and the water solubility, bioconcentration factors of 19 and 5.5, respectively, can be calculated for 2-heptanone using appropriate regression equations [25]. These values indicate that 2-heptanone will not significantly bioconcentrate in fish and aquatic organisms.

Soil Adsorption/Mobility: Based on the log octanol/water partition coefficient and the water solubility, soil adsorption coefficients of 285 and 44, respectively, can be calculated for 2-heptanone using appropriate regression equations [25]. These values indicate that 2-heptanone will display moderate to high mobility in soil [40].

Volatilization from Water/Soil: The vapor pressure of 2-heptanone indicates that it may rapidly volatilize from dry soils. The experimental Henry's Law constant indicates rapid volatilization from water and moist soil. Based on this value, the half-life for volatilization from a model river 1 m deep flowing at 1 m/sec with a wind speed of 3 m/sec is 10.1 hr [25].

Water Concentrations: DRINKING WATER: 2-Heptanone has been qualitatively detected in US drinking water supplies [24].

Effluent Concentrations: 2-Heptanone was qualitatively detected in 2 of 46 US industrial effluent samples [4]. 2-Heptanone was detected in the raw effluent from a textile finishing plant, NC, concentration not provided [15]. 2-Heptanone was detected in 4 of 7 effluent samples obtained from energy-related processes at a concentration of 5-224 ppb [32]. 2-Heptanone has been found in 2 of 63 industrial effluent samples obtained from a wide range of chemical manufacturers in areas across the US at a concn of <10 ug/L [33]. It was detected in the effluent from municipal landfills at a concn of 600-800 mg/L [3]. 2-Heptanone has been detected in household waste and was indicated to be a biodegradation product [44].

2-Heptanone

Sediment/Soil Concentrations:

Atmospheric Concentrations: SOURCE DOMINATED: 2-Heptanone was qualitatively detected in air samples obtained in the industrialized Kanawha Valley, WV, 1977 [12]. RURAL: 2-Heptanone was detected in air above Whitaker's Forest, Sierra Nevada Mt. CA [18].

Food Survey Values: 2-Heptanone was detected as a volatile flavor component of roasted filberts [22], baked potatoes [7], blue cheese [8], Beaufort (Gruyere) cheese [11], fried bacon [20], clove essential oil [29], chickpeas [36], fried chicken [41], sweet corn [5], peanut oil [6], duck meat and roasted duck [47] and in food products processed from the Cassava root [10]. 2-Heptanone has been identified in swiss cheese (0.45 ppm) evaporated milk, butter, milk, cream (0.004-0.007 ppm), milk fat (16 ppm), white bread, soybeans (1 ppm) peaches and orange juice [23], and meat [21,35,38].

Plant Concentrations: 2-Heptanone was identified as a volatile constituent of kiwi fruit flowers [42].

Fish/Seafood Concentrations:

Animal Concentrations:

Milk Concentrations:

Other Environmental Concentrations:

Probable Routes of Human Exposure: Occupational exposure to 2-heptanone may occur by inhalation or dermal contact during its production, formulation or transport. Exposure to the general population may occur by ingestion of food in which it occurs naturally, by inhalation during the use of commercial products in which it is used as a solvent or by the ingestion of contaminated drinking water.

Average Daily Intake:

2-Heptanone

Occupational Exposure: NIOSH (NOES 1981-83) has statistically estimated that 76,382 workers are exposed to 2-heptanone in the US [30].

Body Burdens: 2-Heptanone was found in 36 of 46 samples analyzed for the National Human Adipose Tissue Survey, FY 1982 [31]. 2-Heptanone was qualitatively detected in 6 out of 8 breast milk samples [13].

REFERENCES

1. Atkinson R; Chem Rev 85: 69-201 (1985)
2. Atkinson R; Atmos Environ 24A: 1-41 (1990)
3. Brown KW, Donnelly KC; Haz Wast Haz Water 5: 1-30 (1988)
4. Bursey JT, Pellizzari ED; Analysis of Industrial Wastewater for Organic Pollutants in Consent Decree Survey Res Triangle Park, NC: USEPA (1982)
5. Buttery RG et al; J Agr Food Chem 42:791-95 (1994)
6. Chung TY et al; J Agr Food Chem 41:1467-70 (1993)
7. Coleman EC et al; J Agric Food Chem 29: 42-8 (1981)
8. Day EA, Anderson DF; J Agr Food Chem 13: 2-4 (1965)
9. Dore M et al; La Tribune du Cebedeau 28: 3-11 (1975)
10. Dougan J et al; J Sci Food Agric 34: 874-84 (1983)
11. Dumont JP, Adda J; J Agr Food Chem 26: 364-7 (1978)
12. Erickson MD, Pellizzari ED; Analysis of Organic Air Pollutants in the Kanawha Valley, WV and the Shenandoah Valley VA USEPA-903/9-78-007 (1978)
13. Erickson MD et al; Acquisition and Chemical Analysis of Mother's Milk for Selected Toxic Substances, EPA-560/13-80-029 (1980)
14. Gerhold RM, Malaney GW; J Wat Pollut Control Fed 38: 562-79 (1966)
15. Gordon WA, Gordon, M; Trans Ky Acad Sci 42: 149-57 (1981)
16. Guenther A et al; Atm Environ 28:1197-1210 (1994)
17. Hansch C, Leo AJ; Medchem Project Issue No.26 Pomona College, Claremont CA (1985)
18. Helmig D, Arey J; Sci Total Environ 112:233-250 (1992)
19. Heukelekian H, Rand MC; J Water Pollut Contr Assoc 29: 1040-53 (1955)
20. Ho CT et al; J Agric Food Chem 31: 336-42 (1983)
21. King M-F et al; J Agr Food Chem 41:1974-1961 (1993)
22. Kinlin TE et al; J Agr Food Chem 20: 1021-8 (1972)
23. Lande SS et al; Investigation of Selected Environmental Contaminants: Ketonic Solvents USEPA-560/2-76-003 (1976)
24. Lucas SV; GC/MS Analysis of Organics in Drinking Water Concentrates and Advanced Waste Treatment Concentrates: Vol 3 USEPA-600/1-84-020 (NTIS PB85-128247), Columbus, OH (1984)
25. Lyman WJ et al; Handbook of Chemical Property Estimation Methods NY: McGraw-Hill pp. 15-15 to 15-29 (1982)
26. Mackay D et al; Volatilization of Organic Pollutant from Water. USEPA-600/53-82-019 (1982)

27. Malaney GE, Gerhold RM; Proc 17th Ind Waste Conf Eng Bull Purdue Univ Eng Ext Ser 112: 249-57 (1962)
28. Mills EVJR, Stack VTJR; pp. 492-517 in Proc 8th Ind Waste Conf. Eng Bull Purdue Univ Eng Ext Ser (1962)
29. Muchalal M, Crouzet J; Agric Biol Chem 49: 1583-9 (1985)
30. NIOSH; National Occupational Exposure Survey (NOES) (1989)
31. Onstot et al; Characterization of HRGC/MS Unidentified Peaks from the Broad Scan Analysis of the FY82 NHATS Composites, Draft Final Report, MRI Project No. 8823-A01, Midwest research Institute (1987)
32. Pellizzari ED et al; Amer Soc Test Mat Spec Tech Publ STP 685: 256-74 (1979)
33. Perry DL et al; Identification of Organic Compound in Industrial Effluent Discharges USEPA-600/4-79-016 (1979)
34. Perry JJ; Antonie Van Leeuwenhoek 34: 27-36 (1968)
35. Ramarathnam N et al; J Agr Food Chem 39:1839-47 (1991)
36. Rembold H et al; J Agric Food Chem 37: 659-62 (1989)
37. Riddick JA et al; Organic Solvents 4th ed. NY: Wiley: Interscience (1986)
38. Shahidi R et al; CRC Crit Rev Food Sci Nature 24:141-243 (1986)
39. Sax NI, Lewis RJSR; Hawley's Condensed Chemical Dictionary. NY: Van Nostrand Reinhold Co 11th ed. p. 758 (1987)
40. Swann RL et al; Res Rev 85: 17-28 (1983)
41. Tang J et al; J Agric Food Chem 31: 1287-92 (1983)
42. Tatsuka K et al; J Agric Food Chem 38: 2176-80 (1990)
43. Wallington TJ, Kurylo MJ; J Phys Chem 91: 5050-4 (1987)
44. Wilkins K; Chemosphere 29:47-53 (1994)
45. Windholz M et al; The Merck Index. Rahway, NJ: Merck & Co Inc 10th ed. (1983)
46. Winer AM et al; Atm Environ 26A: 2647-59 (1992)
47. Wu C-M, Liou S-E; J Agr Food Chem 40:838-841 (1992)

Hexabromobenzene

Synonyms:

Structure:

CAS Registry Number: 87-82-1

Molecular Formula: C_6Br_6

SMILES Notation: c(c(c(c(c1Br)Br)Br)Br)(c1Br)Br

CHEMICAL AND PHYSICAL PROPERTIES

Boiling Point:

Melting Point: 327 °C

Molecular Weight: 551.49

Dissociation Constants:

Log Octanol/Water Partition Coefficient: 6.07 [4]

Water Solubility: 0.16 ug/L at 22 °C [12]

Vapor Pressure: 2.35E-6 mm Hg at 25 °C (estimated) [8]

Henry's Law Constant: 2.8×10^{-5} atm-m^3/mol at 25 °C [9]

ENVIRONMENTAL FATE/EXPOSURE POTENTIAL

Summary: Hexabromobenzene, which has been used as a flame retardant, may enter the environment as a fugitive emission during its manufacture, formulation, or use. If released to the soil, hexabromobenzene is expected to adsorb strongly. Hexabromobenzene is not expected to undergo microbial degradation in soil, based on limited data. Volatilization from the soil surface to the atmosphere is not expected to be a significant process. If released to water, hexabromobenzene is expected to adsorb strongly to sediment and suspended organic matter. Volatilization of hexabromobenzene from water to the atmosphere may occur, although its adsorption to sediment and suspended matter is expected to attenuate the rate of this process considerably. The estimated volatilization half-life from a model river is approximately 3 days; however, the volatilization half-life from a model pond, which takes into account adsorptive processes, is 114 months. Neither hydrolysis, direct photochemical degradation, nor chemical oxidation of hexabromobenzene are expected to occur. Conflicting data on the bioaccumulation of hexabromobenzene in fish appear in the open literature. In short-term studies, hexabromobenzene was found to be nonaccumulative in fish. In a 96 day study, a mean bioconcentration factor of 1,100 was obtained. Bioconcentration in fish and aquatic organism would be expected based on its water solubility and octanol/water partition coefficient. Microbial degradation of hexabromobenzene in environmental waters is not expected to occur, based on limited data. In the atmosphere, hexabromobenzene is expected to occur predominately in the particulate form. Neither direct photochemical degradation nor destruction by the vapor-phase reaction with photochemically produced hydroxyl radicals is expected to occur. Hexabromobenzene may undergo dry deposition to the surface. Occupational exposure to hexabromobenzene may occur by dermal contact during its production and use as a flame retardant. Exposure to the general population may occur by dermal contact with commercial products in which it is contained.

Natural Sources: Hexabromobenzene is of anthropogenic origin, and it is not believed to occur naturally.

Artificial Sources: Hexabromobenzene may enter the environment as a fugitive emission from its use as a flame retardant [7,16]. Hexabromobenzene may also enter the environment as a result of the high temperature breakdown of octabromodiphenyl ether, decabromodiphenyl ether, pentabromophenol, and hexabromobiphenyl, all of which are used as flame retardants [2].

Hexabromobenzene

Terrestrial Fate: If released to soil, hexabromobenzene is expected to adsorb strongly to soil [8,12]. Volatilization from the soil surface to the atmosphere is not expected to be an important process based upon the vapor pressure and Henry's Law constant. Based on limited data, hexabromobenzene is not expected to undergo microbial biodegradation in soil [13].

Aquatic Fate: If hexabromobenzene is released to water, the estimated Henry's Law constant suggests that volatilization from water to the atmosphere may be a significant process. The estimated half-life for volatilization from a model river is 3.3 days [8]; however, it is expected to strongly adsorb to sediment and suspended organic matter [12], which may attenuate the rate of volatilization considerably. The estimated volatilization half-life from a model pond, which takes into account the process of adsorption, is 114 months [15]. In water, hexabromobenzene is not expected to undergo direct photochemical degradation, hydrolysis, or chemical-induced oxidation [5]. Conflicting data on the bioconcentration of hexabromobenzene in fish and aquatic organisms have appeared in the literature. In short-term studies (4-16 days), hexabromobenzene was found to be nonaccumulative in fish [12,17]; however, in a 96 day study, a mean bioconcentration factor of 1,100 was obtained [11]. Based on limited data, hexabromobenzene is not expected to biodegrade in aquatic systems [13].

Atmospheric Fate: In the atmosphere, hexabromobenzene is expected to exist predominately in the particulate form [3]. Hexabromobenzene may undergo direct deposition to the surface, or dry deposition while adsorbed to particulate matter. Degradation by direct photolysis [5] or from the reaction with photochemically produced hydroxyl radicals [1] is not expected to occur.

Biodegradation: Hexabromobenzene was listed as being confirmed to be nonbiodegradable in a Japanese MITI screening BOD test [13].

Abiotic Degradation: Hexabromobenzene is not expected to be degraded by direct photolysis, hydrolysis, or chemical oxidation [5].

Bioconcentration: Conflicting data on the bioconcentration of hexabromobenzene have appeared in the literature. Juvenile Atlantic salmon

(*Salmo salar*) did not uptake any hexabromobenzene from water, or from contaminated food [17], and it did not bioconcentrate in 1 year old male guppies (*Poecilia reticulat*) [12]. These experimental studies were performed over a 4 and 16 day period, respectively. Hexabromobenzene was listed as a chemical compound confirmed to be nonaccumulative in screening studies [13], and hexabromobenzene has been reported to have a bioconcentration factor (BCF) of approximately 10 [6]. In a 96 day experiment in a flow-through tank, hexabromobenzene had a mean BCF in rainbow trout (*Salmo gairdneri*) of 1,100 [11]. In the first 7 days of the latter experiment, no hexabromobenzene was detected in the whole fish samples [11]. Both the water solubility and the estimated log octanol/water partition coefficient imply that bioaccumulation in fish and aquatic organisms would occur [8].

Soil Adsorption/Mobility: From the water solubility of hexabromobenzene and the estimated octanol/water partition coefficient, soil adsorption coefficients of 28,000 and 48,000, respectively, can be calculated using appropriate regression equations [8]. The magnitude of these values suggest that hexabromobenzene will be essentially immobile in soil [14].

Volatilization from Water/Soil: The estimated Henry's Law constant for hexabromobenzene suggests that volatilization from water to the atmosphere may be significant. The estimated half-life for volatilization from a model river 1 m deep, flowing at 1 m/sec, with a wind velocity of 3 m/sec is 3.3 days [8]; however, the expected strong adsorption of hexabromobenzene to sediment and suspended organic matter may attenuate the rate of this process considerably. The estimated volatilization half-life from a model pond, which takes into account adsorption processes, is 114 months [15]. Similarly, volatilization from moist soil is not expected to occur. The estimated vapor pressure suggests that volatilization from dry soil to the atmosphere will not be a significant process.

Water Concentrations:

Effluent Concentrations:

Sediment/Soil Concentrations:

Hexabromobenzene

Atmospheric Concentrations:

Food Survey Values:

Plant Concentrations:

Fish/Seafood Concentrations:

Animal Concentrations:

Milk Concentrations:

Other Environmental Concentrations:

Probable Routes of Human Exposure: Occupational exposure to hexabromobenzene may occur through inhalation and dermal contact during its production, formulation, and use as a flame retardant. A probable route of exposure for the general population is through dermal contact with commercial products in which hexabromobenzene is contained.

Average Daily Intake:

Occupational Exposure: NIOSH (NOES 1981-1984) has statistically estimated that 28 workers are exposed to hexabromobenzene in the US [10].

Body Burdens: Hexabromobenzene was detected in human adipose tissue in Japan at a concn range of 0.35 to 0.65 ng/g (wet weight) [16].

REFERENCES

1. Atkinson R et al; Chem Rev 85: 69-201 (1985)
2. Buser HR; Environ Sci Tech 20: 404-8 (1986)
3. Eisenreich SJ et al; Environ Sci Tech 15: 30-38 (1981)
4. Hansch C et al; Exploring QSAR Hydrophobic, Electronic, and Steric Constants, ACS, Washington, DC (1995)
5. Jaber HM et al; Data Acquisition for Environmental Transport and Fate Screening for Compounds of Interest to the Office of Solid Waste: SRI International USEPA-600/6-84-010 (1984)
6. Kawasaki M; Ecotox Environ Safety 4: 444-54 (1980)

Hexabromobenzene

7. Larson ER; Kirk-Othmer Encycl Chem Tech 3rd Ed. NY: John-Wiley 10: 373-95 (1980)
8. Lyman WJ et al; Handbook of Chemical Property Estimation Methods NY: McGraw-Hill pp. 5-1 to 5-30 (1982)
9. Meylan W, Howard PH; Environ Toxicol Chem 10: 1283-93 (1991)
10. NIOSH; National Occupational Exposure Survey (NOES) (1984)
11. Oliver BG, Niimi AJ; Environ Sci Tech 19: 842-9 (1985)
12. Opperhuizen A; ASTM Spec Tech Publ 921, Aquatic Tox Environ Fate 9: 305-15 (1986)
13. Sasaki S; pp 298-98 in Aquatic Pollutants. Hutzinger O et al Ed Oxford: Pergamon Press (1978)
14. Swann RL et al; Res Rev 85: 17-28 (1983)
15. USEPA; Exams II Computer Simulation (1987)
16. Yamaguchi et al; Chemosphere 17: 703-7 (1988)
17. Zitko V, Hutzinger O; Bull Environ Contam Toxicol 16: 665-73 (1976)

Hexachloroacetone

SUBSTANCE IDENTIFICATION

Synonyms:

Structure:

CAS Registry Number: 116-16-5

Molecular Formula: C_3Cl_6O

SMILES Notation: O=C(C(Cl)(Cl)Cl)C(Cl)(Cl)Cl

CHEMICAL AND PHYSICAL PROPERTIES

Boiling Point: 202-204 °C

Melting Point: -2 °C

Molecular Weight: 264.75

Dissociation Constants:

Log Octanol/Water Partition Coefficient: 2.48 (estimated) [8]

Water Solubility: 150 mg/L at 25 °C (estimated from Kow) [5]

Vapor Pressure: 0.376 mm Hg at 20 °C [9]

Henry's Law Constant: 8.73×10^{-4} atm-m³/mol (calculated from vapor pressure and water solubility)

ENVIRONMENTAL FATE/EXPOSURE POTENTIAL

Summary: Hexachloroacetone naturally appears as a component of the edible Hawaiian seaweed *Asparagopsis taxiformis*. It can be artificially

236

released to the environment through its application as a weedkiller or through chlorination processes involved in sewage treatment and pulp mill effluents. If released to water, hexachloroacetone is expected to evaporate into the atmosphere with a volatilization half-life of 6.5 hours from a model river. In water, some hexachloroacetone is expected to partition to sediment. In soil, hexachloroacetone should evaporate from dry soil surfaces, but its estimated Koc value suggests moderate soil mobility. In the ambient atmosphere, hexachloroacetone is not expected to react with photochemically produced hydroxyl radicals. There are no data that suggest that the general population is exposed to hexachloroacetone, but workers involved in the previous application of hexachloroacetone as a weedkiller were probably exposed through inhalation or through dermal contact.

Natural Sources: Hexachloroacetone is a constituent of the edible Hawaiian (US) seaweed, *Asparagopsis taxiformis* (Limu kohu) [13].

Artificial Sources: Hexachloroacetone was detected after the chlorination process in pulp mill effluents with a concentration of approximately 30 g hexachloroacetone per ton pulp [10] and a concentration of approximately 0.2 g hexachloroacetone in the spent chlorination liquor per ton pulp [3]. Hexachloroacetone was found in chlorinated municipal sewage effluents at a concentration of 30 ppb [2]. Hexachloroacetone was once used as a weedkiller, but it is currently believed to be of little commercial interest [12].

Terrestrial Fate: Hexachloroacetone's vapor pressure indicates that the chemical would evaporate from dry surfaces. The estimated Koc of 530 [5] suggests that hexachloroacetone will have moderate mobility in soil [11]. No data are available regarding biodegradation or other chemical processes in soil.

Aquatic Fate: If released to water, hexachloroacetone is expected to evaporate into the atmosphere with a volatilization half-life of 6.5 hours from a model river [5]. Bioconcentration is not expected to be important. No data are available on other potential degradation processes in water. Based on the estimated Koc of 530 [5], hexachloroacetone may partition from the water column to the sediment.

Atmospheric Fate: Based on the vapor pressure, hexachloroacetone will exist almost entirely in the vapor phase in the ambient atmosphere [4]. Hexachloroacetone will not react with photochemically produced hydroxyl radicals in the ambient atmosphere [1,7].

Biodegradation:

Abiotic Degradation: Hexachloroacetone will not react with photochemically produced hydroxyl radicals in the ambient atmosphere based on the fact that hexachloroacetone has no extractable protons [1,7].

Bioconcentration: Based on the estimated log Kow, the BCF for hexachloroacetone can be estimated to be 45 using a recommended regression-derived equation [5]. This value would suggest that bioconcentration in aquatic organisms will not be important [5].

Soil Adsorption/Mobility: Based on the estimated log Kow and a regression-derived equation [5], the Koc for hexachloroacetone can be estimated to be 530, indicating that hexachloroacetone would have moderate mobility in soil [11].

Volatilization from Water/Soil: The value of the Henry's Law constant suggests that hexachloroacetone will volatilize significantly, but not rapidly, from water. Based on this value, the volatilization half-life of hexachloroacetone from a model river 1 m deep flowing 1 m/sec with a wind velocity of 3 m/sec has been estimated to be approximately 6.5 hours [5].

Water Concentrations:

Effluent Concentrations: Concentration of hexachloroacetone in chlorination stage effluent from a bleached kraft pulp mill in the interior of British Columbia was approximately 1.1 mg/L [6]. Hexachloroacetone was detected in pulp mill effluents with a concentration of approximately 30 g hexachloroacetone per ton pulp [10] and a concentration of approximately 0.2 g hexachloroacetone in the spent chlorination liquor per ton pulp [3] after the chlorination process. Hexachloroacetone was found in chlorinated municipal sewage effluents at a concentration of 30 ppb [2].

Sediment/Soil Concentrations:

Atmospheric Concentrations:

Food Survey Values:

Plant Concentrations:

Fish/Seafood Concentrations:

Animal Concentrations:

Milk Concentrations:

Other Environmental Concentrations:

Probable Routes of Human Exposure: Hexachloroacetone was once used as a weedkiller, but it is currently believed to be of little commercial interest [12]. Because hexachloroacetone was previously used as an herbicide, workers involved in the application of the chemical were probably exposed through inhalation or through dermal contact.

Average Daily Intake:

Occupational Exposure:

Body Burdens:

REFERENCES

1. Atkinson R; Intern J Chem Kinet 19: 799-828 (1987)
2. Bourquin AW, Gibson DT; Science 2: 253-64 (1978)
3. Carlberg GE et al; Sci Total Environ 48: 157-67 (1986)
4. Eisenreich SJ et al; Environ Sci Technol 15: 30-8 (1981)
5. Lyman WJ et al; Handbook of Chemical Estimation Methods NY: McGraw-Hill (1982)
6. McKague AB et al; Mutat Res 91 (4-5): 301-6 (1981)
7. Meylan WM, Howard PH; Chemosphere 26:2293-2299 (1993)
8. Meylan W, Howard PH; J Pharm Sci 84: 83-92 (1995)
9. Sunshine I; Handbook of Analytical Toxicology p. 518 (1969)

Hexachloroacetone

10. Suntio LR et al; Chemosphere 17: 1249-90 (1988)
11. Swann RL et al; Res Rev 85: 17-28 (1983)
12. Worthing CR, Walker SB; The Pesticide Manual 8th ed Laneham, England: British Crop Protection Council p. 887 (1987)
13. Zochlinski H, Mower H; Mutat Res 89 (2): 137-44 (1981)

4-Hydroxy-4-methyl-2-pentanone

SUBSTANCE IDENTIFICATION

Synonyms:

Structure:

CAS Registry Number: 123-42-2

Molecular Formula: $C_6H_{12}O_2$

SMILES Notation: O=C(CC(O)(C)C)C

CHEMICAL AND PHYSICAL PROPERTIES

Boiling Point: 167.9 °C at 760 mm Hg

Melting Point: -44 °C

Molecular Weight: 116.16

Dissociation Constants:

Log Octanol/Water Partition Coefficient: -0.34 (estimated) [21]

Water Solubility: Miscible [26]

Vapor Pressure: 1.71 mm Hg at 25 °C [6]

Henry's Law Constant: 4.24×10^{-9} atm-m³/mol at 25 °C [20]

ENVIRONMENTAL FATE/EXPOSURE POTENTIAL

Summary: 4-Hydroxy-4-methyl-2-pentanone may be released to the environment as a result of its manufacture and use as a solvent, additive or synthetic intermediate for many materials. If released to soil, it will be expected to exhibit very high mobility, based upon the reported infinite solubility of the compound in water and an estimated Koc of 16. Although no data were located regarding its biodegradation in environmental media, the compound may be subject to biodegradation in soil based upon results observed in laboratory biodegradation aqueous aerobic screening tests. It should not be subject to volatilization from moist near-surface soil based upon its estimated Henry's Law constant. However, it may volatilize from dry near-surface soil and other dry surfaces based upon its vapor pressure. If released to water, it will not be expected to adsorb to sediment or suspended particulate matter or bioconcentrate in aquatic organisms based upon its estimated Koc and estimated BCF of 0.3, respectively. The compound may be subject to biodegradation in natural waters based upon results observed in laboratory aqueous aerobic biodegradation screening tests using acclimated mixed microbial cultures as inoculum. It should not be subject to volatilization from surface waters based upon the estimated Henry's Law constant. Hydrolysis should not be an important removal process since aliphatic alcohols and ketones (the two functional groups that 4-hydroxyl-4-methyl-2-pentanone contains) generally are resistant to hydrolysis. If released to the atmosphere, it can be expected to exist mainly in the vapor phase in the ambient atmosphere based upon its vapor pressure. The estimated atmospheric half-life for vapor-phase reaction with photochemically produced hydroxyl radicals is 4 days at an atmospheric concentration of $5 \times 10^{+5}$ hydroxyl radicals per cm^3. 4-Hydroxy-4-methyl-2-pentanone may be susceptible to direct photolysis in the atmosphere based upon its possible absorption of light at wavelengths > 290 nm. Based upon its high water solubility, the compound may be susceptible to removal from the atmosphere by washout. Human exposure will occur via ingestion of drinking water contaminated with the compound and ingestion of foods contaminated with the compound due to contact with acetone containing paints.

Natural Sources: It is the toxic principle in Stipa Vaseyi, commonly known as sleepy grass [25].

Artificial Sources: 4-Hydroxy-4-methyl-2-pentanone may be released as a result of its manufacture and use as a solvent for many materials including but not limited to lacquers, nitrocellulose, cellulose acetate, epoxy resins, vinyl chloride-vinyl acetate, dyestuffs, inks, various oils, waxes, fats, tars, wood preservatives, and pesticides [1-4]; and its use in textile printing, fuel additives, antifreeze mixtures, and ink removal [14,16,24,34]. It has been used in hydraulic fluids [24], as a pharmaceutical preparation preservative [34], and as an intermediate in the synthesis of dyes, inhibitors, pharmaceuticals, and insecticides [16].

Terrestrial Fate: If 4-hydroxy-4-methyl-2-pentanone is released to soil, it will be expected to exhibit very high mobility [30], based upon the reported high water solubility and an estimated Koc of 16. It may be subject to biodegradation in soil based upon results observed in laboratory biodegradation aqueous aerobic screening tests using acclimated mixed microbial cultures as inocula [2,31] provided suitably acclimated microbial populations are present [3]. It should not be subject to volatilization from moist near-surface soil based upon the estimated Henry's Law constant. However, it may volatilize from dry near-surface soil and other dry surfaces based upon its vapor pressure.

Aquatic Fate: If 4-hydroxy-4-methyl-2-pentanone is released to water, it will not be expected to adsorb to sediment or suspended particulate matter or to bioconcentrate in aquatic organisms based upon its estimated Koc and BCF, respectively. Although no data were found regarding biodegradation in environmental media, the compound may be subject to biodegradation in natural waters based upon results observed in laboratory aqueous aerobic biodegradation screening tests using acclimated mixed microbial cultures as inocula [2,31] provided suitably acclimated microbial populations are present [3]. It should not be subject to volatilization from surface waters based upon the estimated Henry's Law constant. Hydrolysis should not be an important removal process since aliphatic alcohols and ketones (the two functional groups that 4-hydroxy-4-methyl-2-pentanone contains) generally are resistant to hydrolysis [19].

Atmospheric Fate: If 4-hydroxy-4-methyl-2-pentanone is released to the atmosphere, it can be expected to exist mainly in the vapor phase in the ambient atmosphere [8] based upon the experimental vapor pressure. The estimated rate constant for vapor-phase reaction with photochemically produced hydroxyl radicals is 4×10^{-12} cm^3/molecule-sec at 25 °C [22], which corresponds to an atmospheric half-life of 4 days at an atmospheric concentration of $5 \times 10^{+5}$ hydroxyl radicals per cm^3. 4-Hydroxy-4-methyl-2-pentanone may be susceptible to direct photolysis in the atmosphere based upon its possible absorption of light at wavelengths >290 nm [33]. Based upon its high water solubility, the compound may be susceptible to removal from the atmosphere by washout.

Biodegradation: Two sets of tests using acclimated mixed microbial cultures as inoculum gave percent theoretical BOD of 47% [2] and 46% [31] after 5 days for 4-hydroxy-4-methyl-2-pentanone under aerobic conditions [2,31]. A percent theoretical BOD of 3% was observed after 5 days in screening tests using the standard dilution technique under aerobic conditions and effluent sewage from a biological sanitary waste treatment plant as inoculum; 31% theoretical BOD was observed after 5 days using adapted effluent sewage [3]. No information regarding biodegradation in natural media was found.

Abiotic Degradation: The rate constant for the vapor-phase reaction of 4-hydroxy-4-methyl-2-pentanone with photochemically produced hydroxyl radicals has been estimated to be 4 x 10^{-12} cm^3/molecule-sec at 25 °C [1], which corresponds to an atmospheric half-life of 4 days at an atmospheric concentration of 5 x 10^{+5} hydroxyl radicals per cm^3. 4-Hydroxy-4-methyl-2-pentanone may be susceptible to direct sunlight photolysis because it may absorb light at wavelengths >290 nm; the compound has a weak absorption maximum of light at a wavelength of 282 nm in ethyl alcohol solution [33], which may extend into the environmentally important wavelengths (>290 nm). For certain ketones, however, this absorption maximum is shifted to lower wavelengths in water solutions (e.g., acetone has an absorption maximum at 279 nm in hexane solution and at 264.5 nm in water solution [29]). This suggests that 4-hydroxy-4-methyl-2-pentanone in water solution may not absorb light at wavelengths >290 and may not, therefore, be susceptible to direct photolysis in surface waters. Photochemical smog studies results placed 4-hydroxy-4-methyl-2-pentanone in the high reactivity category in relation to maximum oxidant formation, intermediate reactivity category in relation to NO_2 maximum times, and low reactivity category in relation to formaldehyde formation and eye response times [17]. Hydrolysis should not be an important removal process since aliphatic alcohols and ketones (the two functional groups that 4-hydroxy-4-methyl-2-pentanone contains) generally are resistant to hydrolysis [19].

Bioconcentration: An estimated BCF of 0.3 can be calculated from an estimated log Kow using a recommended regression equation [19]. Based upon the estimated BCF and the reported infinite solubility of the compound in water, 4-hydroxy-4-methyl-2-pentanone will not be expected to significantly bioconcentrate in aquatic organisms.

Soil Adsorption/Mobility: An estimated Koc of 16 can be calculated from an estimated log Kow using a recommended regression equation [19]. Based upon the estimated Koc and the reported infinite solubility of the compound in water, 4-hydroxy-4-methyl-2-pentanone will not be expected to strongly adsorb to sediment or suspended particulate matter [30]. It will be expected to exhibit very high mobility in soil [30].

Volatilization from Water/Soil: Based upon the value for the Henry's Law constant, volatilization of 4-hydroxy-4-methyl-2-pentanone from surface water will not be expected to be an important transport process [19].

Water Concentrations: DRINKING WATER: 4-Hydroxy-4-methyl-2-pentanone was qualitatively detected in 3 of 16 samples of drinking water concentrate derived from large volume (>400 gallons) samples from 3 of 7 cities (New Orleans, LA, January 1976; Philadelphia, PA, February 1976; and Seattle, WA, November 1976 [18]. It was qualitatively detected in one raw drinking water source (a river) out of 13 raw water sources in the U.K. (including 3 ground water supplies, 7 rivers, and 3 reservoirs); it was not detected (detection limit not specified) in the treated drinking water from this or any of the other 12 U.K. water sources studied [10]. SURFACE WATER: 4-Hydroxy-4-methyl-2-pentanone was qualitatively detected in samples of surface water taken from the St. Joseph River, a tributary of Lake Michigan [13]. Trace amounts (defined as generally <1 ng/L) have been detected in samples of water from the lower reaches of the River Lee, a water supply for North London [32].

Effluent Concentrations: 4-Hydroxy-4-methyl-2-pentanone was qualitatively detected in 2 out of >4000 samples of effluent from 2 of 46 industrial categories (organics and plastics manufacture and rubber processing) [4]. It was qualitatively detected in 4 of 16 samples of concentrate derived from large volume (>400 gallons) samples of advanced treatment concentrate from 4 of 6 cities (Lake Tahoe, CA, October 1974; Pomona, CA, September 1974; Orange County, CA, February 1976; and Blue Plains, Washington, DC, May 1975) [18]. The compound was detected in samples of water from a landfill monitoring well in New Castle, DE, at a concn of 2,900 ppm [16]. It was qualitatively detected in a petrochemical plant wastewater stream which emptied into the Calcasieu River, LA [15]. The compound was detected at an estimated concn of 10.9 ppb in ground water from a landfill well near Norman, OK (concn was estimated by comparing gas chromatography peak heights of samples with that obtained for known concn of the compound) [7]. It was detected at a concn of 20 ppb in the effluent for the Los Angeles County wastewater treatment plant between November 1980 and August 1981 [12]. The compound was qualitatively detected in the final effluent from the Addison, IL, publicly owned treatment works sampled in April 1980; it was not detected (detection limit not specified) in the final effluent from seven other publicly owned treatment works, a refinery and an industrial plant in Illinois sampled between February and June 1980 [9]. It was qualitatively found in trench leachate from low level radioactive waste sites at Maxey Flats, KY, and/or West Valley, NY [11]. The compound was found in spent chlorination liquor from bleaching of sulfite pulp at normal lignin content after oxygen treatment at a concn of 1 g/ton of pulp, at 9 g/ton pulp in spent chlorination liquor from bleaching of sulfite pulp at normal lignin content after oxygen and alkali treatments and at 0.3 g/ton in spent chlorination liquor from bleaching of sulfite pulp at high lignin content [5].

Sediment/Soil Concentrations: 4-Hydroxy-4-methyl-2-pentanone was qualitatively detected in sediment from one of two sites located midchannel in Tobin Lake, Nipawin, Saskatchewan, Canada [28].

Atmospheric Concentrations:

Food Survey Values: 4-Hydroxy-4-methyl-2-pentanone has been qualitatively detected in foodstuffs exposed to acetone: meat carcasses exposed to surfaces that were freshly painted with acetone containing paints and vegetables and seeds that were extracted with acetone [16].

Plant Concentrations:

Fish/Seafood Concentrations:

Animal Concentrations:

4-Hydroxy-4-methyl-2-pentanone

Milk Concentrations:

Other Environmental Concentrations:

Probable Routes of Human Exposure: Exposure to 4-hydroxy-4-methyl-2-pentanone may occur via ingestion of drinking water contaminated with the compound [10,18] and ingestion of foods contaminated with the compound due to contact with surfaces that were freshly painted with acetone containing paints [16].

Average Daily Intake:

Occupational Exposure: NIOSH (NOES 1981-1983) has statistically estimated that 225,328 workers are potentially exposed to 4-hydroxy-4-methyl-2-pentanone in the US [23]. The time weighted avg concns of the compound found in the breathing zone of workers in a screen printing plant in various job sites were as follows (avg of 12-19 samples each): 14 ppm in printing press, 3.5 ppm in automatic dryer, 12 ppm in manual drying, 2.8 ppm in paint mixing, and 6.8 ppm in screen wash [27]. The concns were somewhat lower in the general air of most of the work areas than in the breathing zone air except in the automatic and manual dryer areas [27]. The time weighted avg concns of the compound found in the general area of workers in the screen printing plant were as follows (avg of 6-10 samples each): 9 ppm in printing press area, 4.8 ppm in automatic dryer area, 12.5 ppm in manual drying area, 1.8 ppm in paint mixing area, and 4.5 ppm in screen wash area [27]. The overall avg concn found in the general air of the plant was 3.2 ppm [27].

Body Burdens:

REFERENCES

1. Atkinson R; Environ Toxicol Chem 7: 435-42 (1988)
2. Babeu L, Vaishnav DD; J Indust Microbiol 2: 107-15 (1987)
3. Bridie AL et al; Water Res 13: 627-30 (1979)
4. Bursey JT, Pellizzari ED; Analysis of Industrial Wastewater for Organic Pollutants in Consent Decree Survey. Contract No. 68-03-2867. Athens, GA: USEPA Environ Res Lab p. 98 (1982)
5. Carlberg GE et al; Sci Total Environ 48: 157-67 (1986)

6. Daubert TE, Danner RP; Physical and Thermodynamic Properties of Pure Compounds Am Inst Chem Eng (1989)
7. Dunlap WJ et al; pp. 453-77 in Identif Anal Organic Pollut Water. Keith LH ed (1976)
8. Eisenreich SJ et al; Environ Sci Technol 15: 30-8 (1981)
9. Ellis DD et al; Arch Environ Contam Toxicol 11: 373-82 (1982)
10. Fielding M et al; Organic Pollutants in Drinking Water, TR-159 Water Res Cent p 19 (1981)
11. Francis AJ et al; Nuclear Technol 50: 158-63 (1980)
12. Gossett RW et al; Mar Pollut Bull 14: 387-92 (1983)
13. Great Lakes Water Quality Board; Inventory Chem Subst Id Great Lakes Ecos pp. 195 (1983)
14. Hawley GG; Condensed Chemical Dictionary 10th ed Van Nostrand Reinhold NY p. 316-7 (1981)
15. Keith LH; Sci Total Environ 3: 87-102 (1974)
16. Lande SS et al; Investigation of Selected Potential Environmental Contaminants: Ketonic Solvents USEPA-560/2-76-007 (1976)
17. Levy A; Photochemical Smog Reactivity . In: Solvent Theory and Practices. p 70-94 Washington DC: Amer Chem Soc (1973)
18. Lucas SV; GC/MS Anal of Org in Drinking Water Concentrates and Advanced Treatment Concentrates Vol 1 USEPA-600/1-84-020a NTIS PB85-128221 p. 170 (1984)
19. Lyman WJ et al; Handbook of Chem Property Estimation Methods NY: McGraw-Hill pp. 15-16 to 15-29 (1982)
20. Meylan W, Howard PH; Environ Toxicol Chem 10: 1283-93 (1991)
21. Meylan W, Howard PH; J Pharm Sci 84: 83-92 (1995)
22. Meylan WM, Howard PH; Chemosphere 26:2293-2299 (1993)
23. NIOSH; The National Occupational Exposure Survey (NOES) (1983)
24. Papa AJ; Kirk-Othmer Encycl Chem Tech 3rd NY, NY: Wiley 13: 912-5 (1981)
25. Patty; Industrial Hyg & Tox 3rd ed p 4659 (1982)
26. Riddick JA et al; Organic Solvents NY,NY: John Wiley & Sons Inc (1984)
27. Samimi B; Amer Indust Hyg Assoc J 43: 43-8 (1982)
28. Samoiloff MR et al; Environ Sci Technol 17: 329-34 (1983)
29. Silverstein RM et al; Spectrometric Id of Org Cmpd NY, NY: J Wiley & Sons Inc 3rd ed p. 242 (1974)
30. Swann RL et al; Res Rev 85: 17-28 (1983)
31. Vaishnav DD et al; Chemosphere 16: 695-703 (1987)
32. Waggott A; Chem Water Reuse 2: 55-9 (1981)
33. Weast RC et al; Hnbk Chem & Phys 53rd ed p. C-415 Boca Raton,FL: CRC Press (1972)
34. Windholz M et al; Merck Index 10th ed. Rahway, NJ p. 428-9 (1983)

Isobutyric Acid

SUBSTANCE IDENTIFICATION

Synonyms: 2-Methylpropanoic acid

Structure:

$$CH_3$$
$$H_3C \quad \diagdown \quad O$$
$$HO$$

CAS Registry Number: 79-31-2

Molecular Formula: $C_4H_8O_2$

SMILES Notation: O=CC(C)C

CHEMICAL AND PHYSICAL PROPERTIES

Boiling Point: 152-155 °C at 760 mm Hg

Melting Point: -47 °C

Molecular Weight: 88.1

Dissociation Constants: pKa = 4.8 [10]

Log Octanol/Water Partition Coefficient: 0.94 [19]

Water Solubility: 2.0 x 10^{+5} mg/L at 20 °C [18]

Vapor Pressure: 1.5 mm Hg at 20 °C [22]

Henry's Law Constant: 9.09 x 10^{-7} atm-m^3/mol at 25 °C [14]

ENVIRONMENTAL FATE/EXPOSURE POTENTIAL

Summary: Isobutyric acid may be released to the environment in emissions and effluents from manufacturing and use facilities, in leachate from disposed consumer and industrial products in which this compound is contained, in automobile gas exhaust and as a biodegradation product of more complex organic compounds. Isobutyric acid is also a naturally occurring component of food (cheese, butter, milk protein, vinegar and beer) and feedstuffs and is reportedly found in several essential oils. If released to soil, isobutyric acid is expected to be highly mobile. This compound should volatilize fairly rapidly from dry soil surfaces; however, volatilization from moist soils should not be significant. Removal by chemical hydrolysis should not be important. If released to water, isobutyric acid is not expected to chemically hydrolyze, oxidize (estimated half-life 1.7 years), volatilize, bioaccumulate or adsorb to suspended solids or sediment. The chemical structure would suggest that biodegradation of isobutyric acid may be important. If released to the atmosphere, isobutyric acid is expected to exist entirely in the vapor phase and has the potential to be transported long distances before it is removed by washout in precipitation or reaction with photochemically generated hydroxyl radicals (estimated half-life 3.6 days). The most probable routes of exposure to isobutyric acid by the general population are inhalation and ingestion. Workers involved in the manufacture, handling or use of this compound may also be exposed by inhalation and dermal contact.

Natural Sources: Isobutyric acid is a naturally occurring component of food (cheese, butter, milk protein, vinegar, and beer) and feedstuffs and is produced during the intermediary hepatic and microbial metabolism of valine. Isobutyric acid is present in human feces, presumably due to action of intestinal microflora.

Artificial Sources: Isobutyric acid may be released to the environment in emissions and effluents from manufacturing and use facilities. This compound may also enter the environment from the disposal of consumer and industrial products in which this compound is contained [6,8,9], it may be emitted in automobile gas exhaust [11] and it may be formed as a biodegradation product of more complex organic compounds [7].

Terrestrial Fate: If released to soil, isobutyric acid is expected to be highly mobile. This compound should volatilize fairly rapidly from dry soil surfaces; however, volatilization from moist soils should not be significant. Removal by chemical hydrolysis should not be important. The chemical structure would suggest that in soil isobutyric acid will biodegrade under aerobic conditions and there is experimental data to suggest that anaerobic biodegradation will be important.

Aquatic Fate: If released to water, isobutyric acid is not expected to chemically hydrolyze, oxidize (estimated half-life 1.7 years), volatilize, bioaccumulate or adsorb to suspended solids or sediments. The chemical structure would suggest that in water isobutyric acid will biodegrade under aerobic conditions and there is experimental data to suggest that anaerobic biodegradation will be important.

Atmospheric Fate: If released to the atmosphere, isobutyric acid is expected to exist entirely in the vapor phase and has the potential to be transported long distances before it is removed by washout in precipitation or reaction with photochemically generated hydroxyl radicals (estimated half-life 3.6 days).

Biodegradation: Isobutyric acid is reported to be susceptible to anaerobic biodegradation [20]. Under anaerobic conditions, isobutyric acid was metabolized by an enriched acetate culture cross-acclimated with isobutyric acid at a rate of 250 mg/L following a 3 day lag period [2].

Abiotic Degradation: Chemical hydrolysis of isobutyric acid is not expected to be significant since this compound contains no hydrolyzable functional groups [13]. The half-life for isobutyric acid reacting with photochemically generated hydroxyl radicals in water has been calculated to be 1.7 years based on a reaction rate constant of $1.26 \times 10^{+9}$ 1/mol-sec [5] and a hydroxyl radical concn of 1×10^{-17} mol/L [16]. The half-life for isobutyric acid vapor reacting with photochemically generated hydroxyl radicals in the atmosphere has been calculated to be 3.6 days based on a reaction rate constant of 2.8×10^{-12} cm^3/molecule-sec at 25 °C and an ambient hydroxyl radical concn of $8.0 \times 10^{+5}$ molecules/cm^3 [14].

Bioconcentration: The bioconcentration factor (BCF) of isobutyric acid has been calculated to be <2 using the water solubility and the octanol/water partition coefficient [13]. This BCF value indicates that isobutyric acid should not bioaccumulate in aquatic organisms.

Soil Adsorption/Mobility: Using linear regression analysis and the water solubility, a soil adsorption coefficient (Koc) of 5 has been calculated for isobutyric acid [13]; using a method of structural contributions, a Koc of 1.9 has also been calculated [2]. Based on these BCF values adsorption of isobutyric acid to sediments and suspended solids in water is not expected to be significant and isobutyric acid should be highly mobile in soil [21].

Volatilization from Water/Soil: Ionization of isobutyric acid under environmental conditions (pH 5-9), as indicated by the pKa and its relatively low Henry's Law constant, suggest that volatilization would be insignificant from water and moist soil surfaces. Considering the vapor pressure of isobutyric acid, volatilization from dry soil surfaces would probably be fairly rapid.

Water Concentrations: DRINKING WATER: Isobutyric acid was tentatively identified in drinking water from Cincinnati, OH, during 1980 and Philadelphia, PA, during 1976 [12]. GROUND WATER: Isobutyric acid was detected in the ground water at the site of a wood treatment facility in Pensacola, FL, at a maximum concn of 4.13 ppm [9]. Isobutyric acid was found in ground water below a municipal landfill in Norman, OK, at an estimated concentration of 48.7 ppb [6].

Effluent Concentrations: During 1974, isobutyric acid was tentatively identified in water at an advanced waste treatment plant in Pomona, CA [12]. Isobutyric acid was detected in retort water from Australian oil shales at a concn of 31 ppm [4]. Isobutyric acid has been qualitatively identified in effluents from the following industries: paint and ink, textile mills, auto and other laundries and organic chemicals [1].

Sediment/Soil Concentrations:

Atmospheric Concentrations: During 1984, isobutyric acid was monitored in the atmosphere of Los Angeles at concns ranging from 0.003 to 0.056 ppb [11].

Food Survey Values: Isobutyric acid has been identified as a volatile flavor component of Idaho Russet Burbank baked potatoes [3].

Plant Concentrations:

Fish/Seafood Concentrations:

Animal Concentrations:

Milk Concentrations:

Other Environmental Concentrations: Isobutyric acid was found in trench leachate samples taken from low level radioactive waste disposal sites in Maxey Flats, KY, and West Valley, NY, at a concn ranging from 0.40 to 3.6 ppm and 12 to 21 ppm, respectively [8]. Isobutyric acid was detected in used automobile motor oil at a concn of 88 ppb and automobile gas exhaust at a concn of 0.056 ppb [11].

Probable Routes of Human Exposure: The most probable routes of exposure to isobutyric acid by the general population are inhalation and ingestion. Workers involved in the manufacture, handling or use of this compound may also be exposed by inhalation and dermal contact.

Average Daily Intake:

Occupational Exposure: A National Occupational Exposure Survey (NOES) estimates that 3,180 workers are exposed to isobutyric acid; however, this figure does not include the number of workers exposed to trade name products containing isobutyric acid [17].

Body Burdens:

REFERENCES

1. Bursey JT, Pellizzari ED; Analysis of Industrial Wastewater for Organic Pollutants in Consent Decree Survey Contract No. 68-03-2867 USEPA Environ Res Lab, Athens, GA (1982)
2. Chou WL et al; Biotech Bioeng Symp 8: 391 (1979)
3. Coleman EC et al; J Agric Food Chem 29: 42 (1981)
4. Dobson KR et al; Water Res 19: 849 (1985)
5. Dorfman LM, Adams GE; Reactivity of the Hydroxyl Radical in Aqueous Solution NSRD-NBS-46, NTIS COM-73-50623 (1973)

6. Dunlap WJ et al; Organic Pollutants Contributed to Groundwater by a Landfill USEPA 600/9-76-004 (1976)
7. Francis AJ et al; Appl Environ Microbiol 40: 108 (1980)
8. Francis AJ et al; Nuclear Tech 50: 158 (1980)
9. Goerlitz DF et al; Environ Sci Tech 19: 955 (1985)
10. Hemphill L, Swanson WS; Proc of the 18th Industrial Waste Conf 18: 204 (1964)
11. Kawamura K et al; Environ Sci Tech 19: 1082-6 (1985)
12. Lucas SV; GC/MS Analysis of Organics in Drinking Water Concentrates and Advanced Waste Treatment Concentrates, Vol.1 USEPA-600/1-84-020A (1984)
13. Lyman WJ et al; Handbook of Chemical Property Estimation Methods. Environmental Behavior of Organic Compounds. McGraw-Hill NY (1982)
14. Meylan W, Howard PH; Environ Toxicol Chem 10: 1283-93 (1991)
15. Meylan WM et al; Environ Sci Technol 11:1560-1567 (1992)
16. Mill T et al; Science 207: 886 (1980)
17. NIOSH; National Occupational Exposure Survey (NOES) Sept 20 (1985)
18. Perry RH, Green D; Perry's Chemical Engineer's Handbook 6th ed McGraw-Hill Book Co. New York (1983)
19. Sangster J LOGKOW Database, Sangster Research Lab., Quebec, Canada (1993)
20. Speece RE; Environ Sci Tech 17: 416A (1983)
21. Swann RL et al; Res Rev 85: 17 (1983)
22. Weber RC et al; Vapor Pressure Distribution of Selected Organic Chemicals USEPA-600/2-81-021 (1981)

Iso-octane

SUBSTANCE IDENTIFICATION

Synonyms: Pentane, 2,2,4-trimethyl-

Structure:

CAS Registry Number: 540-84-1

Molecular Formula: C_8H_{18}

SMILES Notation: C(CC(C)C)(C)(C)C

CHEMICAL AND PHYSICAL PROPERTIES

Boiling Point: 99.238 °C

Melting Point: -107.45 °C

Molecular Weight: 114.22

Dissociation Constants:

Log Octanol/Water Partition Coefficient: 4.09 [31]

Water Solubility: 2.44 mg/L at 25 °C [51]

Vapor Pressure: 49.3 mm Hg at 25°C [8]

Henry's Law Constant: 3.01 atm-m^3/mol at 25°C [32]

ENVIRONMENTAL FATE/EXPOSURE POTENTIAL

Summary: Iso-octane (2,2,4-trimethylpentane) is a volatile constituent of petroleum products and natural gas. Although the compound occurs naturally, it is principally released to the environment via the manufacture, use and disposal of products associated with the petroleum and gasoline industry. Photolysis and hydrolysis of iso-octane are not expected to be important environmental fate processes in any environmental media. Based upon limited data, biodegradation of the compound may occur slowly in soil and water, but probably only where the compound has been preexposed. Volatilization of the compound from water and soil surfaces may be the most important fate process. The volatilization half-life of iso-octane from a model river 1 m deep flowing at 1 m/sec with a wind speed of 3 m/sec has been estimated to be 3.1 hr. The Koc value of iso-octane indicates that it will not be very mobile in soil and may partition from the water column to suspended solids and sediments in water. The estimated bioconcentration factor indicates that bioconcentration may be important in aquatic organisms. The most important fate process for this compound in air may be its reaction with photochemically produced hydroxyl radicals. This reaction has an estimated half-life of 4.4 days. The most probable route of human exposure to iso-octane is by inhalation. Both general population and occupational exposure to this compound are likely.

Natural Sources: Iso-octane is naturally found in crude petroleum [9] and in small amounts in natural gas [17].

Artificial Sources: Since iso-octane is a constituent of petroleum and gasoline, it is released to the environment by the petroleum industries during refining processes and during the use of gasoline [37]. Automotive exhaust and automotive evaporative emissions are the most important sources of iso-octane in the atmosphere [15,35,44,53,54]. Products, such as polyethylene pipes used for distribution of drinking water, can release this compound in water and the atmosphere [1]. Hazardous wastes sites [23], landfills [23,52] and emissions from wood combustion [54] also release iso-octane into the environment. JP-4 (jet fuel) samples contain iso-octane [7], which may be released to the environment.

Terrestrial Fate: Photolysis and hydrolysis of iso-octane are not expected to be important in soil based upon the lack of hydrolyzable functional groups [27] and the lack of UV absorption [45]. Although iso-octane may undergo slow biodegradation in soil [20], volatilization from dry and wet soil surfaces is expected to be a more important fate process. The vapor pressure and Henry's Law constant indicate high volatility from dry and moist soil surfaces. In subsurface soil, adsorption of iso-octane is expected to be important, as indicated by an average log Koc value of 4.35 in three sediments [29] and an estimated log Koc value of 3.43 in soil [27].

Aquatic Fate: The hydrolysis of iso-octane in water is not expected to be important because the compound does not contain any hydrolyzable group [27]. The photolysis of the compound in water is also expected to be unimportant because iso-octane is transparent to wavelengths available in sunlight [45]. Although slow biodegradation may occur in aquatic media [20], volatilization from water is expected to be the dominant process as indicated by the value of the Henry's Law constant. The average log Koc value of 4.35 in sediments [29] indicates that most of the compound may remain adsorbed to sediment and suspended solids in aquatic media. A log bioconcentration factor of 2.57 estimated [27] from the water solubility indicates that bioconcentration in aquatic organisms may be important.

Atmospheric Fate: The rate constant for the reaction of iso-octane with atomic oxygen (3P) is $5.5 \times 10^{+10}$ cm^3/molecule-sec [16]. Based on this rate constant and the concentration of atomic oxygen of $2.5 \times 10^{+4}$ molecules/cm^3 in a typical atmosphere [11], this reaction will not be important in the atmosphere. The gas-phase reactions of alkanes with ozone and nitrate radicals are of negligible importance as atmospheric loss processes [3]. The rate constant for the reaction of iso-octane with OH radicals at 25 °C is given as 3.66-3.68×10^{-12} cm^3/molecule-sec [3,18]. Based on an average 24-hr atmospheric OH radical concentration of $5 \times 10^{+5}$ molecules/cm^3 [4], the half-live of iso-octane due to this reaction is 4.4 days.

Biodegradation: When a well water (previously contaminated by gasoline spill, but contained no detectable gasoline at the time of the experiment) with added iso-octane to a concentration of 48 mg/L was incubated in the dark at 13 °C, the natural flora in the ground water biodegraded 13% of the compound in 8 days [20]. Therefore, it is likely that iso-octane may undergo slow biodegradation with microorganisms acclimated to this compound.

Abiotic Degradation: Alkanes do not contain any hydrolyzable group and are generally resistant to hydrolysis [27]. Therefore, hydrolysis of iso-octane in water and soil is not expected to be important. The compound is transparent to tropospheric sunlight [45] indicating that direct photolysis in ambient water and soil should not be important. The rate constants for the gas phase reaction of iso-octane with oxygen atoms (3P) and hydroxyl radicals are $5.5 \times 10^{+10}$ cm^3/molecule-sec [16] and $3.66\text{-}3.68 \times 10^{-12}$ cm^3/molecule-sec [3,18], respectively. Based on typical 24-hr average atmospheric atomic oxygen concentrations of $2.5 \times 10^{+4}$ molecules/cm^3 [11] and OH radical concentration of $5 \times 10^{+5}$ radicals/cm^3 [4], the reaction with atomic oxygen can be estimated to be unimportant and the half-life for reaction with OH radicals can be estimated to be 4.4 days. The rate constant for the reaction of OH radicals with iso-octane in aqueous solution is $6.023 \times 10^{+9}$ M-sec^{-1} [50]. If it is assumed that the concentration of OH radicals in a typical natural eutrophic water is 3×10^{-17} M [28], the half-life for the oxidation of iso-octane by photochemically produced OH radicals in water is expected to be 44 days.

Bioconcentration: Based upon the water solubility, the log bioconcentration factor for iso-octane has been estimated to be 2.57 from a regression equation [27]. This value indicates bioconcentration may be important in aquatic organisms.

Soil Adsorption/Mobility: The adsorption of several hydrocarbons including iso-octane in five types of soil was studied by measuring the retention volumes [5]. Hydrocarbon retention by dry soils increased with molecular weight and unsaturation and decreased with branching. The retention volume of iso-octane was found to be high, but the authors did not provide any Koc values. The log Koc for iso-octane in soil has been estimated to be 3.43 from the water solubility and a regression equation [27]. The average log Koc for this compound in sediments from a salt marsh, a pond and a river was 4.35. Therefore, iso-octane is expected to generally remain strongly adsorbed to soil and sediments [46].

Volatilization from Water/Soil: Based on water and vapor phase mass transfer coefficients, the volatilization half-life of iso-octane at 25 °C from a still body of water at a depth of 1 m was estimated to be 5.6 hr [30]. Using more realistic winding conditions, the evaporation half-life of the compound from a model river 1 m deep, flowing at 1 m/sec with a wind speed of 3 m/sec, was estimated to be 3.1 hr [27]. However, neither estimation method considers the effect of adsorption on the rate of volatility. Using the EXAMS model, which considers various input parameters including the effect of adsorption, the volatilization half-life of iso-octane from a model pond has been estimated to be 15 days [48].

Water Concentrations: GROUND WATER: Iso-octane was detected in one on-site and one off-site well water sample from a waste site of a nonlubricating automotive fluids production plant in Michigan at concentrations of 0.005 mg/L and 0.008 mg/L, respectively [49].

Effluent Concentrations: The average exhaust from 67 gasoline fueled vehicles driven through an Australian urban driving cycle on a chassis dynamometer contained 1% (w/w) iso-octane of the total nonmethane hydrocarbon emitted [35]. The average emission of iso-octane from 46 in-use emission-controlled US passenger cars made between 1975 and 1982 was 2.4% of the total emission (both exhaust and evaporative) under a federal test procedure (FTP) driving cycle [44]. On the average, 0.25% of the total nonmethane hydrocarbon emitted from a wood combustion stack was due to iso-octane [54]. Iso-octane has been detected in emitted gases from waste and landfill sites [1,5].

Sediment/Soil Concentrations:

Atmospheric Concentrations: URBAN/SUBURBAN: Iso-octane has been qualitatively detected in the ambient air of many cities around the world including air in Pullman, WA [36], Pretoria, South Africa [26], Johannesburg, South Africa [26], Durban, South Africa [26], Zurich, Switzerland [12] and Leningrad, USSR [19]. The concentration range of iso-octane in Los Angeles air in 1981 was 3-15 ppb [13]. The average and maximum concentrations of this compound in Houston, TX, air obtained from 679 sampling points were 6 and 101 ppbC, respectively, in 1977 [34]. In Houston, TX, the level of iso-octane ranged from none detected to 416.3 ppbC in September 1983 and 39.1 to 146.8 ppbC in January 1984 [25]. The background concentration of iso-octane in Janesville, WI, air in 1978 was <0.5 ug/m^3, but the concns in atmospheric plume at distances of 10, 15 and 40 miles above Janesville were 1.5, 2.5 and 2.0 ug/m^3, respectively [43]. The concn level of iso-octane in Tulsa, OK, air in 1978 ranged 2.1-8.5% of the total hydrocarbon, although its composition was 17.1% in the air over a petroleum refinery [2]. The median and max concns of iso-octane in 39 US cities during 1984-1986 were 6.8 and 106 ppbC [42]. The ground level concn of iso-octane on a fall day in 1968 in the Los Angeles Basin was 1.1-

10.2 ppb, while the concn 1,500 ft above ground was 18.9 ppb [39]. The concn of the compound in urban plumes over Lake Michigan (in the vicinity of Chicago and Milwaukee) in 1976 was 6.8-10.0 ppb [33]. The average concn over Tokyo, Japan, in 1980 was 0.2 ppb [47]. SOURCE DOMINATED: The average concn of iso-octane inside Lincoln Tunnel was 56.6 ppbC in 1982 compared to 275.3 ppbC in 1970 [24]. This 5-fold decrease in concn was attributed to the use of catalyst-equipped vehicles [24]. The amount of this compound detected along US Highway 70 near Raleigh, NC, amounted to 1.07-1.48% of the total nonmethane hydrocarbon concn [53]. The in-vehicle concn of iso-octane in two four-door sedans averaged 21.1 ug/m³ in urban, 12.0 ug/m³ in interstate and 2.4 ug/m³ in rural areas for these driving routes. The corresponding concns along the sidewalks were 3.8 ug/m³, 1.6 and 0.5 ug/m³ [6]. RURAL/REMOTE: The average concentration of iso-octane in two ground level air samples collected in an orange grove near Dunedin, FL, in 1976 was 1.3 ppbC [24]. The concn of the compound at Jones State Forest, 38 miles north of Houston, TX, in 1978 ranged from 0.3 to 16.7 ppbC [2]. The median iso-octane concn in four remote sites in Northwestern North Carolina in 1981-1982 was 0.2 ppbC [41]. The compound constituted 0.5% of the total hydrocarbon composition in Smoky Mountain air in 1978 [2].

Food Survey Values: Iso-octane has been detected in the volatile components of Chickpea (*Cicer arietinum* L) [38].

Plant Concentrations:

Fish/Seafood Concentrations: Unspecified trimethylpentane isomers (probably a mixture) were detected at a concn of 0.2 ng/g (wet wt) in oysters and 1.3 ng/g (wet wt) in clams from Lake Pontchartrain, LA [10].

Animal Concentrations:

Milk Concentrations:

Other Environmental Concentrations: Iso-octane was detected at a concentration of 0.25 mg/L in water when 10 g of sliced polyethylene pipes used for distribution of potable water was allowed to soak in a liter of mineral water for 48 hr [1].

Probable Routes of Human Exposure: The most probable route of human exposure to iso-octane is by inhalation as indicated by the monitoring data in the gasoline industry [14,21,37]. The detection of the compound in the expired air of "normal" humans (not occupationally exposed) [22] indicates inhalation exposure to the general population may be important.

Average Daily Intake:

Occupational Exposure: The occupational exposures of truck drivers and terminal loading operators to gasoline vapors containing iso-octane were studied [14]. The mean air concns of iso-octane for service station attendants, transport drivers and outside operators in the petroleum industry were 0.670 mg/m³, 1.495 mg/m³ and 0.824 4 mg/m³, respectively [37]. The concn of iso-octane in the personal air of a high volume service station attendant was 0.2 ppm [21].

Body Burdens: Iso-octane was detected in the expired air of 29.7% of selected normal people not occupationally exposed to the compound [22]. The geometric mean concentration of iso-octane in the expired air of this general population was 0.604 ng/L [22].

REFERENCES

1.   Anselme C et al; Sci Total Environ 47: 371-84 (1985)
2.   Arnts RR, Meeks SA; Atmos Environ 15: 1643-51 (1981)
3.   Atkinson R; Atmos Environ 24A: 1-41 (1990)
4.   Atkinson R; Chem Rev 85: 69-201 (1985)
5.   Bohn HL et al; J Environ Qual 9: 563-5 (1980)
6.   Chan CC et al; Environ Sci Technol 25: 964-72 (1991)
7.   Cooper RC et al; Environmental Quality Research - Fate of Toxic Jet Fuel Components in Aquatic Systems; Air Force Aerosp Med Res Lab, AFAMRL-TR (US), ISS AFAMRL-TR-82-64, 102 (1982)
8.   Daubert TE, Danner RP; Physical & Thermodynamic Properties of Pure Chemicals. Hemisphere Publ. Corp., NY Vol 3 (1991)
9.   Elliott JJ, Melchoir MT; Kirk-Othmer Encycl Chem Technol NY: John Wiley & Sons 17: 122 (1982)
10.   Ferrario JB et al; Bull Environ Contam Toxicol 34: 246-55 (1985)
11.   Graedel TE; Chemical Compounds in the Atmosphere NY: Academic Press p. 7 (1978)
12.   Grob K, Grob G; J Chromatogr 62: 1-13 (1971)
13.   Grosjean D, Fung K; J Air Pollut Control Assoc 34: 537-43 (1984)

14. Halder CA et al; Am Ind Hyg Assoc J 47: 164-72 (1986)
15. Hampton CV et al; Environ Sci Technol 16: 287-98 (1982)
16. Herron JT, Huie RE; J Phys Chem Ref Data 2: 467-518 (1973)
17. Hillard JH; Kirk-Othmer Encycl Chemical Technol NY: John Wiley & Sons 11: 637 (1980)
18. Hodson J; Chemosphere 17: 2339-48 (1988)
19. Ioffe BV et al; J Chromatogr 142: 787-95 (1977)
20. Jamison VW et al; pp. 187-96 in Proc Third Int Biodeg Symp Shapley JM, Kaplan AM (eds), Essex, England (1976)
21. Kearney CA, Dunham DB; Am Ind Hyg Assoc J 47: 535-9 (1986)
22. Krotoszynski BK et al; J Anal Toxicol 3: 225-34 (1979)
23. LaRegina J et al; Environ Prog 5: 18-27 (1986)
24. Lonneman WA et al; Environ Sci Technol 20: 790-6 (1986)
25. Lonneman WA et al; Hydrocarbon in Houston Air, USEPA 600/3-79-018, Environ Sci Res Lab, Office of Res. Dev., USEPA, Research Triangle Park, NC (1979)
26. Louw CW et al; Atmos Environ 11: 703-17 (1977)
27. Lyman WJ et al; Handbook of Chemical Property Estimation Methods NY: McGraw-Hill (1982)
28. Mabey T, Mill W; p. 207 in Environmental Exposure from Chemicals Vol I Neely WB, Blau GE (eds) Boca Raton, FL: CRC Press (1985)
29. MacIntyre WG et al; Hydrocarbon Fuel Chemistry Virginia Inst of Marine Science, Gloucester Point, VA. NTIS AD-A117928, Springfield, VA p. 53 (1982)
30. Mackay D, Leinonen PJ; Environ Sci Technol 9: 1178-80 (1975)
31. Meylan W, Howard PH; J Pharm Sci 84: 83-92 (1995)
32. Meylan W, Howard PH; Environ Toxicol Chem 10: 1283-93 (1991)
33. Miller DF, Alkezweeny AJ; Ann NY Acad Sci 338: 219-32 (1980)
34. Monson PR et al; Houston Oxidant Field Study: Summer 1977. 71st Annual Meeting of the Air Pollution Control Assoc, Houston, TX, June 25-30, Paper No. 78-50.4 (1978)
35. Nelson PF, Quigley SM; Atmos Environ 18: 79-87 (1984)
36. Nutmagui W et al; Anal Chem 55: 2160-4 (1983)
37. Rappaport SM et al; Appl Ind Hyg 2: 148-54 (1987)
38. Rembold H et al; J Agric Food Chem 37: 659-62 (1989)
39. Scott Res Labs Inc; Atmospheric Reaction Studies in the Los Angeles Basin, Phase I, Vol I, Plumsteadville, PA, NTIS PB-194058, Springfield, VA p. 86 (1969)
40. Seila RL; Non-urban Hydrocarbon Concentrations in Ambient Air North of Houston, Texas, USEPA-600/3-79-010 (1979)
41. Seila RL et al; Atmospheric Volatile Hydrocarbon Composition at Five Remote Sites in Northwestern North Carolina. USEPA-600/D-84-092, Research Triangle Park, NC, NTIS PB-84-177930, Springfield, VA (1984)
42. Seila RL et al; Determination of C2 to C12 Ambient Air Hydrocarbons in 39 U.S. Cities, from 1984 Through 1986, Atmospheric Research and Exposure Assessment Lab, USEPA, Research Triangle Park, NC, NTIS PB-89-214142/AS, Springfield, VA (1989)
43. Sexton K; Environ Sci Technol 17: 402-7 (1983)

44. Sigsby JE et al; Environ Sci Technol 21: 466-75 (1987)
45. Silverstein RM, Bassler GC; Spectrometric Identification of Organic Compounds NY: John Wiley & Sons p. 154 (1963)
46. Swann RL et al; Res Rev 85: 17-28 (1983)
47. Uno I et al; Atmos Environ 19: 1283-93 (1985)
48. USEPA; EXAMS II Computer Simulation, Athens, GA (1987)
49. USEPA; Superfund Record of Decision (EPA Region 5): U.S. Aviex, MI (First Remedial Action), Sept. 1988, NTIS PB 89-225502, Springfield, VA (1989)
50. Wallington TJ et al; J Phys Chem 92: 5024-8 (1988)
51. Yalkowsky SH; Arizona Database of Aqueous Solubility, College of Pharmacy Univ of Arizona, Tuscon, AZ (1989)
52. Zimmerman RE et al; pp. 320-29 in Proc Int Gas Res Conf, Hirsch LH (ed), Government Institutes, Inc. Rockville, MD (1983)
53. Zweidinger RB et al; Environ Sci Technol 22: 956-62 (1988)
54. Zweidinger RB et al; Environ Sci Technol 24: 538-42 (1990)

Isooctyl Alcohol

SUBSTANCE IDENTIFICATION

Synonyms: 6-Methylheptanol

Structure:

CAS Registry Number: 26952-21-6

Molecular Formula: $C_8H_{18}O$

SMILES Notation: CC(C)CCCCCO

CHEMICAL AND PHYSICAL PROPERTIES

Boiling Point: 186 °C

Melting Point: -15 °C (estimated)

Molecular Weight: 130.22

Dissociation Constants:

Log Octanol/Water Partition Coefficient: 2.73 [14]

Water Solubility: 640 mg/L at 25 °C [16]

Vapor Pressure: 1.24 mm Hg at 25 °C (estimated) [24]

Henry's Law Constant: 3.1×10^{-5} atm-m³/mol (estimated) [15]

ENVIRONMENTAL FATE/EXPOSURE POTENTIAL

Summary: Isooctyl alcohol may be released to the environment through effluents at sites where it is produced or used. Photolysis or hydrolysis of isooctyl alcohol is not expected to be environmentally important. Based on information about structurally analogous compounds, isooctyl alcohol may biodegrade rapidly in acclimated soil and water. Estimated Koc values for isooctyl alcohol indicate that it may partition from the water column to organic matter in sediments and suspended solids. The potential for bioconcentration of isooctyl alcohol in aquatic organisms is low. The Henry's Law constant for isooctyl alcohol suggests that volatilization of isooctyl alcohol from natural waters may be an important fate process. The volatilization half-lives from a model river and a model pond, the latter considers the effect adsorption, have been estimated to be about 1.2 and 54 days, respectively. Isooctyl alcohol is expected to exist entirely in the vapor phase in ambient air. It will react with photochemically produced hydroxyl radicals, resulting in an estimated atmospheric half-life of 1.3 days. The most probable human exposure to isooctyl alcohol would be occupational exposure, which may occur through contact or inhalation at places where it is produced or used.

Natural Sources:

Artificial Sources: Isooctyl alcohol may be released to the environment via effluents at sites where it is produced or used.

Terrestrial Fate: Based on analogy to n-octanol isomers, isooctyl alcohol may biodegrade rapidly in acclimated terrestrial environments. Estimated Koc values for isooctyl alcohol of 24 and 125 suggest that isooctyl alcohol would have high to very high mobility in soil. Based on its estimated Henry's Law constant, the volatilization of isooctyl alcohol from moist surface soils may be an important transport process.

Aquatic Fate: Photolysis and hydrolysis of isooctyl alcohol in aquatic systems are not expected to be important fate processes [11]. Based on analogy to n-octanol isomers, isooctyl alcohol may biodegrade rapidly in acclimated aquatic systems. The bioconcentration factor (BCF) has been estimated to be 70 [11] indicating that the potential for isooctyl alcohol to bioconcentrate in aquatic organisms is low. Estimated Koc values for isooctyl alcohol of 24 and 125 suggest that isooctyl alcohol may partition from the water column to organic matter contained in sediments and suspended solids. Its estimated Henry's Law constant suggests that volatilization of isooctyl alcohol from natural waters may be an important fate process. Based on this Henry's Law constant, the volatilization half-life from a model river has been estimated to be 1.2 days [11]. The volatilization half-life from a model pond, which considers the effect of adsorption, has been estimated to be about 54 days [23].

Atmospheric Fate: Isooctyl alcohol is expected to exist entirely in the vapor phase in ambient air [6]. Vapor phase isopropyl alcohol reacts with photochemically produced hydroxyl radicals in the atmosphere by H-atom abstraction [1]. The rate constant for this reaction is estimated to be 1.20×10^{-11} cm^3/molecule-sec at 25 °C [1,12], which corresponds to an atmospheric half-life of about 1.3 days assuming an atmospheric concn of 5×10^5 hydroxyl radicals per cm^3. Isooctyl alcohol in air is not expected to undergo direct photolysis because it does not absorb UV light in the environmentally significant range [21].

Biodegradation: Data regarding the biodegradation of isooctyl alcohol were not available. However, a number of aerobic and anaerobic biological screening studies, which utilized settled waste water, sewage, or activated sludge for inocula, have demonstrated that n-octanol is biodegradable [1-9]. Since isooctyl alcohol is a mixture of isomers where most of the molecule is a straight chain alcohol, the results of the studies on n-octanol would suggest rapid biodegradation where acclimation has occurred.

Abiotic Degradation: Alcohols are generally resistant to hydrolysis [11]. Alcohols absorb UV light at wavelengths <185 nm, which is not in the environmentally significant range, >290 nm [21]. Likewise, alcohols are commonly used as solvents for obtaining UV spectra [21]. Therefore, isooctyl alcohol would not be expected to undergo hydrolysis or direct photolysis in the environment. The rate constant for the vapor-phase reaction of isooctyl alcohol with photochemically produced hydroxyl radicals in air has been estimated to be 1.20×10^{-11} cm^3/molecule-sec at 25 °C, which corresponds to an atmospheric half-life of about 1.3 days, assuming an atmospheric concn of 5×10^5 hydroxyl radicals per cm^3 [12].

Bioconcentration: Using its log Kow, the bioconcentration factor for isooctyl alcohol has been calculated to be 70, from recommended regression-derived equations [11]. This BCF values indicate that the potential for isooctyl alcohol to bioconcentrate in aquatic organisms is low.

Soil Adsorption/Mobility: Based on its water solubility, a Koc of 125 can be estimated for isooctyl alcohol using a regression-derived equation [11]. A Koc calculated from molecular structure is 24 [13]. According to a suggested classification scheme [22], these Koc values indicate that isooctyl alcohol would have high to very high mobility in soil.

Volatilization from Water/Soil: Using the estimated Henry's Law constant, the volatilization half-life of isooctyl alcohol from a model river (1 m deep flowing 1 m/sec with a wind speed of 3 m/sec) has been estimated to be 1.5 days [11]. The volatilization half-life from a model pond, which considers the effect of adsorption, has been estimated to be about 54 days [23].

Water Concentrations:

Effluent Concentrations:

Sediment/Soil Concentrations: Isooctyl alcohol was qualitatively detected in sediments from Lake Tobin of the Saskatchewan River, Canada [18].

Atmospheric Concentrations:

Food Survey Values:

Plant Concentrations:

Fish/Seafood Concentrations:

Animal Concentrations:

Milk Concentrations:

Other Environmental Concentrations:

Probable Routes of Human Exposure:

Average Daily Intake:

Isooctyl Alcohol

Occupational Exposure: The most probable human exposure to isooctyl alcohol would be occupational exposure, which may occur through dermal contact or inhalation at places where it is produced or used. NIOSH (NOES 1981-1983) has estimated that 11,583 workers are potentially exposed to isooctyl alcohol in the US [17].

Body Burdens: The air expired from humans contained isooctyl alcohol in 15.8% of the 387 samples collected from 54 subjects [9,10]. The average isooctyl alcohol concn of 1.3 ng/L was expressed as the geometric mean with upper and lower limits of 0.33 and 5.1, respectively [10].

REFERENCES

1. Atkinson R; Intern J Chem Kin 19: 799-828 (1987)
2. Babeu L, Vaishnav D; J Indust Microb 2: 107-15 (1987)
3. Day EA, Anderson J Agr Food Chem 13: 2-4 (1965)
4. Dore M et al; Trib Cebedeau 28: 3-11 (1975)
5. Dumont JP, Adda J; J Agric Food Chem 26: 364-7 (1978)
6. Eisenreich SJ et al; Environ Sci Technol 15: 30-8 (1981)
7. Gerhold RM, Malaney GW; J Water Pollut Contr Fed 38: 562-79 (1966)
8. Heukelekian H, Rand MC; J Water Pollut Control Assoc 29: 1040-53 (1955)
9. Krotoszynski BK et al; J Chromat Sci 15: 239-44 (1977)
10. Krotoszynski BK et al; J Anal Toxicol 3: 225-34 (1979)
11. Lyman WJ et al; Handbook of Chemical Property Estimation Methods NY: McGraw-Hill Chapt. 4, 5, 7, 8, 15 (1982)
12. Meylan WM, Howard PH; Chemosphere 26: 2293-9 (1993)
13. Meylan WM et al; Environ Sci Technol 26:1560-7 (1992)
14. Meylan WM, Howard PH; Group Contribution Method for Estimating Octanol-Water Partition Coefficents. SETAC Meeting Cincinnati, OH. Nov 8-12, (1992)
15. Meylan WM, Howard PH; Environ Toxicol Chem 10: 1283-93 (1991)
16. Neely WB; Chemosphere 13: 813-9 (1984)
17. NIOSH; National Occupational Exposure Survey (NOES) (1989)
18. Samoiloff MR et al; Environ Sci Tech 17: 329-34 (1983)
19. Shelton DR, Tiedje JM; Development of Tests for Determining Anaerobic Biodegradation Potential USEPA 560/5-81-013 (NTIS PB84-166495) pp. 92 (1981)
20. Shelton DR, Tiedje JM; Appl Environ Microbiol 47: 850-7 (1981)
21. Silverstein RM, Bassler GC; Spectrometric Id of Org Cmpd, NY: J Wiley & Sons Inc pp. 148-169 (1963)
22. Swann RL et al; Res Rev 85: 16-28 (1983)
23. USEPA; EXAMS II Computer Simulation (1987)
24. USEPA; PCGEMS (Graphical Estimation Modeling System); PCCCHEM (1987)
25. Wagner R; Vom Wasser 42: 271-305 (1974)

26.   Yonezawa Y, Urushigawa Y; Chemosphere 8: 139-42 (1979)
27.   Yonezawa Y et al; Kogai Shigen Kenkyusho Iho 11: 77-82 (1981)

Isopentane

SUBSTANCE IDENTIFICATION

Synonyms:

Structure:

CAS Registry Number: 78-78-4

Molecular Formula: C_5H_{12}

SMILES Notation: C(CC)(C)C

CHEMICAL AND PHYSICAL PROPERTIES

Boiling Point: 27.8 °C AT 760 mm Hg

Melting Point: -159.9 °C

Molecular Weight: 72.15

Dissociation Constants:

Log Octanol/Water Partition Coefficient: 2.30 [23]

Water Solubility: 48 mg/L at 25 °C [53]

Vapor Pressure: 689 mm Hg at 25 °C [18]

Henry's Law Constant: 1.4 atm-m^3/mol (calculated from vapor pressure and water solubility)

ENVIRONMENTAL FATE/EXPOSURE POTENTIAL

Summary: Isopentane is both a naturally occurring and an anthropogenic compound which has a wide array of commercial uses. Isopentane may be released to the environment as a fugitive emission during its production and use, as a volatile emission from gasoline, in motor vehicle exhaust or in incinerator stack emissions. If released to soil, estimated soil adsorption coefficients for isopentane indicate low to moderate mobility in soil. Isopentane is expected to rapidly volatilize from both dry and moist soil to the atmosphere. Limited data indicate that isopentane has the potential to biodegrade in soil. If released to water, isopentane is expected to rapidly volatilize from water to the atmosphere. The half-life for volatilization of isopentane from a model river is 2.5 hr. Estimated bioconcentration factors ranging from 33-70 suggest that isopentane will not significantly bioconcentrate in fish and aquatic organisms. Isopentane is not expected to significantly adsorb to sediment and suspended organic matter. Limited data indicate that it has the potential to biodegrade in water. If released to the atmosphere, isopentane is expected to undergo gas-phase oxidation reactions with photochemically produced hydroxyl radicals; the half-life for this process is estimated at 4.2 days. Occupational exposure to isopentane may occur by inhalation and dermal contact during its production, formulation or transport. Workers in the petroleum field are likely to be exposed to isopentane. Exposure to the general population may occur by inhalation of isopentane due to its emission from gasoline, motor vehicle exhaust, and natural sources.

Natural Sources: Isopentane is a naturally occurring compound that is emitted from bay-leaved willows, aspens, balsam poplars, European oaks, European larches, European firs, sorbs, silver fir trees, red bilberry shrubs, biberry shrubs, ferns and deciduous mosses [30] and a variety of arboreous plants [66]. Emissions of isopentane are also associated with natural gas [3]. Isopentane may be produced naturally in ocean sediments [76].

Artificial Sources: Isopentane is used as a solvent, in the manufacture of chlorinated derivatives, and as a blowing agent for polystyrene [57]. In 1991, 4 US facilities manufactured isopentane [67]. Isopentane may be released to the environment as a fugitive emission during its production or use. It may also be released to the environment as a volatile emission from gasoline [21,32], in the exhaust of motor vehicles [22,70,71,77] or in incinerator stack emissions [16].

Terrestrial Fate: If released to soil, estimated soil adsorption coefficients for isopentane ranging from 424-519 [42], calculated from the water solubility and experimental log octanol/water partition coefficient, indicate low to moderate mobility [72]. Limited data indicate that isopentane has the potential to biodegrade in soil [31]. The vapor pressure of isopentane and its estimated Henry's Law constant, indicate that it is expected to rapidly volatilize from both dry soil and moist soil, respectively.

Aquatic Fate: If released to water, isopentane is expected to rapidly volatilize from water to the atmosphere. Based on the Henry's Law constant, the half-life for volatilization of isopentane from a model river 1 m deep, flowing at 1 m/sec, with a wind speed of 3 m/sec, is 2.5 hr [42]. Estimated bioconcentration factors for isopentane ranging from 33-70, obtained from its water solubility and its log octanol/water partition coefficient, suggest that isopentane will not significantly bioconcentrate in fish and aquatic organisms. Estimated soil adsorption coefficients ranging from 424-519 [42] indicate that isopentane may not significantly adsorb to sediment and suspended organic matter. Limited data indicate that isopentane may have the potential to biodegrade in water [31].

Atmospheric Fate: If released to the atmosphere, isopentane is expected to undergo gas-phase oxidation reactions with photochemically produced hydroxyl radicals. A recommended value for the reaction of 3.78×10^{-12} cm^3/molecule-sec [11] translates to a half-life of 4.2 days. Laboratory experiments in which air samples containing isopentane were irradiated with either natural or artificial sunlight for approximately 6 hrs resulted in 24% to 50% oxidation of isopentane [34,68,78].

Biodegradation: Organisms isolated from ground water contaminated by a gasoline spill were found to grow when inoculated with pure samples of isopentane; however, no biodegradation of isopentane was observed when a mixed culture of these organisms was inoculated with gasoline [31].

Abiotic Degradation: Experimental rate constants for the gas-phase reaction of isopentane with photochemically produced hydroxyl radicals of 3.5×10^{-12} cm^3/molecule-sec at room temperature [17], 3.78×10^{-12} cm^3/molecule-sec at 27 °C [12], and 2.0×10^{-9} cm^3/molecule-sec at 32 °C [38] have been reported. Using a recommended value of 3.78×10^{-12} [10] cm^3/molecule-sec and an average atmospheric hydroxyl radical concn of $5 \times 10^{+5}$ molecules/cm^3 [11], a half-life of 4.2 days can be calculated for this process. Isopentane underwent 24% loss when atmospheric samples in large plastic bags were irradiated with natural sunlight (time not provided) [34]. Laboratory irradiation of isopentane in an environmental chamber resulted in 20.9% loss after 5 hr and 22.9% loss after 6 hr [78]. Laboratory irradiation of an ambient air sample containing 27.6 ppb isopentane reduced it to 13.6 ppb after 24 hr [68].

Bioconcentration: Estimated bioconcentration factors for isopentane ranging from 33-70 [42] can be calculated from the experimental log octanol/water partition coefficient and the water solubility, respectively. These values indicated that isopentane is not expected to significantly bioconcentrate in fish and aquatic organisms.

Soil Adsorption/Mobility: Estimated soil adsorption coefficients for isopentane ranging from 424-519 [42] can be calculated from the log octanol/water partition coefficient and the water solubility, respectively. These values suggest that isopentane will display low to moderate mobility in soil [72].

Volatilization from Water/Soil: The vapor pressure of isopentane indicates that it is expected to rapidly volatilize from dry soil. The Henry's Law constant indicates that isopentane is expected to rapidly volatilize from both water and moist soil. The estimated half-life for volatilization of isopentane from a model river 1 m deep, flowing at 1 m/sec with a wind speed of 3 m/sec, is 2.5 hr [42].

Water Concentrations:

Effluent Concentrations: Isopentane has been qualitatively detected as a component of gasoline [65] and in the exhaust of gasoline and diesel engines [22,70]. The average emission of isopentane from 46 in-use passenger cars, as measured using different driving cycles, ranged from 2.5-16.5% of the total non-methane hydrocarbon emissions from 1975-1982 [70]. The estimated emission of isopentane from the exhaust of petroleum run vehicles in the UK, 1983, is 15.76 ktonnes [13,14]. The mean emissions of isopentane from the exhaust of gasoline-powered motor vehicles in the US was found to range from 1.41-4.79% of the total nonmethane hydrocarbon emissions dependent on the gasoline blend, temperature, and vehicle type [71]. The average concn of isopentane in the exhaust of 67 Australian gasoline vehicles was 4.8% w/w of nonmethane hydrocarbon

[48]. Isopentane was identified as both an evaporative and exhaust emission from hydrocarbon and methanol-based fuels [77]. The isopentane wt % in vehicle and refinery emission profiles in Chicago ranged from 4.47 to 30.56% [19]. Isopentane was qualitatively detected in emissions from a municipal refuse incinerator [16]. It was qualitatively identified in the air above 4 or 5 hazardous waste sites monitored in NJ [36]. It has also been detected in municipal landfill gas (anaerobic phase - 0.010 %) in Palos Verdes, CA [15]. Emissions of isopentane in the Tokyo area are associated with vehicular exhaust, gasoline vapor, petroleum refining, vehicle painting shops, ship painting yards, and petrochemical plants [74]. Isopentane was qualitatively detected in water samples and it was detected in gaseous samples at a concn of 270 umol/L at an underwater hydrocarbon vent from an offshore production platform in the Gulf of Mexico, 1976 [56].

Sediment/Soil Concentrations: Isopentane was detected in 9 of 10 near shore sediment samples taken at a depth ranging 48-230 cm at a maximum concn of 0.78 ng/g dry weight in Walvis Bay, South Africa, but it was not detected in samples taken further off shore; the authors concluded that it may arise from natural production [76].

Atmospheric Concentrations: URBAN/SUBURBAN: Isopentane was detected in the air of Tulsa, OK, at concns ranging from 17.3-153.5 ppbC (8 samples) [8,9]. Isopentane was detected in air samples from Los Angeles, CA, Fall of 1981, at concns ranging from 23-83 ppb [20]. The concn of isopentane in 832 samples taken from 39 US cities, 1984-86, ranged from 1.4 ppbC to 3393 ppbC with a median value of 45.3 ppbC [60]. Roadway air samples were found to contain isopentane at 5.52-6.39% of the total nonmethane hydrocarbon concn [79]. The average concn of isopentane in downtown Los Angeles and Azusa, CA, September-November, 1967, was 35 ppb and 16 ppb, respectively [2]. The concn of isopentane in Los Angeles was 0.287 ppbC in 1963-68 and 0.192 ppmC in 1973 [37]. The concn of isopentane measured on the 6th floor of a building in downtown Los Angeles, CA, September-November, 1961, ranged from not detected-0.08 ppm [1]. The concn of isopentane taken from the top of an 82 story building at noon in New York City, 1977, ranged from 25.4-30.3 ug/m³ [5]. The mean concn of isopentane in Manhattan, NY, 1975 and 1978, ranged from 21-214 ppbC [4]. The concn of isopentane at Huntington Park, NJ, October 1968, ranged from 3.0-115 ppbC [58]. The concn in Boone, NC, ranged from 21-48.3 (median 36.8) ppbC downtown and 5.2-9.9 (median

7.0) ppbC on the town outskirts [61]. The ambient air concn of isopentane in Houston, TX, 1973-74, ranged from not detected to 0.06 ppm and from not detected to 1.82 ppm in nearby Pasadena, TX [64]. Air samples in Riverside, CA, 1965, contained isopentane at concns ranging 0.3 ppbC to >11.4 ppbC [69]. The concn of isopentane in downtown Albany, and Troy, NY, ranged from 20.2-26.6 ug/m^3 and 23.6-14.8 ug/m^3, respectively [6]. The concn of isopentane in Houston, TX, September, 1973, January, 1974, and April, 1974 ranged from 9.4-1227 ppbC (9 samples), 192.8-1076.0 ppbC (7 samples) and 5.2-347.8 ppbC (5 samples), respectively [39]. The average and maximum concns of isopentane measured in 780 samples from Houston, TX, Summer 1977, was 36 ppbC and 59 ppbC, respectively [45]. It was detected in the air of Los Angeles, CA, 1960, at concns ranging from 0.021-0.101 ppm in 17 samples [47]. It was qualitatively detected in the air of Houston, TX, Santa Monica, CA, West Covina, CA, Glendora, CA, Garden Grove, CA, and Anaheim, CA, 1974-5 [50]. The concn of isopentane in the Lincoln Tunnel, NY, was 1297 ppbC in 1970 and 305 ppbC in 1982 [40]. The median, minimum and maximum concns of isopentane in 832 air samples from 39 US cities were 45.3, 1.4, and 3393 ppbC [66]. The estimated annual mean concn of isopentane in London, England, is 8-10 ug/cu-m [13]. The observed annual concn of isopentane in European cities ranged from 5-29 ug/m^3, and the observed background level in Sydney, Australia, was 27.9 ug/m^3 [13]. It was qualitatively detected in the air of Leningrad, 1976, and 5 other Russian cities [27,28,29]. The concn of isopentane in Bangkok, Thailand, ranged from 1-13 ug/m^3 [35]. It was qualitatively detected in the air of South African cities [41]. The mean concn of isopentane in Bombay, India, 1985, ranged from not detected to 7.7 ppb over 12 months [44]. The average concn of isopentane in air samples taken over Tokyo, Japan, was 0.1 ppb in 1980 and 2.1 ppb in 1981 [73]. RURAL/REMOTE: The concn of isopentane in rural areas was measured in 1971: Point Reyes, CA - 0.2 ppbC; Moscow mountain, WA, 0.3 ppbC; Bretway-Hump, VT - 1.0 ppbC [54]. The maximum concns of isopentane at Roan Mt., Grandfather Mt., Linville Gorge, and Rich Mt., NC, were 7.5 ppbc, 6.5 ppbC, 4.2 ppbC, and 2.2 ppbC, respectively, in samples taken 1981-82 [54]. The concn of isopentane in aircraft collected hydrocarbon samples along the San Pablo Bay-Carquinez Strait-Susan Bay Channel, CA, 1975, ranged from <0.5 ug/m^3 to 15.0 ug/m^3 [62]. Isopentane was detected in Rio Blanco County, CO, and the Smokey Mountains, TN, 1978-79, at concns ranging from 1.5-79.3 ppbC (6 samples) and 3.2-6.9 ppbC (9 samples), respectively [8,9]. The average concn of isopentane in the air over Lake Michigan, as determined in airplane flights, August 27 and 28, 1976 was 1.3 and 5.8 ppbC, respectively [43]. The concn of isopentane at Jones State Forest, TX, ranged from 5.9-21.9 ppbC in 15 samples taken in 1978 [59]. The concn of isopentane in rural Deuselbach, Hunnsruck, Germany, 1983, ranged from 0.25-0.37 ppbC, and it was 3.2 ppbC in semi-rural Duren, 1984 [55]. The concn of isopentane in desert areas of upper Egypt, 1982, was 0.09 ppbC [55]. The average concn of isopentane in samples obtained in the Norwegian arctic, Summer of 1982 and Spring of 1983, was <20 ppt and 346 ppt, respectively [26]. Isopentane was qualitatively identified in air samples taken at Homebush Bay, Australia, June 1975 [46]. Background concns of isopentane in a variety of forests ranged from 2.0 to 12 ppbvC [33]. SOURCE DOMINATED: The background concn of isopentane in Janesville, WI, 1978, was 1.0 ug/m^3, yet it was detected at respective concns of 4.0 ug/m^3, 1.5 ug/m^3 and 1.5 ug/m^3 10, 15 and 40 miles downwind of the city [2]. INDOOR AIR: Isopentane was detected at concns of 18.2 to 35.8 ug/m^3 in air in a new office building in Portland, OR, compared to 3.8 ug/m^3 on the roof of the building [25].

Food Survey Values: Isopentane was identified as a volatile component of fried bacon [24].

Plant Concentrations:

Fish/Seafood Concentrations:

Animal Concentrations:

Milk Concentrations:

Other Environmental Concentrations:

Probable Routes of Human Exposure: The probable routes of exposure to isopentane are by inhalation and dermal contact during its production, formulation or transport. Workers in the petroleum field are likely to be exposed to isopentane by inhalation of gasoline fumes during the production, transport or dispensing of motor fuels [21,52]. Dermal exposure is likely to result when gasoline products are spilled on the skin, as isopentane is a component of gasoline [21]. Probable routes of exposure for the general population include inhalation of isopentane due to its presence in gasoline [21,32], motor vehicle exhaust [22,70,71,77], and emission from natural sources [3,30].

Average Daily Intake:

Occupational Exposure: Isopentane was found in 6 of 8 personal air samples and 4 of 12 breath samples of subjects monitored in Bayonne and Elizabeth, NJ, and RTP, NC [75]. Isopentane was detected in 16 of 18 personal air samples of workers at a high volume service station monitored in May 1983, at a concn ranging from 0.1-1.3 ppm, and it was detected, but not measured, in the other 2 samples [32]. Isopentane was detected in 18 of 18 air samples (12 customer exposure and 6 fenceline) at six service stations monitored in October-November 1990, at a concn ranging from 0.03-35 ppm [7]. Personal air samples taken from workers in the petroleum industry indicated that 53 of 56 outside operators were exposed to isopentane at mean concn 4.851 mg/m^3, 49 of 49 transport drivers were exposed at a mean concn of 9.979 mg/m^3 and 48 of 49 service attendants were exposed at a mean concns of 13.426 mg/m^3 [51]. NIOSH (NOES 1981-83) has statistically estimated that 6554 workers are potentially exposed to isopentane in the US [49].

Body Burdens:

REFERENCES

1.   Altshuller AP, Beller TA; J Air Pollut Control Assoc 13: 81-7 (1963)
2.   Altshuller AP et al; Environ Sci Technol 5: 1009-16 (1971)
3.   Altshuller AP; Atmos Environ 17: 2131-65 (1983)
4.   Altwicker ER et al; J Geophys Res 85: 7475-87 (1980)
5.   Altwicker ER, Whiby RA; 72nd Ann Mtg Air Pollut Contr Assoc 79-52.3 (1979)
6.   Altwicker ER et al; pp. 520-3 in Proc The Fourth Internat Clean Air Conf (1977)
7.   API; Laboratory study on solubilities of petroleum hydrocarbons in groundwater; American Petroleum Institute. API Publ No 4395 pp 241 (1985)
8.   Arnts RR, Meeks SA; Atmos Environ 15: 1643-51 (1981)
9.   Arnts RR, Meeks SA; Biogenic Hydrocarbon Contributions to the Ambient Air of Selected Areas. USEPA-600/3-80-023 (1980)
10. Atkinson R, J Chem Phys Ref Data Monographs 1 (1989)
11. Atkinson R; Int J Chem Kinet 19: 799-828 (1987)
12. Atkinson R et al; Adv Photochem 11: 375-488 (1979)
13. Bailey JC et al; Atmos Environ 24A: 34-52 (1990)

14. Bailey JC et al; Sci Total Environ 93: 199-206 (1990)
15. Brosseau J, Heitz M; Atmos Environ 28:285-93 (1994)
16. Carotti AA, Kaiser ER; JAPCA 22: 248-53 (1972)
17. Cox RA et al; Environ Sci Technol 14: 57-61 (1980)
18. Daubert TE, Danner RP; Physical & Thermodynamic Properties of Pure Chemicals NY,NY: Hemisphere Pub Corp (1989)
19. Doskey PV et al; J Air Water Manage Assoc 42:1437-45 (1992)
20. Grosjean D, Fung K; J Air Pollut Control Assoc 34: 537-43 (1984)
21. Halder CA et al; Am Ind Hyg Assoc 47: 164-72 (1968)
22. Hampton CV et al; Environ Sci Technol 16: 287-98 (1982)
23. Hansch C et al; J Org Chem 33: 347 (1968)
24. Ho CT et al; J Agric Food Chem 31: 336-42 (1983)
25. Hodgson AT et al; J Air Waste Manage Assoc 41:1461-8 (1991)
26. Hov O et al; Geophys Res Lett 11: 425-8 (1984)
27. Ioffe BV et al; Environ Sci Technol 13: 864-9 (1979)
28. Ioffe BV et al; Dokl Akad Nauk Sssr. 243: 1186-9 (1978)
29. Ioffe BV et al; J Chromatog 142: 787-95 (1977)
30. Isidorov VA et al; Atmos Environ 19: 1-8 (1985)
31. Jamison VW et al; pp. 187-96 in Proc Int Biodeg Symp 3rd Sharpley JM, Kapalan Am, eds Essex, England (1976)
32. Kearney CA, Dunham DB; Am Ind Hyg Assoc J 47: 535-9 (1986)
33. Khalil MAK, Rasmussen RA; J Air Waste Manage Assoc 42:810-3 (1992)
34. Kopczynski SL et al; Environ Sci Technol 6: 342-7 (1972)
35. Kungskulniti N et al; Chemosphere 20: 673-9 (1990)
36. LaRegina J et al; Environ Prog 5: 18-27 (1986)
37. Leonard MJ et al; J Air Pollut Control Fed 26: 359-63 (1970)
38. Lloyd AC et al; J Phys Chem 80: 789-94 (1976)
39. Lonneman WA et al; Hydrocarbons in Houston Air USEPA-600/3-79/018 (1979)
40. Lonneman WA et al; Environ Sci Technol 20: 790-6 (1986)
41. Louw CW et al; Atmos Environ 11: 703-17 (1977)
42. Lyman WJ et al; Handbook of Chemical Property Estimation Methods NY: McGraw-Hill Chapt 15 (1982)
43. Miller DF, Alkezweeny AJ; Ann NY Acad Sci 338: 219-32 (1980)
44. Mohan Rao AM, Panditt GG; Atmos Environ 2: 395-401 (1988)
45. Monson PR et al; 71st Ann Mtng APCA Paper 78-50.4 (1978)
46. Mulcahy MFR et al; p. 17 in Occurrence Contr Photochem Pollut Proc Symp Workshop Sess Paper IV (1976)
47. Neligan RE; Arch Environ Health 5: 581-91 (1962)
48. Nelson PF, Quigley SM; Environ Sci Technol 18:79-87 (1984)
49. NIOSH; National Occupational Exposure Survey (NOES) (1989)
50. Pellizzari ED; Development of Analytical Techniques for Measuring Ambient Atmospheric Carcinogenic Vapors. USEPA-600/2-75-076 (1975)
51. Rappaport SM et al; Appl Ind Hyg 2: 148-54 (1987)
52. Rapport SM et al; Appl Ind Hyg 2: 148-54 (1987)
53. Riddick JA et al; Organic Solvents 4th ed. NY,NY: Wiley ((1986)

54. Robinson E et al; J Geophysical Res 78: 5345-51 (1973)
55. Rudolph J, Khedim A; Int J Environ Anal Chem 290: 265-82 (1985)
56. Sauer TC Jr; Environ Sci Technol 15: 917-23 (1981)
57. Sax NI, Lewis RJ Jr; Hawley's Condensed Chemical Dictionary 11th ed. NY: Van Nostrand Reinhold Co (1987)
58. Scott Research Labs Inc; Atmospheric Reaction Studies in the Los Angeles Basin NTIS PB-194-058 (1969)
59. Seila RL; Non-urban Hydrocarbons Concentration in Ambient Air North of Houston, TX USEPA-500/3-79-0101 (1979)
60. Seila RL et al; Determination of C2 to C12 Ambient Air Hydrocarbons in 39 US Cities, from 1984 through 1986 USEPA/600/S3-89/058 (1989)
61. Seila RL et al; Atmospheric Volatile Hydrocarbon Composition at Five Remote Sites in NW NC USEPA-600/D-84-092, NTIS PB84-177930 (1984)
62. Sexton K, Westberg H; J Air Pollut Contr Fed 29: 1149-52 (1979)
63. Sexton K; Environ Sci Technol 17: 402-7 (1983)
64. Siddiqi AA, Worley FL; Atmos Environ 11: 131-43 (1977)
65. Sigsby JE et al; Environ Sci Technol 21: 466-75 (1987)
66. Singh HB, Zimmerman PB; Adv Environ Sci Technol 24:177-235 (1992)
67. SRI International; Directory of Chemical Producers USA (1991)
68. Stephens ER; Hydrocarbons in Polluted Air NTIS PB-230 993-/8 (1973)
69. Stephens ER, Burleson FR; J Air Pollut Control Assoc 17: 147-53 (1967)
70. Stump FD, Dropkin DL; Anal Chem 57: 2629-34 (1985)
71. Stump FD et al; Atmos Environ 23: 307-20 (1989)
72. Swann RL et al; Res Rev 85: 17-28 (1983)
73. Uno I et al; Atmos Environ 19: 2183-93 (1985)
74. Wadden RA et al; Environ Sci Tech 20: 473-83 (1986)
75. Wallace LA et al; Environ Res 35: 293-319 (1984)
76. Whelan JK et al; Geochim Cosmochim Acta 44: 1767-85 (1980)
77. Williams RL et al; J Air Waste Manage Assoc 40: 747-56 (1990)
78. Yanagihara S et al; Proc Int Clean Air Cong 4: 472-7 (1977)
79. Zweidinger RB et al; Environ Sci Technol 22: 956-62 (1988)

Isosafrole

SUBSTANCE IDENTIFICATION

Synonyms: 1,2-Methylenedioxy-4-propenyl-benzene

Structure:

CAS Registry Number: 120-58-1

Molecular Formula: $C_{10}H_{10}O_2$

SMILES Notation: O(c(c(c(O1)cc(c2)C=CC)c2)C1

CHEMICAL AND PHYSICAL PROPERTIES

Boiling Point: 252 °C

Melting Point: 6.7-6.8 °C

Molecular Weight: 162.18

Dissociation Constants:

Log Octanol/Water Partition Coefficient: 2.75 (estimate) [3]

Water Solubility: 45 mg/L (estimate) [10]

Vapor Pressure: 9.29 x 10^{-2} mm Hg (estimate) [11]

Henry's Law Constant: 0.036 atm-m^3/mol at 25 °C (estimate) [7]

ENVIRONMENTAL FATE/EXPOSURE POTENTIAL

Summary: Isosafrole occurs naturally as a principal component of the essential oil star anise and also at low quantities in the essential oils of other spices. Isosafrole may be released during its manufacture and use as an intermediate in the production of heliotropin, and in the production of perfumes, flavors, and pesticide synergists. No information was found that indicates that isosafrole is used or is currently produced in the US. If isosafrole is released to soil, it should not hydrolyze and should have a low tendency to leach to ground water. It should be subject to significant volatilization from near-surface soil and other surfaces based upon an estimated Henry's Law constant and an estimated vapor pressure. Isosafrole may biodegrade in soil, although no data are available concerning the biodegradation of isosafrole in culture or in soil and water in the environment. If released to water, it should not hydrolyze, significantly bioconcentrate in aquatic organisms based upon an estimated BCF, or adsorb to sediment or suspended particulate matter based upon an estimated Koc. It should be subject to significant volatilization from surface water with a half-life of 3.8 hr estimated for volatilization of isosafrole from a model river 1 m deep flowing 1 m/sec with a wind velocity of 3 m/sec at 20 °C based on the estimated Henry's Law constant. A half-life of 95.4 hr for volatilization from a model pond can be estimated using a three-compartment EXAMS model, which considers the effect of adsorption. It may be susceptible to direct photolysis in surface water and the atmosphere based upon its absorption of light at wavelengths >290 nm. Isosafrole may biodegrade in natural waters although no data are available. If released to the atmosphere, it can be expected to exist mainly in the vapor phase in the ambient atmosphere based on an estimated vapor pressure. The overall half-lives for the vapor phase reactions of cis- and trans-isosafrole with photochemically produced hydroxyl radicals and ozone have been estimated to be 2.2 and 1.4 hrs. Exposure to isosafrole will likely occur via ingestion of food.

Natural Sources: Isosafrole occurs naturally as a principal component of the essential oil star anise and also at low quantities in the essential oils of other spices [8,9]. Its distribution is generally similar to that of safrole, which is found in several essential oils including the essential oils of sassafras, nutmeg, mace, ginger, cinnamon, and black pepper [8,9].

Artificial Sources: Isosafrole may be released during its manufacture and use [6,8,11]. It has been used as an intermediate in the production of heliotropin and in the production of perfumes, flavors, and pesticide synergists [6,8,11]. No information was found to indicate that isosafrole is used or is currently produced in the US.

Terrestrial Fate: If isosafrole is released to soil, it should not hydrolyze [10]. It should exhibit low mobility in soil [13], based upon an estimated Koc [3,10]. It should be subject to significant volatilization from moist near-surface soil based upon the rapid volatilization predicted from water [3,5,7,10]. Based upon an estimated vapor pressure of 9.29×10^{-2} mm Hg [11], volatilization of isosafrole from dry soil and other surfaces may be an important process. Isosafrole may biodegrade in soil although no data are available concerning the biodegradation of isosafrole in culture or in soil and water in the environment.

Aquatic Fate: If isosafrole is released to water, it should not hydrolyze [10], significantly bioconcentrate in aquatic organisms based upon an estimated BCF [3,10] or adsorb to sediment or suspended particulate matter based upon an estimated Koc [3,10]. It should be subject to significant volatilization from surface water, with a half-life of 3.8 hr estimated for volatilization of isosafrole from a model river 1 m deep flowing 1 m/sec with a wind velocity of 3 m/sec at 20 °C [10] based on a Henry's Law constant of 0.036 atm-m^3/mol, which has been estimated using a structure-based estimation method [7]. It may be susceptible to direct photolysis in surface water based upon its absorption of light at wavelengths >290 nm [12]. Isosafrole may biodegrade in natural waters although no data are available concerning the biodegradation of isosafrole in culture or in soil and water in the environment.

Atmospheric Fate: If isosafrole is released to the atmosphere, it can be expected to exist mainly in the vapor phase in the ambient atmosphere [4] based on an estimated vapor pressure of 9.29×10^{-2} mm Hg at 25 °C [11]. The overall half-lives for the vapor phase reactions of cis- and trans-isosafrole with photochemically produced hydroxyl radicals and ozone have been estimated to be 2.2 and 1.4 hr at atmospheric concentrations of $5 \times 10^{+5}$ hydroxyl radicals per cm^3 and $7 \times 10^{+11}$ ozone molecules per cm^3 [1,2]. It may be susceptible to direct photolysis in the atmosphere based upon its absorption of light at wavelengths >290 nm [12].

Biodegradation: No data are available concerning the biodegradation of isosafrole in culture or in soil and water in the environment.

Abiotic Degradation: Isosafrole may be susceptible to direct photolysis based upon its absorption of light at wavelengths >290 nm [12]. The rate constants for the vapor phase reactions of the cis- and trans- isomers of isosafrole with photochemically produced hydroxyl radicals have been estimated to be 79.97 and 87.57 x 10^{-12} cm^3/molecule-sec, respectively [1]. The rate constants for the vapor phase reactions of the cis- and trans- isomers of isosafrole with ozone have been estimated to be 6.825 and 13.65 x 10^{-17} cm^3/molecule-sec, respectively [2]. The rate constants correspond to atmospheric half-lives for the cis- and trans- isomers of isosafrole of 2.2 and 1.4 hours, respectively, at an atmospheric concentration of 5 x 10^{+5} hydroxyl radicals per cm^3 and 7 x 10^{+11} ozone molecules per cm^3 [1,2].

Bioconcentration: An estimated BCF of 72 can be calculated [10] from an estimated log Kow of 2.75 [3] using a recommended regression equation [10]. Based upon this estimated BCF, isosafrole should not significantly bioconcentrate in aquatic organisms.

Soil Adsorption/Mobility: An estimated Koc of 540 can be calculated [10] from an estimated log Kow of 2.75 [3] using a recommended regression equation [10]. Based upon this estimated Koc, isosafrole will be expected to exhibit low mobility in soil [13].

Volatilization from Water/Soil: The half-life for volatilization of isosafrole from a river 1 m deep flowing 1 m/sec with a wind velocity of 3 m/sec is estimated to be 3.8 hr at 20 °C [10] based on a Henry's Law constant of 0.036 atm-m^3/mol, which has been estimated using a structure-based estimation method [7]. A half-life of approximately 90 hr for volatilization from a model pond can be estimated using a three-compartment EXAMS model, which considers the effect of adsorption [5]. Based upon a vapor pressure of 9.29 x 10^{-2} mm Hg, which can be extrapolated from data at higher temperatures [11], and the estimated Henry's Law constant, volatilization of isosafrole from near-surface soil and other surfaces may be important processes.

Water Concentrations:

Effluent Concentrations:

Sediment/Soil Concentrations:

Isosafrole

Atmospheric Concentrations:

Food Survey Values:

Plant Concentrations:

Fish/Seafood Concentrations:

Animal Concentrations:

Milk Concentrations:

Other Environmental Concentrations:

Probable Routes of Human Exposure: Exposure to isosafrole will likely occur via ingestion of foods that contain small amounts of the compound [8,9].

Average Daily Intake:

Occupational Exposure:

Body Burdens:

REFERENCES

1. Atkinson R et al; Internat J Chem Kin 14: 781-8 (1982)
2. Atkinson R, Carter WP; Chem Rev 84: 437-70 (1984)
3. CLOGP3; PCGEMS Graphical Exposure Modeling System USEPA (1986)
4. Eisenreich SJ et al; Environ Sci Technol 15: 30-8 (1981)
5. EXAMS; Exposure Analysis Modeling System. EXAMS II. USEPA, Environ Res Lab, Athens,GA USEPA/600/3-85/038 (1985)
6. Hawley GG; Condensed Chemical Dictionary 10th ed Van Nostrand Reinhold NY p. 585 (1981)
7. Hine J, Mookerjee PK; J Org Chem 40: 292-8 (1975)
8. IARC; Monograph on the Evaluation of the Carcinogenic Risk of Chemicals to Man 10: 231-5 (1976)
9. IARC; Monograph on the Evaluation of the Carcinogenic Risk of Chemicals to Man 1: 171 (1972)
10. Lyman WJ et al; Handbook of Chem Property Estimation Methods NY: McGraw-Hill pp 4-9, 7-4, 15-16 to 15-29 (1982)

11. Merck; The Merck Index 10th ed Rahway, NJ: Merck & Co p. 751 (1983)
12. Sadtler; UV No. 2308 (1960)
13. Swann RL et al; Res Rev 85: 17-28 (1983)

(D)-Limonene

SUBSTANCE IDENTIFICATION

Synonyms: (+)-p-Mentha-1,8-diene; 4-Isopropenyl-1-methylcyclohexene

Structure:

CAS Registry Number: 5989-27-5

Molecular Formula: $C_{10}H_{16}$

SMILES Notation: C(=CCC(C(=C)C)C1)(C1)C

CHEMICAL AND PHYSICAL PROPERTIES

Boiling Point: 175.5-176 °C at 763 mm Hg

Melting Point: -74.35 °C

Molecular Weight: 136.23

Dissociation Constants:

Log Octanol/Water Partition Coefficient: 4.57 [9]

Water Solubility: 13.8 mg/L [15]

Vapor Pressure: 2.0 mm Hg at 25 °C [8]

Henry's Law Constant: 3.8×10^{-1} atm-m^3/mol (estimated) [13]

ENVIRONMENTAL FATE/EXPOSURE POTENTIAL

Summary: (D)-Limonene is both a naturally occurring and a synthetically produced terpene which is used in flavors and fragrances, as a solvent, and for numerous other commercial uses. If released to soil, (D)-limonene is expected to exhibit low to slight mobility. It should rapidly volatilize from both dry and moist soil to the atmosphere although strong adsorption to soil may attenuate the rate of this process. If released to water, (D)-limonene may bioconcentrate in fish and aquatic organisms and it may significantly adsorb to sediment and suspended organic matter. It is expected to rapidly volatilize from water to the atmosphere. The estimated half-life for volatilization of (D)-limonene from a model river is 3.4 hr, although adsorption to sediment and suspended organic matter may attenuate the rate of this process. If released to the atmosphere, (D)-limonene is expected to rapidly undergo gas-phase oxidation reactions with photochemically produced hydroxyl radicals, ozone, and at night with nitrate radicals. Calculated half-lives for these processes are 2.3-2.6 hr, 25-26 min and 3.1 min, respectively. Occupational exposure to (D)-limonene may occur by inhalation or dermal contact during its production, formulation, transport, or use. Exposure by the general population may occur through inhalation due to its presence in the atmosphere as a result of release from natural sources or by ingestion of contaminated food.

Natural Sources: (D)-Limonene is a naturally occurring compound found in many natural oils including orange, lemon, grapefruit, berry, leaf, caraway, dill, bergamot, peppermint and spearmint oils [5,17,18,23]. (D)-Limonene emissions to the environment are associated with wax myrtle, sweet acacia, oranges, tomatoes, grasses, and California western sagebrush [1].

Artificial Sources: (D)-Limonene is used as a flavor and fragrance [17,18]. It is also used as a solvent, wetting agent, in resins, and as a monomer and copolymer [11,18]. (D)-Limonene may be released to the environment as a fugitive emission during its production, use, or transport. It may also be released to the environment by volatilization during its use as a solvent.

Terrestrial Fate: Based on the solubility of (D)-limonene [15] and its log octanol/water partition coefficient [9], soil adsorption coefficients of 1030-4780 can be calculated using appropriate regression equations [12]. These values indicate that this compound will display low to slight mobility in soil [19]. Its vapor pressure [15] and Henry's Law constant [13] indicate that (D)-limonene will rapidly volatilize from both dry and moist soil surfaces to the atmosphere although its strong adsorption to soil may attenuate the rate of this process.

(D)-Limonene

Aquatic Fate: Based on the water solubility of (D)-limonene [15] and its log octanol/water partition coefficient [9] bioconcentration factors of 246-262 can be calculated using appropriate regression equations [12], indicating that it may bioconcentrate in fish and aquatic organisms. Estimated soil adsorption coefficients ranging from 1030 to 4780 [9,12,15] indicate that (D)-limonene may significantly adsorb to sediment and suspended organic matter. (D)-Limonene's Henry's Law constant [13] suggests that it will rapidly volatilize from water to the atmosphere. The estimated half-life for volatilization of (D)-limonene from a model river 1 m deep flowing at 1 m/sec with a wind speed of 3 m/sec is 3.4 hr [12], although adsorption to sediment and suspended organic matter may attenuate the rate of this process.

Atmospheric Fate: Based on its vapor pressure, (D)-limonene is expected to exist solely in the vapor phase in the ambient environment [8]. If released to the atmosphere, (D)-limonene is expected to rapidly undergo gas-phase oxidation reactions with photochemically produced hydroxyl radicals, ozone, and at night with nitrate radicals. Based on experimental rate constants, calculated half-lives for the gas-phase reaction between limonene and photochemically produced hydroxyl radicals range from 2.3-2.6 hr [4,14,24]. For the gas-phase reaction of (D)-limonene with ozone, half-lives ranging from 25-26 min can be calculated [1,3,4]. A half-life, based on an experimentally determined rate constant, for the night-time reaction of (D)-limonene with nitrate radicals of 3.1 min can also be calculated [2,4]. The atmospheric lifetime of (D)-limonene during the daytime was estimated at 0.2-0.8 hr, dependant on both the local hydroxyl radical and ozone concentration [1].

Biodegradation:

Abiotic Degradation: Experimental rate constants for the gas-phase reaction of (D)-limonene with photochemically produced hydroxyl radicals of 1.70×10^{-10} m^3/molecule-sec at 25 °C [4], 1.63×10^{-10} cm^3/molecule-sec at 25 °C [14] and 1.49×10^{-10} cm^3/molecule-sec at 32 °C [24] correspond to half-lives ranging from 2.3-2.6 hr using an average atmospheric hydroxyl radical concentration of $5 \times 10^{+5}$ molecules/cm^3 [4]. (D)-Limonene was classified as group V in a 5 tiered rating system of relative reactivities towards photochemically produced hydroxyl radicals (methane = 1, (D)-limonene = 18,800), indicating an atmospheric half-life of <0.24 hr [7].

(D)-Limonene

Experimental rate constants for the gas-phase reaction of (D)-limonene with ozone of 6.40 x 10^{-16} cm^3/molecule-sec at 23 °C [1,3,4] and 6.5 x 10^{-16} cm^3/molecule-sec [24] correspond to half-lives ranging from 25-26 min using an average atmospheric ozone concentration of 7 x 10^{+11} molecules/cm^3 [3]. An experimental rate constant for the night-time gas-phase reaction of (D)-limonene with nitrate radicals of 1.3 x 10^{-11} cm^3/molecule-sec at 25 °C [4] corresponds to a half-life of 3.1 min using an average nitrate radical concentration of 2.4 x 10^{+8} molecules/cm^3 [2]. Photolysis of (D)-limonene in the presence of nitrogen oxides produces formaldehyde, formic acid, carbon monoxide, carbon dioxide, acetaldehyde, peroxyacetyl nitrate and acetone [1]. The daytime atmospheric lifetime of (D)-limonene has been estimated at 0.2-0.8 hr depending on both the local hydroxyl radical and ozone concentration [1].

Bioconcentration: Based on the water solubility of (D)-limonene [15] and its log octanol/water partition coefficient [9], bioconcentration factors of 246 and 262, respectively, can be calculated using appropriate regression equations [12]. These values indicate that (D)-limonene may bioconcentrate in fish and aquatic organisms.

Soil Adsorption/Mobility: Based on the water solubility of (D)-limonene [15] and its log octanol/water partition coefficient [9], soil adsorption coefficients of 1030 and 4780, respectively, can be calculated using appropriate regression equations [12]. These values indicate that (D)-limonene is expected to display slight to low mobility in soil [19].

Volatilization from Water/Soil: Based on its vapor pressure [8], (D)-limonene may volatilize rapidly from dry soil to the atmosphere. Its Henry's Law constant [13] indicates that (D)-limonene will rapidly volatilize from both water and moist soil surfaces to the atmosphere [12]. The estimated half-life for volatilization of (D)-limonene from a model river 1 m deep flowing at 1 m/sec with a wind speed of 3 m/sec is 3.4 hr [12]. The estimated half-life for volatilization from a model pond, which takes into account adsorptive processes, is 29 days [21].

Water Concentrations:

Effluent Concentrations: Ten gallons of (D)-limonene were released into Newark Bay and its major tributaries from October 1986-August 1991 [6].

(D)-Limonene

Sediment/Soil Concentrations:

Atmospheric Concentrations: URBAN/SUBURBAN: (D)-Limonene was detected indoors in an office building, 1987, at a concentration ranging from 43-63 ug/m³ [22].

Atmospheric Concentrations: RURAL/REMOTE: The average concentration of (D)-limonene at a rural site in the Rocky Mountains was 0.030 ppb during the day and 0.072 ppb at night, 1982 [16]. (D)-Limonene was qualitatively identified in air samples obtained at a forest in Germany, 1988 [10].

Food Survey Values: (D)-Limonene is a naturally occurring compound found in many natural oils including orange, lemon, grapefruit, berry, leaf, caraway, dill, bergamot, peppermint and spearmint oils [5,17,18,23].

Plant Concentrations: (D)-Limonene was identified as a volatile constituent of kiwi fruit flowers [20].

Fish/Seafood Concentrations:

Animal Concentrations:

Milk Concentrations:

Other Environmental Concentrations:

Probable Routes of Human Exposure: Occupational exposure to (D)-limonene may occur by inhalation or dermal contact during its production, formulation, transport or use. Exposure to the general population may occur by inhalation due to its presence in the atmosphere as a result of its release from natural sources, or by ingestion of food in which it occurs naturally or to which it has been added as a flavor or fragrance.

Average Daily Intake:

Occupational Exposure: (D)-Limonene was detected indoors in an office building, 1987, at a concentration ranging from 43-63 ug/m³ [22].

Body Burdens:

REFERENCES

1. Altshuller AP; Atmos Environ 17: 2131-65 (1983)
2. Atkinson R; Chem Rev 85: 69-201 (1985)
3. Atkinson R, Carter WPL; Chem Rev 84: 437-70 (1984)
4. Atkinson R; Atmos Environ 24A: 1-41 (1990)
5. Bauer K et al; In: Ullmann's Encycl Indust Tech 5th ed. Gerhartz W et al Eds. VCH Publ A11: 141-250 (1988)
6. Crawford DW et al; Ecotoxicol Environ Safety 30: 85-100 (1995)
7. Darnall KR ct al; Environ Sci Tcch 10: 692-6 (1976)
8. Daubert TE, Danner RP; Physical and Thermodynamic Properties of Pure Chemicals: Data Compilation. Design Institute for Physical Property Data, American Institute of Chemical Engineers. Hemisphere Pub Corp, New York, NY. 4 Vol (1989)
9. Hansch C, Leo AJ; Medchem Project, Claremont, CA: Pomona College. Issue #26 (1985)
10. Helmig D et al; Chemosphere 19: 1399-1412 (1989)
11. Holohan SF et al; Kirk-Othmer Encycl Chem Tech NY: Wiley 3rd Ed. 12: 852-69 (1980)
12. Lyman WJ et al; Handbook of Chemical Property Estimation Methods. Washington, DC: Amer Chem Soc (1990)
13. Meylan WM, Howard PH; Environ Toxicol Chem 10: 1283-93 (1991)
14. Mulcahy MFR et al; in Occur Control Photochem Pollut Proc Symp Workshop. SESS, Paper No. IV: 1-7 (1976)
15. Riddick JA et al; Organic Solvents 4th ed NY: Wiley Interscience (1986)
16. Roberts JM et al; Environ Sci Tech 19: 364-9 (1985)
17. Rogers JA Jr; Kirk-Othmer Encycl Chem Tech NY: Wiley 3rd Ed. 16: 307-332 (1981)
18. Sax NI, Lewis RJ Sr; Hawley's Condensed Chemical Dictionary 11th ed. NY: Van Nostrand Reinhold Co pp. 425, 701 (1987)
19. Swann RL et al; Res Rev 85: 17-28 (1983)
20. Tatsuka K et al; J Agric Food Chem 38: 2176-80 (1990)
21. USEPA; EXAMS II (1987)
22. Weschler CJ et al; Am Ind Hyg Assoc J 51: 261-8 (1990)
23. Windholz M et al; The Merck Index. Rahway, NJ: Merck & Co Inc 10th ed: (1983)
24. Winer AM et al; J Phys Chem: 80: 1635-9 (1976)

Limonene

SUBSTANCE IDENTIFICATION

Synonyms: 1,8(9)-p-Menthadiene; 4-Isopropenyl-1-methylcyclohexene

Structure:

CAS Registry Number: 138-86-3

Molecular Formula: $C_{10}H_{16}$

SMILES Notation: C(=CCC(C(=C)C)C1)(C1)C

CHEMICAL AND PHYSICAL PROPERTIES

Boiling Point: 175.5-176.5 °C at 763 mm Hg

Melting Point: -95.5 °C

Molecular Weight: 136.23

Dissociation Constants:

Log Octanol/Water Partition Coefficient: 4.57 [37]

Water Solubility: 13.8 mg/L [40]

Vapor Pressure: 1.51 mm Hg at 25 °C [37]

Henry's Law Constant: 0.38 atm-cm^3/mol (estimated) [37,40]

ENVIRONMENTAL FATE/EXPOSURE POTENTIAL

Summary: Limonene is both a naturally occurring and a synthetic terpene that is used in flavors and fragrances, as a solvent and for numerous other commercial uses. If released to soil, limited data indicate that limonene is expected to be resistant to biodegradation under aerobic conditions. Limonene is expected to exhibit low to slight mobility in soil. It is expected to rapidly volatilize from both dry and moist soil to the atmosphere although adsorption to soil may attenuate the rate of this process. If released to water, limited data indicate that limonene is expected to be resistant to biodegradation under aerobic conditions. Limonene may bioconcentrate in fish and aquatic organisms and it may significantly adsorb to sediment and suspended organic matter. It is expected to rapidly volatilize from water to the atmosphere. The estimated half-life for volatilization of limonene from a model river and lake is 3.5 hrs and 5 days, respectively, although adsorption to sediment and suspended organic matter may attenuate the rate of this process. If released to the atmosphere, limonene is expected to rapidly undergo gas-phase oxidation reactions with photochemically produced hydroxyl radicals, ozone, and at night with nitrate radicals. Calculated lifetimes for these processes in a clean and moderately polluted atmosphere are 2.0 hr and 30 min, 36 min and 11 min, and 9 min and 0.9 min, respectively. Occupational exposure to limonene may occur by inhalation or dermal contact during its production, formulation, transport or use. Exposure to the general population may occur by inhalation due to its presence in the atmosphere as a result of release from natural sources, its presence in household products, or by ingestion of food in which it is contained.

Natural Sources: Limonene is a naturally occurring compound and it is found in many natural oils including orange, lemon, grapefruit, berry, leaf, caraway, dill, bergamot, peppermint and spearmint oils [5,51,66]. Limonene emissions to the environment are associated with wax myrtle, sweet acacia, oranges, tomatoes, grasses, and California western sagebrush [2]. Emissions of limonene are also associated with balsam poplar, European larch, European fir, scots pine, Siberian pine, silver fir, common juniper, zeravshan juniper, pencil cedar, evergreen cypress, northern white cedar, chinese arbor vitae, marsh tea and deciduous moss [28].

Artificial Sources: Limonene is used as a flavor and fragrance [51,52]. It is also used as a solvent, wetting agent, in resins, and as a monomer and copolymer [25,52]. Limonene may be released to the environment as a fugitive emission during its production, use or transport, from volatilization during its use as a solvent, from landfills, and in industrial effluent [9,69]. Limonene emissions have been associated with effluent from the following industries: extraction of pine gum, paper and pulp mills, plastics materials-synthetic resins and nonvulcanizable elastomers, perfumes, cosmetics and other toilet preparations, organic solvents and lubricating oils and greases [1]. Limonene may be emitted to household environments from furniture polishes and room fresheners [59]. The presence of dipentene as 0.2-0.3 of the smoke condensate from both bright (flue-cured) burley tobacco and a commercial blend of bright, burley, turkish and Maryland tobacco was confirmed [42].

Terrestrial Fate: If released to soil, limited data indicate that limonene is expected to be resistant to biodegradation under aerobic conditions [1,48]. Based on limonene's water solubility [40] and its log octanol/water partition coefficient [37], soil adsorption coefficients ranging from 1030-7300 can be calculated using an appropriate regression equation [38], indicating that it will display low to no mobility in soil [55]. Its vapor pressure [37] and Henry's Law constant indicate that limonene will rapidly volatilize from both dry and moist soil to the atmosphere although strong adsorption to soil may significantly attenuate the rate of this process.

Aquatic Fate: If released to water, limited data indicate that limonene is expected to be resistant to biodegradation under aerobic conditions [1,48]. Based on limonene's water solubility [40] and its log octanol/water partition coefficient [37], bioconcentration factors ranging from 140-1750 can be calculated using an appropriate regression equation [38], indicating that it may bioconcentrate in fish and aquatic organisms. Estimated soil adsorption coefficients ranging from 1030 to 7300 [37,38,40] indicate that limonene may significantly adsorb to sediment and suspended organic matter. Limonene's Henry's Law constant [37,40] suggests that limonene will rapidly volatilize from water to the atmosphere. The estimated half-life for the volatilization of limonene from a model river (1 m deep, flowing at 1 m/sec, wind speed of 3 m/sec) is 3.5 hr [38], although adsorption to sediment and suspended matter may attenuate the rate of this process.

Atmospheric Fate: If released to the atmosphere, limonene is expected to rapidly undergo gas-phase oxidation reactions with photochemically produced hydroxyl radicals, ozone, and at night with nitrate radicals. Based on experimental rate constants, calculated lifetimes for the gas-phase reaction between limonene and photochemically produced hydroxyl radicals are 2.0 hr in a clean atmosphere and 30 mins in a moderately polluted atmosphere [67]. Corresponding values for the gas-phase reaction with ozone are 36 min and 11 min, respectively [67]. Calculated lifetimes, based on experimentally determined rate constants, for the night-time reaction of α-pinene with nitrate radicals of 9 min in a clean atmosphere and 0.9 min in a moderately polluted atmosphere have been reported [67]. The atmospheric lifetime of limonene was estimated at 0.2-0.8 hr depending on both the local hydroxyl radical and ozone concentration [2].

Biodegradation: Organisms isolated from soil and water were, in general, unable to oxidize limonene in laboratory experiments [48]. Limonene was listed as a compound difficult to biodegrade and was classified in level 3 (difficult to biodegrade) in a 5 tiered rating system on ease of biodegradability [1]. The concentration of limonene between the influent and effluent of aerated treatment lagoons was found to decrease significantly, which the author ascribed to biological removal processes although complete documentation was not provided [65]. Limonene, at 100 mg/L, reached 41-98% of the theoretical BOD over a period of 2 weeks following inoculation with activated sludge [12].

Abiotic Degradation: An experimental rate constant for the gas-phase reaction of limonene with photochemically produced hydroxyl radicals of 1.49×10^{-10} atm/m^3 molec at 32 °C [68] translates to a half-life of 2.6 hr using an average atmospheric hydroxyl radical concentration of $5 \times 10^{+5}$ molecule/cm^3 [4]. A calculated lifetime for the reaction of limonene with photochemically produced hydroxyl radicals in a clean atmosphere is 2.0 hr or 30 mins in a moderately polluted atmosphere [67]. Limonene was classified as group V in a 5 tiered rating system of relative reactivities towards photochemically produced hydroxyl radicals (methane = 1, limonene = 18,800), indicating an atmospheric half-life of <0.24 hr [17]. An experimental rate constant for the gas-phase reaction of limonene with ozone of 6.5×10^{-16} atm/m^3 molec [68] translates to a half-life of 25 min using an average atmospheric ozone concentration of $7 \times 10^{+11}$ molecules/cm^3 [67]. A calculated lifetime for the reaction of limonene with ozone in a clean atmosphere is 36 min or 11 min in a moderately polluted atmosphere [67]. A calculated lifetime for the reaction of limonene with nitrate radicals in a clean atmosphere is 9 min or 0.9 mins in a moderately polluted atmosphere [67]. Limonene was listed as a compound amenable to direct photochemical degradation in the presence of nitrogen oxides [64]. Photolysis of limonene in the presence of nitrogen oxides produces formaldehyde, formic acid, carbon monoxide, carbon dioxide, acetaldehyde, peroxyacetyl nitrate and acetone [17]. The daytime atmospheric lifetime of limonene was estimated at 0.2-0.8 hr depending on both the local hydroxyl radical and ozone concentration [2].

Bioconcentration: Based on limonene's water solubility [40] and its log octanol/water partition coefficient [37], bioconcentration factors of 140 and 1750, respectively, can be calculated using an appropriate regression equation [38]. These values indicate that limonene may bioconcentrate in fish and aquatic organisms.

Soil Adsorption/Mobility: Based on its water solubility [40] and its log octanol/water partition coefficient [37], soil adsorption coefficients of 1030 and 7300, respectively, can be calculated for limonene using an appropriate regression equation [38]. These values indicate that limonene is expected to display slight to no mobility in soil [55].

Volatilization from Water/Soil: Based on its vapor pressure [37], limonene may volatilize rapidly from dry soil to the atmosphere. Limonene's Henry's Law constant [37,40] indicates that this compound will rapidly volatilize from both water and moist soil to the atmosphere [38]. The estimated half-life for volatilization of limonene from a model river (1 m deep, flowing at 1 m/sec, wind speed of 3 m/sec) is 3.5 hr [38].

Water Concentrations: The concentration of limonene in seawater samples from Resurrection Bay, AK, was 84 ng/L in June 1985 and 0.47 ng/L in June, 1986 [11]. Limonene has been qualitatively detected in the Black Warrior River, near Tuscaloosa, AL, 1975 [6]. Limonene was identified in the River Glatt, Switzerland, 1975 [71]. It was detected, but not quantified, in water samples taken from the River Lee, in the UK, date not given [62]. Limonene was detected in contaminated ground water in the Netherlands at a maximum concentration of 10 ug/L [70]. Limonene was detected in 11 of 11 ground water samples at the site of a former pine-tar manufacturer in Gainesville, FL, at concentrations ranging from 1 ug/L to 130 ug/L [41]. Limonene was listed as a compound identified in US drinking water supplies [34,35]. It was qualitatively detected in treated drinking water supplies in the UK, 1976 [20].

Limonene

Effluent Concentrations: Limonene was detected as a component of landfill gases from sites in the UK at measured concentrations of 21-84 mg/m³ in probes buried underground and 7.4 mg/m³ at above ground vents [69]. It was qualitatively detected in 2 of 46 US industrial effluent samples [9]. Limonene was detected in 6 of 7 samples of kraft pulp mill wastewater at concentrations ranging from 10-220 ppb in 2 Canadian mills monitored in 1973 [65]. It was qualitatively detected in landfill leachate [61]. Limonene was qualitatively identified in the effluent gas from refuse waste obtained from a food center in an experiment designed to determine the gases emitted from decaying waste matter at refuse sites, landfills, and trash transfer sites [33]. Limonene has been associated with effluent from the following industries: extraction of pine gum, paper and pulp mills, plastics materials-synthetic resins and nonvulcanizable elastomers, perfumes, cosmetics and other toilet preparations, organic solvents and lubricating oils and greases [1]. Limonene was identified as a gaseous trace compound in municipal landfill sites at an average concentration of 26,200 ppbV [8].

Sediment/Soil Concentrations: Limonene was detected in soil samples at the site of a former pine-tar manufacturer in Gainesville, FL, at concentrations ranging from not detected to 920 ug/g [41].

Atmospheric Concentrations: Limonene was detected in 97% of 17 indoor air samples taken at residences in Ruston, WA, 1985-86, at concentrations ranging from 1.6-78 ug/m³ (mean and median 18 ug/m³ and 11 ug/m³) (respectively); outdoor concentrations were typically 12 times lower [43]. It was detected in 37 indoor and 12 outdoor samples from 36 houses (50 total measurements) in the Chicago area, concentrations not provided, indoor/outdoor ratio 3.1 [29]. The concentration of limonene in the air above Moscow Mountain, ID, 1976-1977, ranged from <10 ppt to 50 ppt [24]. The mean and maximum concentration of limonene in 40 homes in Oak Ridge/Knoxville, TN, 1982-83, was 16 ug/m³ and 77.5 ug/m³, respectively [23]. The concentration of limonene in Houston, TX, ranged from not detected to 5.7 ppb [7]. Limonene was detected indoors in an office building, 1987, at concentrations ranging from 43-63 ug/m³ [63]. It was listed as a compound typically identified in both indoor and outdoor air [22]. Limonene was qualitatively detected in the air of Leningrad, 1976, and 5 other Russian cities [26,27]. It was detected in suburban air samples in Germany, 1985, at concentrations ranging from not detected to 2.0 ng/m³ [30,31]. Limonene was qualitatively detected in air samples taken at 2

Stockholm preschools, 1981-82 [46]. Limonene was detected in indoor and outdoor air in Northern Italy, 1983-88, at mean concentrations of 140 ug/m^3 and 2 ug/m^3, respectively [18]. The concentration of limonene in the air over a forest in Soviet Georgia, July 1979, ranged from 0.004 ug/m^3 to 0.010 ug/m^3 in 8 samples [53]. The concentration of limonene 1.7 m above a maple forest in Quebec ranged from approximately 100-1750 ppt over a two day period in June 1989 [14]. Limonene was detected in forest air samples in southern black forest region, Germany, 1985, at concentrations ranging from 1.0-89 ng/m^3 [30,31]. Traces of limonene were found in the air over the Landes Forest, France, 1984, which consists mainly of maritime pines [50]. Limonene was detected in forest (Norway spruce, Fir, Scots pine, Larch) air samples in Switzerland, 1988 [3].

Food Survey Values: Limonene has been identified as a volatile component of fried chicken [57], chickpea seed [49], orange juice essence [44], mangos [39], roasted filberts [32], Beaufort (Gruyere) cheese [19], plums [21], apricots [21], and baked potatoes [16]. It has been detected in a headspace analysis of intact, tree-ripened nectarines, but not in an analysis of the blended fruit [56]. Limonene was detected in sweet, cooked corn at concentrations from 2 ppb (fresh kernel) to 10 ppb (frozen kernel) [10]. Limonene was detected in headspace samples of peanut oil that had been heated above 150 °C [13]. Cabernet Sauvignon wines from California contained limonene [54].

Plant Concentrations: Limonene was identified as a volatile constituent of kiwi fruit flowers [58].

Fish/Seafood Concentrations:

Animal Concentrations:

Milk Concentrations: Limonene was qualitatively detected in 8 of 8 samples of mother's milk obtained from residents of urban centers in PA, NJ, and LA [47].

Other Environmental Concentrations:

Probable Routes of Human Exposure: Occupational exposure to limonene may occur by inhalation or dermal contact during its production, formulation, transport or use. Exposure to the general population may occur by inhalation due to its presence in the atmosphere as a result of its release from natural sources [2,28], its presence in household products [59,60], or by ingestion of food in which it occurs either naturally or has been added as a flavor or fragrance.

Average Daily Intake:

Occupational Exposure: NIOSH (NOES 1981-83) has statistically estimated that 94,910 workers are exposed to limonene in the US [45]. Limonene was detected in the air of the vulcanization area of a shoe-sole factory at a concentration of 25-130 ug/m³, and at 5-1700 ug/m ³ in the vulcanization area of a tire retreading factory [15]. Limonene was detected indoors in an office building, 1987, at a concentration ranging from 43-63 ug/m³ [63]. Limonene was qualitatively detected in air samples taken at 2 Stockholm preschools, 1981-82 [46].

Body Burdens: Limonene has been identified in the expired air of urban volunteers [36]. Limonene was qualitatively detected in 8 of 8 samples of mother's milk obtained from residents of urban centers in PA, NJ, and LA [47].

REFERENCES

1. Abrams EF et al; Identification of Organic Compounds in Effluents from Industrial Sources. Washington, DC: USEPA-560/3-75-002 (1975)
2. Altshuller AP; Atmos Environ 17: 2131-65 (1983)
3. Andreani-Aksoyoglu S, Keller J; J Atmos Chem 20: 71-87 (1995)
4. Atkinson R; Chem Rev 85: 69-201 (1985)
5. Bauer K et al; In Ullmann's Encycl Indust Tech, 5th ed. Gerhartz W et al Eds. VCH Publ A11: 141-250 (1988)
6. Bertsch W et al; J Chromatog 112: 701-18 (1975)
7. Bertsch W et al; J Chromatog Sci 12: 175-82 (1974)
8. Brosseau J, Heitz M; Atmos Environ 28: 285-93 (1994)
9. Bursey JT, Pellizzari ED; Analysis of Industrial Wastewater for Organic Pollutants in Consent Decree Survey. Research Triangle Park, NC (1982)
10. Buttery RG et al; J Agric Food Chem 42: 791-75 (1994)
11. Button DK, Juttner F; Marine Chem 26: 57-66 (1989)

12. Chemical Inspection and Testing Institute; Data of Existing Chemicals Based on the CSCL Japan. Japan Chem Indus Ecol-Toxicol & Informat Center, Japan p. 3-119 (1992)
13. Chung TY et al; J Agric Food Chem 41: 1467-70 (1993)
14. Clement B et al; Atmos Environ 24A: 2513-6 (1990)
15. Cocheo V et al; Am Ind Hyg Assoc J 44: 521-7 (1983)
16. Coleman EC et al; J Agric Food Chem 29: 42-8 (1981)
17. Darnall KR et al; Environ Sci Tech 10: 692-6 (1976)
18. DeBortoli M et al; Environ Int 12: 343-50 (1986)
19. Dumont JP, Adda J; J Agr Food Chem 26: 364-7 (1978)
20. Fielding M et al; Organic Micropollutants in Drinking Water TR-159 Medmenham, Eng Water Res Cent 49 pp. (1981)
21. Gomez E et al; J Agric Food Chem 41: 1669-1676 (1993)
22. Harrison RM et al; Environ Tech Lett 9: 521-30 (1988)
23. Hawthorne AR et al; pp. 574-26 in Spec Meas Monit Non-Criter Contam. Frederick, ER ed. Pittsburgh, PA: APCA (1983)
24. Holdren MW et al; J Geophys Res 84: 5083-8 (1979)
25. Holohan SF et al; Kirk-Othmer Encycl Chem Tech, NY: Wiley 3rd ed. 12: 852-69 (1980)
26. Ioffe BV et al; Environ Sci Technol 13: 864-9 (1979)
27. Ioffe BV et al; J Chromatog 142: 787-95 (1977)
28. Isidorov VA et al; Atmos Environ 19: 1-8 (1985)
29. Jarke FH et al; Ashrae Trans 87: 153-6 (1981)
30. Juttner F; Chemosphere 15: 985-92 (1986)
31. Juttner F; Chemosphere 17: 309-17 (1988)
32. Kinlin TE et al; J Agr Food Chem 20: 1021-8 (1972)
33. Koe LC, Ng WJ; Water, Air Soil Pollut 33: 199-204 (1987)
34. Kool HJ et al; CRC Crit Rev Env Control 12: 307-57 (1982)
35. Kopfler FC et al; Adv Environ Sci Technol 8: 419-33 (1977)
36. Krotoszynski B et al; J Chromat Sci 15: 239-44 (1977)
37. Li J, Perdue EM; Physicochemical Properties of Selected Monoterpenes. Preprints of Papers Presented at the 209th ACS National Meeting, Anaheim CA; April 2-7, 1995; 35(1): 134-37 (1995)
38. Lyman WJ et al; Handbook of Chemical Property Estimation Methods. Washington, DC: Amer Chem Soc (1990)
39. MacLeod AJ, Snyder CH; J Agric Food Chem 36: 137-9 (1988)
40. Massaldi HA, King CJ; J Chem Eng Data 18: 393-7 (1973)
41. McCreary JJ et al; Chemosphere 12: 1619-32 (1983)
42. Mold JD et al; Science 144: 1572 (1964)
43. Montgomery DD; Kalman DA; Appl Ind Hyg 4: 17-20 (1989)
44. Moshonas MG, Shaw PE; J Agric Food Chem 38: 2181-4 (1990)
45. NIOSH; National Occupational Exposure Survey (NOES) (1989)
46. Noma E et al; Atmos Environ 22: 451-60 (1988)
47. Pellizzari ED et al; Bull Environ Contam Toxicol 28: 322-8 (1982)
48. Perry JJ; The Role of Co-Oxidation and Commensalism in the Biodegradation of

Recalcitrant Molecules. US Army Res Off DAAG-29-76-G0159 (1980)

49. Rembold H et al; J Agric Food Chem 37: 659-62 (1989)
50. Riba ML et al; Atmos Environ 21: 191-3 (1987)
51. Rogers JA Jr: Kirk-Othmer Encycl Chem Tech, NY: Wiley 3rd ed. 16: 307-32 (1981)
52. Sax NI, Lewis RJ Sr; Hawley's Condensed Chemical Dictionary 11th ed NY: Van Nostrand Reinhold Co. pp. 425, 701 (1987)
53. Shaw RW Jr et al; Environ Sci Tech 17: 389-95 (1983)
54. Shimoda M et al; J Agric Food Chem 41: 1664-68 (1993)
55. Swann RL et al; Res Rev 85: 17-28 (1983)
56. Takeoka GR et al; J Agric Food Chem 36: 553-60 (1988)
57. Tang J et al; J Agric Food Chem 31: 1287-92 (1983)
58. Tatsuka K et al; J Agric Food Chem 38: 2176-80 (1990)
59. Tichenor BA, Mason MA; JAPCA 38: 264-8 (1988)
60. Tichenor BA; Organic Emission Measurements via Small Chamber Testing. USEPA, Research Triangle Park, NC. PB87-199154 (1987)
61. Venkataramani ES, Ahlert RC; J Water Purif Contr Fed 56: 1178-84 (1984)
62. Waggott A et al; Chem Water Reuse 2: 55-9 (1981)
63. Weschler CJ et al; Am Ind Hyg Assoc J 51: 261-8 (1990)
64. Westberg HH, Rasmussen RA; Chemosphere 4: 163-8 (1972)
65. Wilson D, Hrutfiord B; Pulp and Paper Canada 76: 91-3 (1975)
66. Windholz M et al; The Merck Index. Rahway, NJ: Merck & Co Inc 10th ed (1983)
67. Winer AM et al; Science 224: 156-9 (1984)
68. Winer AM et al; J Phys Chem: 80: 1635-9 (1976)
69. Young P, Parker A; Vapors, pp. 24-41 in Hazardous and Industrial Waste Management and Testing, 3rd Symp. Amer Soc Test Mater (1984)
70. Zoeteman BCJ et al; Sci Total Environ 21: 187-202 (1981)
71. Zurcher F, Giger W; Vom Wasser 47: 37-55 (1976)

2-Mercaptoethanol

SUBSTANCE IDENTIFICATION

Synonyms: 1-Hydroxy-2-mercaptoethane; Thioglycol

Structure:

CAS Registry Number: 60-24-2

Molecular Formula: C_2H_6OS

SMILES Notation: OCCS

CHEMICAL AND PHYSICAL PROPERTIES

Boiling Point: Decomposes at boiling point 157-158 °C, 742 mm Hg

Melting Point:

Molecular Weight: 78.13

Dissociation Constants:

Log Octanol/Water Partition Coefficient: -0.20 (estimated) [11]

Water Solubility: Miscible in water [15]

Vapor Pressure: 1.00 mm Hg at 20 °C [15]

Henry's Law Constant: 3.4×10^{-8} atm-cm^3/mol (estimated) [8]

ENVIRONMENTAL FATE/EXPOSURE POTENTIAL

Summary: The decomposition of naturally occurring products such as swine manure and proteins (produced by marine algae and other marine plants) forms 2-mercaptoethanol. Human sources of releases may include solvent evaporation. If released to air, 2-mercaptoethanol will degrade relatively rapidly by reaction with photochemically produced hydroxyl radicals (estimated half-life of 4.0 hr). Physical removal from air via wet deposition is possible since it is miscible in water. If released to water or soil, 2-mercaptoethanol may biodegrade. Although the results from a soil degradation study indicate that 2-mercaptoethanol is biodegradable, insufficient data are available to predict the relative importance or rate of microbial degradation. Leaching in soil is expected as 2-mercaptoethanol has an estimated Koc of 18 and is miscible in water. 2-Mercaptoethanol can evaporate to air from solid surfaces. Occupational exposure to 2-mercaptoethanol may occur through inhalation of vapor and dermal contact.

Natural Sources: 2-Mercaptoethanol has been identified as a volatile substance evolved during aerobic and anaerobic microbial decomposition of liquid swine manure [7].

Artificial Sources: 2-Mercaptoethanol's use as a solvent and as a chemical intermediate [15] could release the compound to the environment through waste effluents. Evaporation from solvent applications [15] could release 2-mercaptoethanol directly to air.

Terrestrial Fate: 2-Mercaptoethanol may leach readily in soil based upon an estimated Koc value of 16 and its miscibility in water [5,6]. Although the results of one soil degradation study indicate that 2-mercaptoethanol is biodegradable [5], insufficient data are available to predict the relative importance or rate of microbial degradation in soil.

Aquatic Fate: Although the results of one soil degradation study indicate that 2-mercaptoethanol is biodegradable [5], insufficient data are available to predict the relative importance or rate of microbial degradation in water. Aquatic volatilization, bioconcentration, and adsorption to sediment are not expected to be important environmental processes.

Atmospheric Fate: Based upon a vapor pressure of 1.0 mm Hg at 20 °C [15], 2-mercaptoethanol is expected to exist almost entirely in the vapor phase in the ambient atmosphere [4]. It will degrade relatively rapidly in an average ambient atmosphere by reaction with photochemically produced hydroxyl radicals (estimated half-life of 4.0 hr) [1]. Since 2-

mercaptoethanol is miscible in water [15], physical removal from air via wet deposition is likely to occur.

Biodegradation: The biodegradation rate of 2-mercaptoethanol added to a soil at 10,000 ppm has been reported as 40 kg/wk [5].

Abiotic Degradation: The rate constant for the vapor phase reaction of 2-mercaptoethanol with photochemically produced hydroxyl radicals has been estimated to be 4.4×10^{-11} cm^3/molecule-sec at 25 °C, which corresponds to an atmospheric half-life of about 4.0 hr at an atmospheric concentration of $5 \times 10^{+5}$ hydroxyl radicals per cm^3 [2]. The rate constant for the reaction between hydroxyl radicals and 2-mercaptoethanol in aqueous solution at pH 6.5 has been experimentally determined to be $6.8 \times 10^{+9}$ 1/M-sec [3]. Assuming a hydroxyl radical concentration of 1×10^{-17} M in brightly sunlit natural water [10], the half-life for this reaction would be 118 days.

Bioconcentration: Based on the log Kow [9], the BCF for 2-mercaptoethanol can be estimated to be 0.4 from a regression-derived equation [6]. This BCF value suggests that 2-mercaptoethanol will not bioconcentrate significantly in aquatic organisms. 2-Mercaptoethanol is miscible in water [15], which also suggests that it will not biconcentrate in aquatic organisms.

Soil Adsorption/Mobility: Based on the log Kow [9], the Koc for 2-mercaptoethanol can be estimated to be 18 from a regression-derived equation [6]. This Koc value suggests that 2-mercaptoethanol has very high soil mobility [16]. 2-Mercaptoethanol is miscible in water [15], which indicates that it may leach in soil.

Volatilization from Water/Soil: The estimated Henry's Law constant [8] indicates that 2-mercaptoethanol is essentially nonvolatile from water [6].

Water Concentrations:

Effluent Concentrations:

Sediment/Soil Concentrations: 2-Mercaptoethanol and other thiol concentrations of less than 100 uM were detected in intertidal marine

2-Mercaptoethanol

sediments from Biscayne Bay, FL [11]; the presence of the thiols was attributed to protein degradation, where the protein source was marine algae and other higher plants [11]; this degradation may be an important environmental source of thiols [11].

Atmospheric Concentrations:

Food Survey Values: 2-Mercaptoethanol concentrations of 0.01 ppb or less were detected in samples of German wines [14].

Plant Concentrations:

Fish/Seafood Concentrations:

Animal Concentrations:

Milk Concentrations:

Other Environmental Concentrations:

Probable Routes of Human Exposure: 2-Mercaptoethanol's use as a solvent for dyestuffs and as a chemical intermediate for a wide variety of products [15] could expose occupational workers to this compound through inhalation of vapor and dermal contact.

Average Daily Intake:

Occupational Exposure: NIOSH (NOES 1981-1983) has statistically estimated that 7,155 workers are potentially exposed to 2-mercaptoethanol in the US [12]. NIOSH (NOHS 1972-1974) has statistically estimated that 38,676 workers are potentially exposed to 2-mercaptoethanol in the US [13].

Body Burdens:

REFERENCES

1. Atkinson R; J Phys Chem Ref Data, Monograph 1, p. 145 (1989)
2. Atkinson R; J Inter Chem Kinet 19: 799-828 (1987)

3. Buxton GV et al; J Phys Chem Ref Data 17: 733 (1988)
4. Eisenreich SJ et al; Environ Sci Technol 15: 30-8 (1981)
5. Huddleston RL et al; pp. 41-61 in Water Resour Symp 13 (Land Treat: Hazard Waste Manage Altern) (1986)
6. Lyman WJ et al; Handbook of Chemical Property Estimation Methods. Washington DC: Amer Chem Soc (1990)
7. Mehlorn G et al; Proc Int Kongr Tarhyg, 5th 2: 725-30 (1985)
8. Meylan WM, Howard PH; Environ Toxicol Chem 10: 1283-93 (1991)
9. Meylan WM, Howard PH; J Pharm Sci 84: 83-92 (1995)
10. Mill T et al; Science 207: 886-7 (1980)
11. Mooper K, Taylor BF; ACS Symp Ser 305 (Org Mar Geochem): 324-39 (1986)
12. NIOSH; National Occupational Exposure Survey (NOES) (1983)
13. NIOSH; National Occupational Hazard Survey (NOHS) (1974)
14. Rapp A et al; Am J Enol Vitic 36: 219-21 (1985)
15. Sax NI, Lewis RJ Sr; Hawley's Condensed Chemical Dictionary 11th ed NY; Van Nostrand Reinhold Co. p. 740 (1987)
16. Swann RL et al; Res Rev 85: 23 (1983)

4-Methoxy-4-methyl-2-pentanone

SUBSTANCE IDENTIFICATION

Synonyms:

Structure:

CAS Registry Number: 107-70-0

Molecular Formula: $C_7H_{14}O_2$

SMILES Notation: O=C(CC(OC)(C)C)C

CHEMICAL AND PHYSICAL PROPERTIES

Boiling Point: 147-163 °C

Melting Point:

Molecular Weight: 130.21

Dissociation Constants:

Log Octanol/Water Partition Coefficient: 0.36 [7]

Water Solubility: 280,000 mg/L at 25 °C [9]

Vapor Pressure: 3.16 mm Hg at 25 °C [3]

Henry's Law Constant: 1.93 x 10^{-6} atm-m^3/mol (calculated from vapor pressure and water solubility)

ENVIRONMENTAL FATE/EXPOSURE POTENTIAL

Summary: 4-Methoxy-4-methyl-2-pentanone is released to the atmosphere by evaporation from its use as a solvent in a variety of resin coatings. The time required to evaporate about 50% of applied 4-methoxy-4-methyl-2-pentanone from a typical solvent application in lacquers, thinners, and paints is about 20 minutes. If released to the atmosphere, it will degrade by reaction with photochemically produced hydroxyl radicals (estimated half-life of 3.1 days). Direct photolysis may also contribute to its atmospheric degradation. Physical removal from air via wet deposition is possible since 4-methoxy-4-methyl-2-pentanone is very water soluble. If released to soil surfaces or water, 4-methoxy-4-methyl-2-pentanone can degrade through direct photolysis (exact environmental photolysis rates are not available). Insufficient data are available to predict the importance of biodegradation. Evaporation from dry surfaces is an important environmental transport process, but the Henry's Law constant value would suggest slow evaporation from moist soils and water. When released to soil, 4-methoxy-4-methyl-2-pentanone may leach readily based upon an estimated Koc value of 4.4. Occupational exposure to 4-methoxy-4-methyl-2-pentanone can occur through inhalation of vapor and dermal contact. The general population may be exposed through the same routes when handling consumer products (such as paints, thinners, lacquers, etc) that contain 4-methoxy-4-methyl-2-pentanone.

Natural Sources:

Artificial Sources: 4-Methoxy-4-methyl-2-pentanone's use as a solvent in a variety of resin coatings [10] will release the compound directly to air through evaporation.

Terrestrial Fate: Evaporation from dry surfaces, such as soil, is an important environmental transport process for 4-methoxy-4-methyl-2-pentanone. The time required to evaporate about 50% of applied 4-methoxy-4-methyl-2-pentanone from a typical solvent application in lacquers, thinners, and paints is about 20 minutes [9]. When released to soil, 4-methoxy-4-methyl-2-pentanone may leach readily based upon an estimated Koc value of 4.4 [5]. Since 4-methoxy-4-methyl-2-pentanone can be degraded by direct photolysis at environmental wavelengths [2], photodegradation may occur on sunlit surfaces. Insufficient data are available to predict the importance of biodegradation in soil.

Aquatic Fate: 4-Methoxy-4-methyl-2-pentanone can be degraded by direct photolysis at environmentally relevant wavelengths [2]; although the photolysis rate is not certain. This process may contribute to its removal from sunlit water. Volatilization from water is slow. The volatilization half-lives from a model environmental river (1 m deep) and model pond have been estimated to be 22 days and 236 days, respectively [5,12]. Insufficient data are available to predict the importance of biodegradation. Chemical hydrolysis, bioconcentration, and adsorption to sediment are not expected to be important in water.

Atmospheric Fate: Based upon the vapor pressure, 4-methoxy-4-methyl-2-pentanone is expected to exist almost entirely in the vapor phase in the ambient atmosphere [4]. It is expected to degrade in an average ambient atmosphere by reaction with photochemically produced hydroxyl radicals with an estimated half-life of about 3.1 days [6]. 4-Methoxy-4-methyl-2-pentanone can be degraded by direct photolysis at environmentally relevant wavelengths [2]; although the photolysis rate in the environment is not certain, this process may contribute to atmospheric removal. Physical removal via wet deposition is likely since 4-methoxy-4-methyl-2-pentanone is very soluble in water.

Biodegradation:

Abiotic Degradation: The rate constant for the vapor-phase reaction of 4-methoxy-4-methyl-2-pentanone with photochemically produced hydroxyl radicals has been estimated to be 5.15×10^{-12} cm^3/molecule-sec [6] at room temperature, which corresponds to an atmospheric half-life of about 3.1 days at an atmospheric concn of $5 \times 10^{+5}$ hydroxyl radicals per cm^3 [1]. Ketone, ether, and alkane functional groups are generally not susceptible to significant aqueous hydrolysis in the environment [5]; therefore, 4-methoxy-4-methyl-2-pentanone is not expected to hydrolyze in environmental waters. Laboratory experiments demonstrated that 4-methoxy-4-methyl-2-pentanone is decomposed by direct photolysis [2]; irradiation in hexane solution at 313 nm resulted in a quantum yield measurement of 0.46 and a projected degradation half-life of about 6 hr [2]; irradiation in ethanol and allyl alcohol solution yielded similar results [2]; the photolysis products included mesityl oxide, methanol, a hydroxyfuran derivative, and small amounts of acetone and methyl isopropenyl ether [2].

Bioconcentration: Based upon the water solubility, the BCF for 4-methoxy-4-methyl-2-pentanone can be estimated to be 0.5 from a regression-derived equation [5]. This BCF value suggests that 4-methoxy-4-methyl-2-pentanone will not bioconcentrate significantly in aquatic organisms.

Soil Adsorption/Mobility: Based upon the water solubility, the Koc for 4-methoxy-4-methyl-2-pentanone can be estimated to be 4.4 from a regression-derived equation [5]. This Koc value suggests that 4-methoxy-4-methyl-2-pentanone has very high soil mobility [11].

Volatilization from Water/Soil: From the value of the Henry's Law constant, the volatilization from environmental waters is slow with the possible exception of very shallow rivers [5]. Based on the Henry's Law constant, the volatilization half-life from a model river (1 m deep flowing 1 m/sec with a wind velocity of 3 m/sec) can be estimated to be about 22 days [5]. Volatilization half-life from a model environmental pond can be estimated to be about 236 days [12].

Water Concentrations:

Effluent Concentrations:

Sediment/Soil Concentrations:

Atmospheric Concentrations:

Food Survey Values:

Plant Concentrations:

Fish/Seafood Concentrations:

Animal Concentrations:

Milk Concentrations:

Other Environmental Concentrations:

Probable Routes of Human Exposure: Occupational exposure to 4-

methoxy-4-methyl-2-pentanone can occur through inhalation of vapor and dermal contact. The general population may be exposed through the same routes when handling consumer products (such as paints, thinners, lacquers, etc.) that contain 4-methoxy-4-methyl-2-pentanone.

Average Daily Intake:

Occupational Exposure: NIOSH (NOES 1981-1983) has statistically estimated that 4,101 workers are potentially exposed to 4-methoxy-4-methyl-2-pentanone in the US [8].

Body Burdens:

REFERENCES

1. Atkinson R; J Inter Chem Kinet 19: 799-828 (1987)
2. Coyle DJ et al; J Amer Chem Soc 86: 3850-54 (1964)
3. Dykyj JJ, Repas M; Petrochemia 18: 179-98 (1973)
4. Eisenreich SJ et al; Environ Sci Technol 15: 0-8 (1981)
5. Lyman WJ et al; Handbook of Chemical Property Estimation Methods NY: McGraw-Hill p. 4-9 (1982)
6. Meylan WM, Howard PH; Chemosphere 26:2293-2299 (1993)
7. Meylan W, Howard PH; J Pharm Sci 84: 83-92 (1995)
8. NIOSH; National Occupational Exposure Survey (NOES) (1983)
9. Norton LC, Scherzinger RA; Paint Varnish Prod 51: 25-31 (1961)
10. Sax NI, Lewis RJ Jr; Hawley's Condensed Chemical Dictionary 11th ed. NY: Van Nostrand Reinhold Co p. 754 (1987)
11. Swann RL et al; Res Rev 85: 23 (1983)
12. USEPA; EXAMS II Computer Simulation (1987)

2-Methyl-5-ethylpyridine

SUBSTANCE IDENTIFICATION

Synonyms:

Structure:

CAS Registry Number: 104-90-5

Molecular Formula: $C_8H_{11}N$

SMILES Notation: n(c(ccc1CC)C)c

CHEMICAL AND PHYSICAL PROPERTIES

Boiling Point: 177.8 °C at 747 mm Hg

Melting Point: -70.3 °C

Molecular Weight: 121.18

Dissociation Constants: pKa = 6.40 [6]

Log Octanol/Water Partition Coefficient: 2.39 [11]

Water Solubility: 12,000 mg/L at 25 °C [6]

Vapor Pressure: 1.4 mm Hg at 25 °C [4]

Henry's Law Constant: 1.38×10^{-5} atm-m^3/mol at 25 °C [10]

ENVIRONMENTAL FATE/EXPOSURE POTENTIAL

Summary: 2-Methyl-5-ethylpyridine may be released to the environment via effluents at sites where it is produced or used as a chemical intermediate in medicine, agriculture and industry. It may also be released to the

312

environment via effluents from oil processing facilities. Information pertaining to the biodegradation of 2-methyl-5-ethylpyridine in soil and water was not located in the available literature. 2-Methyl-5-ethylpyridine and its conjugate acid will exist in most environmental media in varying proportions that are pH dependent. Ions generally do not volatilize. A Henry's Law constant of 1.38×10^{-5} atm-m^3/mol at 25 °C indicates that volatilization of 2-methyl-5-ethylpyridine from environmental waters and moist soil should generally be slow; however, volatilization from shallow, rapid waters may be significant. The volatilization half-lives from a model river and a model pond, the latter considers the effect of adsorption, have been estimated to be 3 days and 34 days, respectively. 2-Methyl-5-ethylpyridine should evaporate from dry surfaces, especially when present in high concn such as in spill situations. In aquatic systems, the bioconcentration of 2-methyl-5-ethylpyridine is not expected to be an important fate process. A low Koc indicates 2-methyl-5-ethylpyridine should not partition from the water column to organic matter contained in sediments and suspended solids. 2-Methyl-5-ethylpyridine should be highly mobilized in soil. In the atmosphere, 2-methyl-5-ethylpyridine is expected to exist entirely in the vapor phase and reactions with photochemically produced hydroxyl radicals should be important (estimated half-life of 6 days). The high water solubility of 2-methyl-5-ethylpyridine suggests that physical removal from air by wet deposition (rainfall and dissolution in clouds, etc.) may occur. The most probable human exposure would be occupational exposure, which may occur through dermal contact or inhalation at workplaces where 2-methyl-5-ethylpyridine is produced or used. The most common nonoccupational exposure is likely to result from the ingestion of contaminated drinking water or certain foods.

Natural Sources: 2-Methyl-5-ethylpyridine was found in peppermint oils [13].

Artificial Sources: Pyridine bases are produced from the pyrolysis of coal or synthetically by reactions between aldehydes and ketones with ammonia [6]. 2-Methyl-5-ethylpyridine is a commercially important derivative that is utilized in medicine, agriculture and industry [6]. Consequently, 2-methyl-5-ethylpyridine may be released to the environment via effluents at sites where it is produced or used. 2-Methyl-5-ethylpyridine is also released to the environment via effluents from oil processing facilities [1,3].

2-Methyl-5-ethylpyridine

Terrestrial Fate: Information pertaining to the biodegradation of 2-methyl-5-ethylpyridine in soil was not located in the available literature. Based upon the pKa, 2-methyl-5-ethylpyridine and its conjugate acid will exist in most soils in varying proportions that are pH dependent. Ions generally do not volatilize. The Henry's Law constant indicates that volatilization of 2-methyl-5-ethylpyridine from moist soil should be slow [9]. Yet, 2-methyl-5-ethylpyridine should evaporate from dry surfaces, especially when present in high concn such as in spill situations. An estimated Koc range of 25 to 540 [9] indicates that 2-methyl-5-ethylpyridine should be highly mobile in soil [14].

Aquatic Fate: Information pertaining to the biodegradation of 2-methyl-5-ethylpyridine in aquatic systems was not located in the available literature. Based upon the pKa, 2-methyl-5-ethylpyridine and its conjugate acid will exist among most environmental waters in varying proportions that are pH dependent. Ions are not expected to volatilize from water. The Henry's Law constant indicates that volatilization of 2-methyl-5-ethylpyridine from environmental waters should generally be slow [9]. Volatilization from shallow, rapid waters may be significant. Based on the Henry's Law constant, the volatilization half-life from a model river has been estimated to be 3 days [9]. The volatilization half-life from a model pond, which considers the effect of adsorption, has been estimated to be about 34 days [16]. An estimated Koc range of 25 to 540 [9] indicates 2-methyl-5-ethylpyridine should not partition from the water column to organic matter [14] contained in sediments and suspended solids; and estimated bioconcentration factors (log BCF) of 0.49 and 1.66 [9] indicate that 2-methyl-5-ethylpyridine should not bioconcentrate in aquatic organisms.

Atmospheric Fate: Based upon the vapor pressure, 2-methyl-5-ethylpyridine is expected to exist entirely in the vapor phase in ambient air [5]. In the atmosphere, vapor-phase reactions with photochemically produced hydroxyl radicals may be an important fate process. The rate constant for the vapor-phase reaction of 2-methyl-5-ethylpyridine with photochemically produced hydroxyl radicals has been estimated to be 2.74 x 10^{-12} cm^3/molecule-sec at 25 °C [12], which corresponds to an atmospheric half-life of about 6 days at an atmospheric concn of 5 x 10^{+5} hydroxyl radicals per cm^3 [2]. The high water solubility of 2-methyl-5-ethylpyridine suggests that physical removal from air by wet deposition (rainfall and dissolution in clouds, etc.) may occur.

2-Methyl-5-ethylpyridine

Biodegradation:

Abiotic Degradation: The rate constant for the vapor-phase reaction of 2-methyl-5-ethylpyridine with photochemically produced hydroxyl radicals has been estimated to be 2.74 x 10^{-12} cm^3/molecule-sec at 25 °C [12], which corresponds to an atmospheric half-life of about 6 days at an atmospheric concn of 5 x 10^{+5} hydroxyl radicals per cm^3 [2].

Bioconcentration: Based on the water solubility and an estimated log Kow, respective bioconcentration factors (log BCF) of 0.49 and 39 for 2-methyl-5-ethylpyridine have been calculated using recommended regression-derived equations [9]. These BCF values indicate that 2-methyl-5-ethylpyridine should not bioconcentrate among aquatic organisms.

Soil Adsorption/Mobility: Based on the water solubility and an estimated log Kow, the Koc of 2-methyl-5-ethylpyridine has been calculated using regression-derived equations to range from 25 to 478 [9]. These Koc values indicate 2-methyl-5-ethylpyridine will be highly mobile in soil [14], and it should not partition from the water column to organic matter contained in sediments and suspended solids. However, based upon the pKa, 2-methyl-5-ethylpyridine and its conjugate acid will exist among most environmental media in varying proportions that are pH dependent.

Volatilization from Water/Soil: Based upon the pKa, 2-methyl-5-ethylpyridine and its conjugate acid will exist among most environmental media in varying proportions that are pH dependent. Ions are not expected to volatilize from water. Based upon the Henry's Law constant, volatilization of 2-methyl-5-ethylpyridine from natural bodies of water and moist soil should generally be slow [9]. Volatilization from shallow, rapid waters may be significant [9]. The volatilization half-life from a model river (1 m deep flowing 1 m/sec with a wind speed of 3 m/sec) has been estimated to be about 3 days [9]. The volatilization half-life from a model pond, which considers the effect of adsorption, has been estimated to be approximately 34 days [16]. Based upon the vapor pressure, 2-methyl-5-ethylpyridine should evaporate from dry surfaces, especially when present in high concn such as in spill situations.

Water Concentrations: DRINKING WATER: 2-Methyl-5-ethylpyridine was listed as a contaminant found in drinking water [1,7] for a survey of US cities including Pomona, Escondido, Lake Tahoe and Orange Co, CA and Dallas, Washington, DC, Cincinnati, Philadelphia, Miami, New Orleans, Ottumwa, IA, and Seattle [8].

Effluent Concentrations: 2-Methyl-5-ethylpyridine was detected in 1 of 21 industrial categories of wastewater effluents [3]. Extract from the wastewater of an organic chemical manufacturer contained 2-methyl-5-ethylpyridine at an average concn of 3021 mg/L [1,3].

Sediment/Soil Concentrations:

Atmospheric Concentrations:

Food Survey Values: 2-Methyl-5-ethylpyridine was qualitatively identified as a volatile component of fried chicken [15].

Plant Concentrations:

Fish/Seafood Concentrations:

Animal Concentrations:

Milk Concentrations:

Other Environmental Concentrations:

Probable Routes of Human Exposure: The most probable route of human exposure to 2-methyl-5-ethylpyridine is by inhalation, dermal contact and ingestion. Drinking water supplies [1,7,8] and certain foods have been shown to contain 2-methyl-5-ethylpyridine [15].

Average Daily Intake:

Occupational Exposure:

Body Burdens:

REFERENCES

1. Abrams EF et al; Identification of Organic Compounds in Effluents from Industrial Sources USEPA-560/3-75-002 (1975)
2. Atkinson R; Intern J Chem Kin 19: 799-828 (1987)
3. Bursey JT, Pellizzari ED; Analysis of Industrial Wastewater for Organic Pollutants in Consent Decree Survey. Contract No 68-03-2867. Athens, GA: USEPA Environ Res Lab (1982)
4. Chao J et al; J Phys Chem Ref Data 12: 1033-63 (1983)
5. Eisenreich SJ et al; Environ Sci Technol 15: 30-8 (1981)
6. Goe GL; Kirk-Othmer Encycl Chem Tech 3rd NY: Wiley Interscience 19: 454-83 (1982)
7. Kopfler FC et al; Adv Environ Sci Technol 8: 419-33 (1977)
8. Lucas SV; GC/MS Anal of Org in Drinking Water Concentrates and Advanced Treatment Concentrates Vol 1 USEPA-600/1-84--020A (NTIS PB85-128239) p. 397 (1984)
9. Lyman WJ et al; Handbook of Chemical Property Estimation Methods NY: McGraw-Hill (1982)
10. Meylan W, Howard PH; Environ Toxicol Chem 10: 1283-93 (1991)
11. Meylan W, Howard PH; J Pharm Sci 84: 83-92 (1995)
12. Meylan WM, Howard PH; Chemosphere 26:2293-2299 (1993)
13. Sakurai K et al; Agric Biol Chem 47 (19): 2307-12 (1983)
14. Swann RL et al; Res Rev 85: 16-28 (1983)
15. Tang J et al; J Agric Food Chem 31: 1287-92 (1983)
16. USEPA; EXAMS II Computer Simulation (1987)

2-Methylpentane

Synonyms:

Structure:

$$CH_3$$

H₃C — CH₃

CAS Registry Number: 107-83-5

Molecular Formula: C_6H_{14}

SMILES Notation: C(CCC)(C)C

CHEMICAL AND PHYSICAL PROPERTIES

Boiling Point: 60.3 °C

Melting Point: -153.7 °C

Molecular Weight: 86.18

Dissociation Constants:

Log Octanol/Water Partition Coefficient: 3.74 (estimated) [38]

Water Solubility: 14 mg/L [30]

Vapor Pressure: 211 mm Hg at 25 °C [8]

Henry's Law Constant: 1.71 atm-m^3/mol (calculated from vapor pressure and water solubility)

ENVIRONMENTAL FATE/EXPOSURE POTENTIAL

Summary: 2-Methylpentane is a highly volatile chemical that occurs naturally in petroleum and may be released to the environment from natural

318

petroleum seepages or as fugitive emissions or spills wherever petroleum products are refined, stored, transferred or used. Another large source of release is from motor vehicle exhaust, and as a result, the atmospheric concn of 2-methylpentane is strongly related to traffic. Other anthropogenic releases of 2-methylpentane are related to its use as a solvent. It is also a plant volatile. If released to land, 2-methylpentane will be lost primarily by volatilization due to its high vapor pressure and may slowly leach in soil. Volatilization from water (estimated half-life 2.7 hr in a model river) should be the most important fate process in aquatic systems. Adsorption to sediment is expected to be moderate and bioconcentration in aquatic organisms low to moderate. In the atmosphere, 2-methylpentane will degrade by reaction with photochemically produced hydroxyl radicals (half-life 2.9 days). Degradation is much faster under photochemical smog conditions. Human exposure will be primarily via inhalation, especially in areas of high traffic or where petroleum products are used.

Natural Sources: 2-Methylpentane occurs naturally in petroleum and natural gas and as a plant volatile [1,2]. Emissions would therefore occur as a result of natural oil seepages or gas vents and emissions from vegetation. Forest fires are another emission source [10].

Artificial Sources: 2-Methylpentane is found in sources associated with petroleum products such as petroleum manufacture, natural gas, turbines, and automobiles [10]. Emissions will occur from the evaporation of petroleum during transfer, transport and storage of petroleum products and evaporation of gasoline from carburetors, gas tanks, and during refueling of motor vehicles [12]. The fact that 2-methylpentane in cities is associated with traffic is clearly demonstrated by the similarity of auto exhaust profiles compared with ambient air profiles in areas of high traffic in Bangkok City [19]. 2-Methylpentane is also used as a solvent [14] and will be released to the atmosphere as a result of solvent evaporation. 2-Methylpentane was a volatile component of a stainless steel polish cleaner and carpet glue [33].

Terrestrial Fate: If released on soil, 2-methylpentane will be lost primarily by volatilization due to its high vapor pressure. It is estimated to have low mobility in soil. 2-Methylpentane is oxidized by microorganisms isolated from soil and ground water and therefore may biodegrade in soil, especially where populations of suitable microorganisms have become established as a result of petroleum-related spills. However, there are no data establishing biodegradation rates of 2-methylpentane in environmental media.

Aquatic Fate: 2-Methylpentane is estimated to have a very high Henry's Law constant and therefore will volatilize rapidly from water. Its volatilization half-life from a model river is 2.7 hr [23]. It is estimated to have a moderately high adsorptivity and therefore may adsorb to sediment and particulate matter in the water column. Bioconcentration in fish is estimated to be low to moderate.

Atmospheric Fate: In the atmosphere, 2-methylpentane will degrade by reaction with photochemically produced hydroxyl radicals (half-life 2.9 days) [2]. Degradation is much faster under photochemical smog conditions with a 26.4% loss in 5 hr being reported in one study [41].

Biodegradation: A quarter of the microorganisms, most notably of the *Nocardia* sp., isolated from gasoline-contaminated ground water supported growth of 2-methylpentane [17]. Generally isoalkanes are significantly more resistant to microbial attack than n-alkanes [5]. However, a soil microorganism, *Corynebacterium* sp., oxidized 2-methylpentane, at about the same rate as n-pentane, although at a significantly lower rate than n-hexane [5].

Abiotic Degradation: Aliphatic hydrocarbons do not absorb radiation >290 nm and therefore 2-methylpentane will not directly photolyze in the environment [23]. 2-Methylpentane reacts with photochemically produced hydroxyl radicals in the atmosphere by H-atom abstraction. The rate constant for this reaction is 5.57×10^{-12} cm^3/molecule-sec [2]. Assuming a OH concn of 5×10^5 radicals/cm^3, the half-life of 2-methylpentane in the atmosphere is 2.9 days. 2-Methylpentane has low to moderate photochemical reactivity [41]. Smog chamber studies of the NOX-air photooxidation of 2-methylpentane resulted in a 26.4% depletion of 2-methylpentane in 5 hr [41]. In the Houston Oxidant Field Study, the diurnal patterns of NO$_2$, O$_3$, and 2-methylpentane concns suggested that the photochemical reaction of 2-methylpentane and NO$_2$ contribute to elevated ozone levels [24]. The order-of-magnitude lower concn of 2-methylpentane in the arctic summer compared with the spring is explained by differences in levels of sunlight and photochemical loss [16]. Alkanes are resistant to hydrolysis [23].

Bioconcentration: Using the estimated log octanol/water partition coefficient and a water solubility for 2-methylpentane, BCFs are estimated for 2 methylpentane ranging from 100 to 408 using six different regression equations [23]. Therefore 2-methylpentane is estimated to have a moderately low bioconcentration potential.

Soil Adsorption/Mobility: Using its water solubility, one estimates a Koc of 1022 for 2-methylpentane using a recommended regression equation [23]. Therefore 2-methylpentane would be expected to have a low mobility in soil [35].

Volatilization from Water/Soil: Using its calculated Henry's Law constant, one estimates a half-life for volatilization of 2-methylpentane, from a 1 m deep river having a 1 m/sec current with a 3 m/sec wind speed, of 2.7 hr [23]. Volatilization is therefore rapid and is controlled by the rate of diffusion of 2-methylpentane in water [23]. Because of its high vapor pressure, volatilization from dry soil should be rapid.

Water Concentrations: DRINKING WATER: 2-Methylpentane was detected in tap water in New Jersey during a pilot broad spectrum analysis of exposure to chemicals conducted by the EPA's Total Exposure Assessment Methodology (TEAM) Study [40]. SURFACE WATER: 2-Methylpentane was measured in eight samples of sea water taken during a cruise on the Indian Ocean; concns ranged from 0.03-0.79 nanoliters of vapor per liter of water at 30 °C [3].

Effluent Concentrations: Methylpentane (isomer unspecified) was identified in air at four to five locations in 4 New Jersey hazardous waste sites on the National Priorities list and one landfill; concns were not reported [21]. It was identified in 12 of 63 industrial waste water effluents, all at concn levels <10 μg/L [28].

Sediment/Soil Concentrations:

2-Methylpentane

Atmospheric Concentrations: The median 2-methylpentane concn at rural/remote (10 samples), urban/suburban (46 samples), and source dominated (59 samples) sites in the United States was 1.1, 4.2, and 3.1 ppb, respectively [4]. The concn for the 6 site studies ranged from 0.8 to 45 ppb [4]. RURAL/REMOTE: Norwegian arctic 0.182 ppb spring, <0.020 ppb summer [16]. Four remote sites in northwest North Carolina sampled over 13 months 0.24-0.72 ppb, median; 0.12-2.4 ppb, range [32]. Detected at one of four rural sites in the US at a mean level of 0.8 ppb [9]. Smokey Mountains, TN (9 samples) 1.0-5.2 ppb [1]; Rio Blanco County, CO (5 sites) 1.0-3.0 ppb [4]; rural sites in NW England, <0.6 ppb, mean [7]. Detected, not quantified, in a spruce forest in Germany [15]. URBAN/SUBURBAN: 39 US cities from 1984 through 1986 (836 samples) 1.4-647 ppb, 25th percentile 10.2, median 17.9, 75th percentile 28.2 ppb [31]. The concn at a residential site in Houston, TX, removed from highways (684 data points) was 28 and 272 ppb, mean and maximum, respectively [24]. The average concn between 6 and 9 am (72 data points) was 43 ppb [24]. Tulsa, OK (6 samples) 7.7-42.5 ppb [1]. Los Angeles (23 days, morning samples) 8-28 ppb [11]. 2-Methylpentane was identified, but not quantified, in the air of Pretoria, Johannesburg and Durban, South Africa [22]. Concentration measured in aircraft 350 to 500 m over Tokyo (258 samples) 0.8 ppb, mean [37]. NW England, urban samples 25.4 ppb, mean [6]. Vienna, Austria, 22.9 ppb - the concn on top of a 52 m building was 47% that at ground level [20]. Sydney, Australia (140 samples) 2.6 ppb, mean [25]. SOURCE DOMINATED: 2-Methylpentane was detected, but not quantitated, in ambient roadside air [34] and in the Allegheny Mountain Tunnel on the Pennsylvania Turnpike [13]. Natural gas facility, Rio Blanco County, CO, 59.4 ppb [1]. Refinery, Tulsa, OK (2 samples), 19.0-47.8 ppb [1]. PERSONAL AIR: 2-Methylpentane was detected in 3 of 8 personal air samples of New Jersey residents during a pilot broad spectrum analysis of exposure to chemicals conducted by the EPA's Total Exposure Assessment Methodology (TEAM) Study [40].

Food Survey Values: 2-Methylpentane has been detected as a volatile in nectarines [36].

Plant Concentrations:

Fish/Seafood Concentrations:

322

2-Methylpentane

Animal Concentrations:

Milk Concentrations:

Other Environmental Concentrations: The average (standard deviation) composition of 2-methylpentane in the exhaust of 67 motor vehicles in Sydney, Australia, was 2.3% (0.4%) by weight [26]. The percent 2-methylpentane of vehicle exhaust, gasoline vapor, petroleum refinery air, and petrochemical plant air in the Tokyo region was 3.8, 6.3, 3.7, and 1.5% [39].

Probable Routes of Human Exposure: The general population will be exposed to 2-methylpentane in ambient air, especially in areas of high traffic and at filling stations. Exposure via inhalation and dermal contact may also result from the use of glues and other products in which 2-methylpentane is contained as a solvent.

Average Daily Intake: AIR INTAKE (assume typical urban concn 17.9 ppb [31]): 1.3 mg. WATER INTAKE: Insufficient data. FOOD INTAKE: Insufficient data.

Occupational Exposure: NIOSH (NOES 1981-1983) has statistically estimated that 42 workers are potentially exposed to 2-methylpentane in the US [27]. These estimated exposures are to the pure chemical and therefore do not include exposures to petroleum products or mixed solvents. The gasoline vapor that workers are exposed to at bulk terminals and marine loading facilities contained an average of 3.5 and 4.9% 2-methylpentane by weight [12]. A survey of a high volume service station in eastern PA, whose site represented a maximum exposure to gasoline vapors found that all 15 personal, long-term samples contained 2-methylpentane; the 3 that were quantifiable ranged from 0.1 to 0.3 ppm [18]. Air concns (95% confidence limits) of 2-methylpentane measured by eight petroleum companies for service station attendants (49 measurements), transport drivers (49 measurements), and outside operators (56 measurements) were 2.01 (1.55-2.47), 1.69 (0.80-2.57), and 0.61 (0.37-0.84) mg/m^3 [29], respectively.

Body Burdens: 2-Methylpentane was detected in 3 of 12 breath samples collected in New Jersey during a pilot broad spectrum analysis of exposure to chemicals conducted by the EPA's Total Exposure Assessment Methodology (TEAM) Study [40].

REFERENCES

1. Arnts RR, Meeks SA; Atmos Environ 15: 1643-51 (1981)
2. Atkinson R; Atmos Environ 24A:1-41 (1990)
3. Bonsang B et al; J Atmos Chem 6:3-20 (1988)
4. Brodzinsky R, Singh HB; Volatile Organic Chemicals in the Atmosphere Menlo Park, CA: Atmos Sci Ctr SRI 68-02-3452 (1982)
5. Buswell JA, Jurtshuk P; Arch Microbiol 64: 215-22 (1969)
6. Colbeck I, Harrison RM; Atmos Environ 19: 1899-904 (1985)
7. Colbeck I, Harrison RM; Atmos Environ 19: 1899-904 (1985)
8. Daubert TE, Danner RP; Data Compilation Tables of Properties of Pure Compounds NY: Amer Inst for Phys Prop Data (1989)
9. Duce RA et al; Rev Geophys Space Phys 21: 921-52 (1983)
10. Graedel TE; Chemical Compounds In The Atmosphere NY, NY: Academic Press (1978)
11. Grosjean D, Fung K; J Air Pollut Control Assoc 34: 537-43 (1984)
12. Halder CA et al; Am Ind Hyg Assoc J 47: 164-72 (1986)
13. Hampton CV et al; Environ Sci Technol 16: 287-98 (1982)
14. Hawley CG; The Condensed Chemical Dictionary 10th ed NY: Van Nostrand Reinhold Co (1981)
15. Helmig D et al; Chemosphere 19: 1339-1412 (1989)
16. Hov O et al; Geophys Res Lett 11: 425-8 (1984)
17. Jamison VW et al; pp. 187-96 in Proc Int Biodeg Symp 3rd Sharpley JM, Kaplan AM, eds Essex, England (1976)
18. Kearney CA, Dunham DB; Am Ind Hyg Assoc J 47: 535-9 (1986)
19. Kungskulniti N, Edgerton SA; Chemosphere 20: 673-9 (1990)
20. Lanzerstorfer C, Puxbaum H; Water Air Soil Pollut 51: 345-55 (1990)
21. LaRegina J et al; Environ Prog 5: 18-27 (1986)
22. Louw CW et al; Atmos Environ 11: 703-17 (1977)
23. Lyman WJ et al; Handbook of Chemical Property Estimation Methods NY: McGraw-Hill Chapt 4, 5, 7, 8, 15 (1982)
24. Monson PR et al; 71st Ann Mtng APCA Paper 78-50.4 (1978)
25. Nelson PF, Quigley SM; Environ Sci Technol 16: 650-5 (1982)
26. Nelson PF, Quigley SM; Atmos Environ 18: 79-87 (1984)
27. NIOSH; National Occupational Exposure Survey (NOES) (1989)
28. Perry DL et al; Identification of Organic Compound in Industrial Effluent Discharges USEPA-600/4-79-016, NTIS PB-294794 (1979)
29. Rappaport SM et al; Appl Ind Hyg 2: 148-54 (1987)
30. Riddick JA et al; Organic Solvents 4th ed. NY: Wiley (1986)
31. Seila RL et al; Determination of C2 to C12 Ambient Air Hydrocarbons in 39 US Cities, from 1984 through 1986 USEPA/600/S3-89/058 (1989)

32. Seila RL et al; Atmospheric Volatile Hydrocarbon Composition at Five Remote Sites in NW NC USEPA-600/D-84-092, NTIS PB84-177930 (1984)

33. Sheldon LS et al; Indoor Air Quality in Public Buildings, Vol 1 USEPA-600/6-88-009a (1988)

34. Stump FD, Dropkin DL; Anal Chem 57: 2629-34 (1985)

35. Swann RL et al; Res Rev 85: 17-28 (1983)

36. Takeoka GR et al; J Agric Food Chem 36: 553-60 (1988)

37. Uno I et al; Atmos Environ 19: 1283-93 (1985)

38. USEPA; PCGEMS (Graphical Estimation Modeling System); CLOGP3 (1988)

39. Wadden RA et al; Environ Sci Technol 20: 473-83 (1986)

40. Wallace LA et al; Environ Res 35: 293-319 (1984)

41. Yanagihara S et al; pp. 472-7 in Proc 4th Int Clean Air Congress (1977)

2-Methylpentanedinitrile

SUBSTANCE IDENTIFICATION

Synonyms:

Structure:

CAS Registry Number: 4553-62-2

Molecular Formula: $C_6H_7N_2$

SMILES Line Notation: C(#N)C(CCC(#N))C

CHEMICAL AND PHYSICAL PROPERTIES

Boiling Point:

Melting Point:

Molecular Weight:

Dissociation Constants:

Log Octanol/Water Partition Coefficient: 0.28 [11]

Water Solubility: 34,810 mg/L at 25 °C [8]

Vapor Pressure: 0.0051 mm Hg at 25 °C [5]

Henry's Law Constant: 2.97 x 10^{-8} atm-m^3/mol (estimated) [9]

ENVIRONMENTAL FATE/EXPOSURE POTENTIAL

Summary: 2-Methylpentanedinitrile may be released to the environment during its industrial production and use as a solvent and intermediate for the production of other compounds. 2-Methylpentanedinitrile may also be released to the environment as a byproduct during adiponitrile production. Hydrolysis of 2-methylpentanedinitrile in water and soil should not be important, although experimental data are unavailable. No biodegradation data on 2-methylpentanedinitrile are available. The low vapor pressure of the compound indicates volatilization from dry soil should not be important. The estimated Henry's Law constant suggests that volatilization of 2-methylpentanedinitrile from water and moist soil should be negligible. Based upon an estimated Koc value, 2-methylpentanedinitrile should have very high mobility in soil and adsorption to suspended solids and sediments in water should not be important. The estimated low log BCF value indicates bioconcentration of 2-methylpentanedinitrile in aquatic organisms should be unimportant. 2-Methylpentanedinitrile should be present predominantly in the vapor phase in air. In the atmosphere in the vapor form, 2-methylpentanedinitrile should react with photochemically produced hydroxyl radicals with an estimated half-life of 13.3 days. Because of its high water solubility, the removal of 2-methylpentanedinitrile from the atmosphere by wet deposition may be important. Workers using 2-methylpentanedinitrile during the production and use of the compound are the most likely people to be exposed to 2-methylpentanedinitrile by dermal contact.

Natural Sources:

Artificial Sources: 2-Methylpentanedinitrile is produced as a byproduct during the production of adiponitrile [12]. Therefore, release of 2-methylpentanedinitrile to the environment may occur during the production of adiponitrile. Since 2-methylpentanedinitrile is used as a solvent and intermediate for the manufacture of other chemicals [4], environmental release may occur during these uses.

Terrestrial Fate: Based on expected slow hydrolysis [3] of 2-methylpentanedinitrile in water, hydrolysis should be unimportant in soil. The log Koc of 1.03 estimated from log Kow indicates that 2-methylpentanedinitrile should be very mobile in soil [13]. The vapor pressure suggests volatilization from dry soil should not be rapid and the low Henry's Law constant suggests low volatilization from moist soil.

Aquatic Fate: Based on hydrolysis rates of a monosubstituted organic nitrile, hydrolysis of 2-methylpentanedinitrile should not be important in environmental waters [3]. From the estimated Henry's Law constant and an estimation method [8], the volatilization half-life of 2-methylpentanedinitrile from a model river has been estimated to be 1259 days [8]. Therefore, volatilization from water should not be important. The estimated log Koc indicates that adsorption of 2-methylpentanedinitrile to suspended solids and sediments in water should not be important [13]. The estimated bioconcentration factor of 0.2 [8] suggests that bioconcentration of 2-methylpentanedinitrile in aquatic organisms should be unimportant.

Atmospheric Fate: Based on the vapor pressure, 2-methylpentanedinitrile should be found predominantly in the vapor phase in the atmosphere [6]. In the atmosphere, vapor phase 2-methylpentanedinitrile should react with photochemically produced hydroxyl radicals. Based on an estimation method [2,10], the rate constant for the reaction of 2-methylpentanedinitrile with hydroxyl radicals has been estimated to be 1.208×10^{-12} cm^3/molecule-sec [2]. Assuming the daily average concn of hydroxyl radicals in the atmosphere as $5 \times 10^{+5}$ per cm^3 [1], the half-life for this reaction has been estimated to be 13.3 days. Based on the high water solubility and a regression equation [8], the removal of atmospheric 2-methylpentanedinitrile by wet deposition should be important.

Biodegradation:

Abiotic Degradation: Organic nitriles are generally susceptible to chemical hydrolysis under acid and basic conditions; but based on very slow hydrolysis of cyano groups in organic nitriles [3], the hydrolysis of 2-methylpentanedinitrile in water and moist soil should not be important. The rate constant for the reaction of vapor phase 2-methylpentanedinitrile with hydroxyl radicals has been estimated to be 1.208×10^{-12} cm^3/molecule-sec by an estimation method [2,10]. Assuming the daily average concn of hydroxyl radicals in the atmosphere of $5 \times 10^{+5}$ per cm^3 [1], the half-life for this reaction has been estimated to be 13.3 days.

Bioconcentration: Based on the estimated log Kow and a regression equation [8], the bioconcentration factor for 2-methylpentanedinitrile has been estimated to be 5.8. Therefore, bioconcentration of 2-methylpentanedinitrile in aquatic organisms should not be important [7].

2-Methylpentanedinitrile

Soil Adsorption/Mobility: Based on the estimated log Kow value and a regression equation [8], a log Koc value of 34 has been estimated for 2-methylpentanedinitrile. This value indicates that 2-methylpentanedinitrile should be highly mobile in soil [13] and the adsorption of 2-methylpentanedinitrile to suspended solids and sediments in water should be unimportant.

Volatilization from Water/Soil: A Henry's Law constant of 2.97×10^{-8} atm-m^3/mol has been estimated for 2-methylpentanedinitrile by the bond contribution method [9]. Based on this value, the volatilization of 2-methylpentanedinitrile from a model river of depth 1 m, flowing at 1 m/sec and a wind speed of 3 m/sec has been estimated to be 1259 days [8]. Therefore, volatilization should not be important in water.

Water Concentrations:

Effluent Concentrations:

Sediment/Soil Concentrations:

Atmospheric Concentrations:

Food Survey Values:

Plant Concentrations:

Fish/Seafood Concentrations:

Animal Concentrations:

Milk Concentrations:

Other Environmental Concentrations:

Probable Routes of Human Exposure: Since 2-methylpentanedinitrile is a liquid at ambient temperatures with low vapor pressure, dermal exposure to this compound may be the most probable route of exposure during its use as a solvent and intermediate in the production of other compounds [4].

2-Methylpentanedinitrile

Average Daily Intake:

Occupational Exposure: Workers using 2-methylpentanedinitrile as solvent and intermediate in the production of other chemicals [4] are the most likely people for exposure to this compound.

Body Burdens:

REFERENCES

1. Atkinson R; Chem Rev 85: 69-201 (1985)
2. Atkinson R; Environ Toxicol Chem 7: 435-42 (1988)
3. Brown SL et al; Research Program on Hazard Priority Ranking of Manufactured Chemicals (Chemicals 41-60) SRI. Menlo Park CA p. 85 NTIS PB-263-163 (1975)
4. Chemocyclopedia 90; The Manual of Commercially Available Chemicals 8: 97 American Chemical Society, Washington DC (1990)
5. Daubert TE, Danner RP; Physical and Thermodynamic Properties of Pure Chemicals Design Inst Phys Prop Data, Amer Inst Chem Eng, Vol 2, NY: Hemisphere Publ Corp (1991)
6. Eisenreich SJ et al; Environ Sci Technol 15: 30-38 (1981)
7. Kenaga EE; Ecotoxicol Environ Safety 4: 26-38 (1980)
8. Lyman WJ et al; Handbook of Chemical Property Estimation Methods NY: McGraw-Hill (1982)
9. Meylan WM, Howard PH; Environ Toxicol Chem 10: 1283-93 (1991)
10. Meylan WM, Howard PH; Chemosphere 26:2293-2299 (1993)
11. Meylan W, Howard PH; J Pharm Sci 84: 83-92 (1995)
12. Offermanns H et Kirk-Othmer Encycl Chem Technol 3rd ed. NY: John Wiley & Sons 24: 65 (1984)
13. Swann RL; Res Rev 85: 17-28 (1983)

Methyl Propionate

SUBSTANCE IDENTIFICATION

Synonyms:

Structure:

CAS Registry Number: 554-12-1

Molecular Formula: $C_4H_8O_2$

SMILES Notation: O=C(OC)CC

CHEMICAL AND PHYSICAL PROPERTIES

Boiling Point: 79.7 °C

Melting Point: -87 °C

Molecular Weight: 88.10

Dissociation Constants:

Log Octanol/Water Partition Coefficient: 0.84 [5]

Water Solubility: 62,370 mg/L at 25 °C [9]

Vapor Pressure: 84.04 mm Hg at 25 °C [6]

Henry's Law Constant: 1.74×10^{-4} atm-m^3/mol at 25 °C [3]

ENVIRONMENTAL FATE/EXPOSURE POTENTIAL

Summary: Methyl propionate occurs naturally in some fruits such as kiwi fruit. It is released directly to the atmosphere by evaporation in its use as a solvent in paints, lacquers, varnishes and other coating compositions. If released to the atmosphere, it will degrade by reaction with photochemically produced hydroxyl radicals (estimated half-life of 16 days). If released to soil or water, methyl propionate is expected to biodegrade. A screening study has demonstrated that methyl propionate is biodegradable. Aqueous hydrolysis will be important only in very alkaline environmental media (pH > 8.5). Methyl propionate may leach readily in soils based upon an estimated Koc of 10. Evaporation from solid surfaces will be an important physical transport mechanism to air. The general population is exposed to methyl propionate through oral consumption since it occurs naturally in some fruits, such as kiwi, and is used as a food flavoring additive. Inhalation exposure will occur from its use as a solvent.

Natural Sources: Methyl propionate has been detected as a volatile component of kiwi fruit [2].

Artificial Sources: Methyl propionate's use as a solvent in paints, lacquers, varnishes and other coating compositions [8] will release it directly to the atmosphere through evaporation.

Terrestrial Fate: Biodegradation is expected to be an important degradation process for methyl propionate in soil. One biodegradation screening study has demonstrated that methyl propionate is readily biodegradable [11]. Based upon an estimated Koc of 10, methyl propionate is expected to leach readily in soil [10]. Methyl propionate's vapor pressure suggests that the compound will evaporate rapidly from solid surfaces.

Aquatic Fate: Biodegradation is expected to be an important degradation process for methyl propionate in water. One biodegradation screening study has demonstrated that methyl propionate is readily biodegradable [11]. Volatilization from water may be an important transport process. The volatilization half-lives from a model environmental river (1 m deep) and model pond have been estimated to be 7.5 hr and 3.5 days, respectively [10,15]. Aquatic bioconcentration and adsorption to sediment are not expected to be important. Aqueous hydrolysis will be important only in very alkaline (pH > 8.5) environmental waters.

Atmospheric Fate: Based upon the vapor pressure, methyl propionate is expected to exist entirely in the vapor phase in the ambient atmosphere [7]. It will degrade in the ambient atmosphere by reaction with photochemically produced hydroxyl radicals with an estimated half-life of about 16 days [1]. Physical removal from air via wet deposition (dissolution into clouds, rainfall) is likely since it is relatively soluble in water.

Biodegradation: In Warburg respirometer studies, methyl propionate was readily biooxidized by activated sludges from three Ohio treatment facilities over 24-hr incubation periods [11].

Abiotic Degradation: The rate constant for the vapor-phase reaction of methyl propionate with photochemically produced hydroxyl radicals has been experimentally determined to be 1.0×10^{-12} cm^3/molecule-sec at 25 °C, which corresponds to an atmospheric half-life of about 16 days at an atmospheric concn of $5 \times 10^{+5}$ hydroxyl radicals per cm^3 [1]. The aqueous hydrolysis rate constant has been estimated to be 0.108 L/mol-sec at 25 °C [12], which corresponds to aqueous hydrolysis half-lives of 2.03 yrs, 74.3 days and 7.43 days at pH 7, 8 and 9, respectively [16]. The rate constant for the reaction between photochemically produced hydroxyl radicals in water and methyl propionate is $4.5 \times 10^{+8}$ 1/mol-sec [4]; assuming that the concn of hydroxyl radicals in brightly sunlit natural water is 1×10^{-17} M [12], the half-life would be about 1780 days of continuous (24 hr/day) sunlight.

Bioconcentration: Based upon the measured water solubility, the BCF for methyl propionate can be estimated to be 1.2 from a recommended regression-derived equation [10]. This indicates that bioconcentration in aquatic organisms is not important.

Soil Adsorption/Mobility: Based upon the measured water solubility, the Koc for methyl propionate can be estimated to be 10 from a linear regression-derived equation [10]. This Koc estimation indicates very high soil mobility [14].

Volatilization from Water/Soil: Methyl propionate's Henry's Law constant indicates that volatilization from environmental waters will occur, but may not be rapid [10]. Based on the Henry's Law constant, the volatilization half-life from a model river (1 m deep flowing 1 m/sec with a wind velocity of 3 m/sec) can be estimated to be about 7.5 hr [10]. The volatilization half-life from a model environmental pond (2 m deep) can be estimated to be about 3.5 days [15]. Methyl propionate has a relatively high vapor pressure, which suggests that rapid evaporation from dry surfaces can occur.

Water Concentrations:

Effluent Concentrations:

Sediment/Soil Concentrations:

Atmospheric Concentrations:

Food Survey Values: Methyl propionate concns of 0.2-0.9% were detected in the volatile components of mature and ripe kiwi fruit collected in New Zealand and Australia [2].

Plant Concentrations: Methyl propionate has been detected as a volatile component of kiwi fruit [2].

Fish/Seafood Concentrations:

Animal Concentrations: A methyl propionate concn of 4.23 ug/g was detected in a mussel (*Mytilus edulis*) sample collected at the Oarai Coast in Ibaraki, Japan, on July 31, 1985 [17].

Milk Concentrations:

Other Environmental Concentrations:

Probable Routes of Human Exposure: The general population is exposed to methyl propionate through oral consumption since it occurs naturally in some fruits, such as kiwi [2], and is used as a food flavoring additive [8]. Inhalation exposure will occur from its use as a solvent in paints, varnishes and other coating compositions [8]. Occupational exposure can occur through inhalation of vapor and dermal contact.

Average Daily Intake:

Occupational Exposure: NIOSH (NOES 1981-1983) has statistically estimated that 1,821 workers are potentially exposed to methyl propionate in the US [13].

Body Burdens:

REFERENCES

1. Atkinson R; J Phys Chem Ref Data. Monograph No. 1 (1989)
2. Bartley JP, Schwede AM; J Agric Food Chem 37: 1023-5 (1989)
3. Buttery RG et al; J Agric Food Chem 17: 385-9 (1969)
4. Buxton GV et al; J Phys Chem Ref Data 17: 704 (1988)
5. Catz P, Friend DR; Int J Pharm 55:1723 (1989)
6. Daubert TE, Danner RP; Physical and Thermodynamic Properties of Pure Chemicals: Data Compilation, NY: Hemisphere Pub Corp (1989)
7. Eisenreich SJ et al; Environ Sci Technol 15: 30-8 (1981)
8. Hawley GG; The Condensed Chemical Dictionary 10th ed. NY: Van Nostrand Reinhold p. 689 (1981)
9. Hine J, Mookerjee PK; J Org Chem 40: 292-8 (1975)
10. Lyman WJ et al; Handbook of Chemical Property Estimation Methods Washington, DC: Amer Chem Soc (1990)
11. Malaney GW, Gerhold RM; pp. 249-57 in Proc 17th Ind Waste Conf, Purdue Univ, Ext Ser 112 (1962)
12. Mill T et al; Science 207: 886-7 (1980)
13. NIOSH; National Occupational Exposure Survey (NOES) (1983)
14. Swann RL et al; Res Rev 85: 23 (1983)
15. USEPA; EXAMS II Computer Simulation (1987)
16. USEPA; PCGEMS Graphical Exposure Modeling System. PCHYDRO (1991)
17. Yasuhara A, Morita M; Chemosphere 16: 2559-65 (1987)

2-Methylpropyl Isobutyrate

SUBSTANCE IDENTIFICATION

Synonyms:

Structure:

CAS Registry Number: 97-85-8

Molecular Formula: $C_8H_{16}O_2$

SMILES: O=C(OCC(C)C)C(C)C

CHEMICAL AND PHYSICAL PROPERTIES

Boiling Point: 148.6 °C

Melting Point: -80.66 °C

Molecular Weight: 144.21

Dissociation Constants:

Log Octanol/Water Partition Coefficient: 2.68 [12]

Water Solubility: 1000 mg/L at 20 °C [6]

Vapor Pressure: 4.33 mm Hg at 25 °C [3]

Henry's Law Constant: 8.22 x 10^{-4} atm-m^3/mol (calculated from vapor pressure and water solubility)

ENVIRONMENTAL FATE/EXPOSURE POTENTIAL

Summary: 2-Methylpropyl isobutyrate occurs naturally in fruits such as strawberries, cantaloupes and muskmelons. It is released directly to the atmosphere by evaporation through its use as a solvent in lacquers, thinners and other coating compositions. If released to the atmosphere, it will degrade by reaction with photochemically produced hydroxyl radicals (estimated half-life of 3.4 days). If released to soil or water, 2-methylpropyl isobutyrate is expected to degrade via biodegradation. Aqueous hydrolysis will be important only in very alkaline environmental media (pH > 8.5). 2-Methylpropyl isobutyrate may leach readily in soils based upon an estimated Koc of 98; however, the importance of leaching will be lessened if rapid biodegradation occurs. Evaporation from solid surfaces will be an important physical transport mechanism to air. The general population is exposed to 2-methylpropyl isobutyrate through oral consumption since it occurs naturally in some fruits. Inhalation exposure will occur from its use as a solvent.

Natural Sources: 2-Methylpropyl isobutyrate has been identified as a volatile constituent of strawberries, cantaloupes, and muskmelons [4,20,21]. It has also been identified in the essential oils of *Heracleum persicum* and *Chamaemelum nobile* plants [2,15].

Artificial Sources: 2-Methylpropyl isobutyrate's use as a solvent in lacquers, thinners and other coating compositions [8,9] will release it directly to the atmosphere through evaporation.

Terrestrial Fate: By analogy to other aliphatic esters [11,14], biodegradation is expected to be an important degradation process for 2-methylpropyl isobutyrate in soil. Based upon an estimated Koc of 98, 2-methylpropyl isobutyrate is expected to leach readily in soil. If rapid biodegradation occurs, the importance of leaching will be lessened. 2-Methylpropyl isobutyrate's vapor pressure suggests that the compound will evaporate from solid surfaces.

Aquatic Fate: By analogy to other aliphatic esters [11,14], biodegradation is expected to be an important degradation process for 2-methylpropyl isobutyrate in water. Volatilization from water may be an important transport process. The volatilization half-lives from a model environmental river (1 m deep) and model pond have been estimated to be 4.8 hr and 3 days, respectively [10,17]. Aquatic bioconcentration and adsorption to sediment are not expected to be important. Aqueous hydrolysis will be important only in very alkaline (pH > 8.5) environmental waters.

2-Methylpropyl Isobutyrate

Atmospheric Fate: Based upon the vapor pressure, 2-methylpropyl isobutyrate is expected to exist almost entirely in the vapor phase in the ambient atmosphere [5]. It will degrade in the ambient atmosphere by reaction with photochemically produced hydroxyl radicals with an estimated half-life of about 3.4 days [1].

Biodegradation: Data specific to the biodegradation of 2-methylpropyl isobutyrate were not located; however, similar aliphatic esters have been shown to biodegrade readily [11,14]; therefore, 2-methylpropyl isobutyrate is expected to biodegrade readily.

Abiotic Degradation: The rate constant for the vapor-phase reaction of 2-methylpropyl isobutyrate with photochemically produced hydroxyl radicals has been experimentally determined to be 4.73×10^{-12} cm^3/molecule-sec at 25 °C, which corresponds to an atmospheric half-life of about 3.4 days at an atmospheric concn of $5 \times 10^{+5}$ hydroxyl radicals per cm^3 [1]. The aqueous hydrolysis rate constant has been estimated to be 0.02383 L/mol-sec at 25 °C [18], which corresponds to aqueous hydrolysis half-lives of 9.22 yrs, 336 days and 33.6 days at pH 7, 8 and 9, respectively [18].

Bioconcentration: Based upon the water solubility, the BCF for 2-methylpropyl isobutyrate can be estimated to be 12.2 from a recommended linear regression-derived equation [10]. This estimated BCF indicates that bioconcentration in aquatic organisms is not important.

Soil Adsorption/Mobility: Based upon the water solubility, the Koc for 2-methylpropyl isobutyrate can be estimated to be 98 from a linear regression-derived equation [10]. This Koc estimation indicates high soil mobility [16].

Volatilization from Water/Soil: Based upon the Henry's Law constant, evaporation from environmental waters is important, but may not be rapid [10]. Based on the Henry's Law constant, the volatilization half-life from a model river (1 m deep flowing 1 m/sec with a wind velocity of 3 m/sec) can be estimated to be about 4.8 hours [10]. The volatilization half-life from a model environmental pond (2 m deep) can be estimated to be about 3 days [17]. 2-Methylpropyl isobutyrate has a relatively high vapor pressure, which suggests that evaporation from dry surfaces will occur.

Water Concentrations: SURFACE WATER: 2-Methylpropyl isobutyrate has been detected in the aquatic ecosystem of the Western Basin of Lake Superior [7]; concns, sampling dates, and sample types (water, whole water or sediment) were not reported.

Effluent Concentrations:

Sediment/Soil Concentrations:

Atmospheric Concentrations:

Food Survey Values: 2-Methylpropyl isobutyrate has been identified as a volatile constituent of strawberries, cantaloupes and muskmelons [4,20,21].

Plant Concentrations: 2-Methylpropyl isobutyrate has been identified as a volatile constituent of strawberries, cantaloupes and muskmelons [4,20,21]. It has also been identified in the essential oils of *Heracleum persicum* and *Chamaemelum nobile* plants [2,15].

Fish/Seafood Concentrations:

Animal Concentrations:

Milk Concentrations:

Other Environmental Concentrations:

Probable Routes of Human Exposure: 2-Methylpropyl isobutyrate has been identified as a volatile constituent of strawberries, cantaloupes and muskmelons [4,20,21]; therefore, the general population will be exposed through ingestion of fruits and other foods that contain 2-methylpropyl isobutyrate as a natural constituent. Inhalation exposure will occur from its use as a solvent in lacquers, thinners and other coating compositions. Occupational exposure can occur through inhalation of vapor and dermal contact.

Average Daily Intake:

2-Methylpropyl Isobutyrate

Occupational Exposure: NIOSH (NOES 1981-1983) has statistically estimated that 47,530 workers are potentially exposed to 2-methylpropyl isobutyrate in the US [13]. Mean air levels of 0.1-0.4 ppm 2-methylpropyl isobutyrate were detected inside three US factories involved in spray painting and spray gluing operations [19]; the highest sample concn, highest time weighted average (TWA) and average TWA were 5.2, 5.0, and 1.7 ppm, respectively [19].

Body Burdens:

REFERENCES

1. Atkinson R; J Inter Chem Kinet 19: 799-828 (1987)
2. Damiani P et al; Fitoterapia 54: 213-22 (1983)
3. Daubert TE, Danner RP; Physical and Thermodynamic Properties of Pure Chemicals: Data Compilation NY: Hemisphere Pub Corp (1989)
4. Dirinck P et al; Anal Volatiles: Methods Appl, Proc - Int Workshop 1983: 381-400 (1984)
5. Eisenreich SJ et al; Environ Sci Technol 15: 30-8 (1981)
6. Flick EW; Industrial Solvents Handbook 4th ed Park Ridge NJ: Noyes Data Corp p. 815 (1991)
7. Great Lakes Water Quality Board; An Inventory of Chemical Substances Identified in the Great Lakes Ecosystem. Volume 1 - Summary. Report to the Great Lakes Water Quality Board. Windsor Ontario, Canada p. 75 (1983)
8. Hawley GG; The Condensed Chemical Dictionary 10th ed. NY: Van Nostrand Reinhold p. 576 (1981)
9. Kuney JH; Chemcyclopedia 1991 Washington DC: Amer Chem Soc p. 83 (1990)
10. Lyman WJ et al; Handbook of Chemical Property Estimation Methods Washington, DC: Amer Chem Soc (1990)
11. Malaney GW, Gerhold RM; pp. 249-57 in Proc 17th Ind Waste Conf, Purdue Univ Ext Ser 112 (1962)
12. Meylan W, Howard PH; J Pharm Sci 84: 83-92 (1995)
13. NIOSH; National Occupational Exposure Survey (NOES) (1983)
14. Price KS et al; J Water Pollut Control Fed 46: 63-77 (1974)
15. Scheffer JJC et al; Planta Med 50: 56-60 (1984)
16. Swann RL et al; Res Rev 85: 23 (1983)
17. USEPA; EXAMS II Computer Simulation (1987)
18. USEPA; PCGEMS Graphical Exposure Modeling System. PCHYDRO (1991)
19. Whitehead LW et al; Amer Ind Hyg Assoc J 45: 767-772 (1984)
20. Yabumoto K et al; J Food Sci 42: 32-7 (1977)
21. Yamashita I et al; Phytochemistry 15: 1633-7 (1976)

Methylpyridine

SUBSTANCE IDENTIFICATION

Synonyms:

Structure:

CAS Registry Number: 1333-41-1

Molecular Formula: C_6H_7N

SMILES Notation: n1c(C)cccc1 (2-isomer), n1cc(C)ccc1 (3-isomer), n1ccc(C)cc1 (4-isomer)

CHEMICAL AND PHYSICAL PROPERTIES

Boiling Point:

Melting Point:

Molecular Weight: 93.10

Dissociation Constants: pKa = 5.63 (3-isomer), 5.98 (4-isomer) [10]

Log Octanol/Water Partition Coefficient: 1.11 (2-isomer), 1.20 (3-isomer) 1.22 (4-isomer) [13]

Water Solubility: miscible [10]

Vapor Pressure: 6.05 (3-isomer), 5.77 (4-isomer) mm Hg at 25 °C [3]

Henry's Law Constant: 7.41 x 10^{-7} (3-isomer), 7.07 x 10^{-7} (4-isomer) atm-m^3/mol at 25 °C [1]

ENVIRONMENTAL FATE/EXPOSURE POTENTIAL

Summary: Methylpyridine may be released to the environment via effluents at sites where it is produced or used as a chemical intermediate in medicine, agriculture and industry. It is also released to the environment with the manufacture and use of coal-derived liquid fuels and during the disposal of coal liquefication and gasification waste byproducts. In addition, oil refineries, landfills and municipal waste incinerators also release methylpyridine to the environment. Based upon pKa's, methylpyridine and its conjugate acids will exist in environmental media in varying proportions that are pH dependent. Ions generally do not volatilize. The Henry's Law constants indicate that volatilization of methylpyridine from environmental waters and moist soil should be extremely slow. Yet, methylpyridine should evaporate from dry surfaces, especially when present in high concns such as in spill situations. In aquatic systems, the bioconcentration of methylpyridine is not expected to be an important fate process. A low Koc indicates methylpyridine should not partition from the water column to organic matter contained in sediments and suspended solids, and it should be highly mobile in soil. Limited monitoring data has shown it can leach to ground water. Biodegradation is likely to be the most important removal mechanism of methylpyridine from aerobic soil and water. Both an aerobic river die-away test and an aerobic soil grab sample study demonstrated rapid biodegradation of methylpyridine after acclimation. In the atmosphere, methylpyridine is expected to exist almost entirely in the vapor phase and reaction with photochemically produced hydroxyl radicals should be important (estimated half-life of 11 days). The complete miscibility of methylpyridine in water suggests that physical removal from air by wet deposition (rainfall, dissolution in clouds) may occur. The most probable human exposure would be occupational exposure, which may occur through dermal contact or inhalation at workplaces where coal-derived or petroleum fuels are produced or used. The most common nonoccupational exposure is likely to result from either passive or active inhalation of cigarette smoke. Limited monitoring data indicates that other nonoccupational exposures can occur from the ingestion of certain foods and contaminated drinking water supplies.

Natural Sources:

Methylpyridine

Artificial Sources: Methylpyridine is produced from the pyrolysis of coal or synthetically by reactions between aldehydes and ketones with ammonia [10]. Methylpyridine may be released to the environment via effluents at places where it is produced or used as an intermediate in medicine, agriculture and industry [10]. Methylpyridine is also released to the environment via effluents from the manufacture and use of coal-derived liquid fuels and the disposal of coal liquefication and gasification waste byproducts [4,9,20,26]. Oil refineries [25], landfills [5,6] and municipal waste incinerators [16] may release methylpyridine to the environment. In addition, methylpyridine can be released to ambient air via cigarette smoke [12].

Terrestrial Fate: Based upon pKa's, methylpyridine and its conjugate acids will exist in soils in varying proportions that are pH dependent. Ions generally do not volatilize. The Henry's Law constants indicate that volatilization of methylpyridine from moist soil should be extremely slow [19]. Yet, methylpyridine should evaporate from dry surfaces, especially when present in high concn such as in spill situations. An estimated Koc of 107 to 110 indicates methylpyridine should be highly mobile in soil [28]; and limited monitoring data has shown it may leach to ground water [5,6,27]. Biodegradation is likely to be the most important removal mechanism of methylpyridine from aerobic soil. An aerobic soil grab sample study demonstrated rapid biodegradation of methylpyridine [23].

Aquatic Fate: Based upon the pKa's, methylpyridine and its conjugate acids will exist among environmental waters in varying proportions that are pH dependent. Under near neutral and acidic conditions, the ratio of methylpyridine to its conjugate acid should increase with increasing pH [19]. The miscibility of methylpyridine in water suggests that volatilization, adsorption and bioconcentration are not important fate processes. This is supported by Henry's Law constants, which indicate that volatilization of methylpyridine from environmental waters should be extremely slow [19]. An estimated Koc of 107 to 110 [19] indicates methylpyridine should not partition from the water column to organic matter [28] contained in sediments and suspended solids; and an estimated bioconcentration factor (log BCF) of 5 indicates methylpyridine should not bioconcentrate in aquatic organisms [19]. Aerobic biodegradation is likely to be the most important removal mechanism of methylpyridine from aquatic systems. An aerobic river die-away test also showed that methylpyridine biodegraded rapidly after acclimation in highly polluted natural waters [8].

Atmospheric Fate: Based on the vapor pressures, methylpyridine is expected to exist almost entirely in the vapor phase in ambient air [7]. In the atmosphere, vapor-phase reactions with photochemically produced hydroxyl radicals may be important. The rate constant for methylpyridine was estimated to be 1.43×10^{-12} cm^3/molecule-sec at 25 °C, which corresponds to an atmospheric half-life of about 11 days at an atmospheric concn of 5 $\times$ 10^{+5} hydroxyl radicals per cm^3 [2]. The complete miscibility of methylpyridine in water suggests that physical removal from air by wet deposition (rainfall, dissolution in clouds) may occur.

Biodegradation: An aerobic biological screening study, which utilized a 10 mg/L yeast extract and an Aeric Ochraqualf soil for inocula, indicates that 3-methylpyridine is not readily biodegraded and 4-methylpyridine is readily biodegradable [24]. At 24 °C and a pH of 7, less than 1% of an initial concn of 12.7 ppm of 3-methylpyridine was mineralized within 30 days, as evidenced via the release of inorganic nitrogen [24]. For 4-methylpyridine, 68% of 15 ppm was mineralized within 24 days [24]. An aerobic soil grab sample study demonstrated rapid biodegradation of both 3- and 4-methylpyridine [23]. Methylpyridines were added to Fincastle silt loam (Aeric Ochraqualf) with a pH of 6.7 and incubated at 25 °C [23]. Within 32 days, 69.3% of the available nitrogen from 3-methylpyridine was released to inorganic forms [23]. Within 16 and 32 days, 71.7% and 100%, respectively, of the available nitrogen from 4-methylpyridine was released to inorganic forms [23]. Respective sterilized controls lost 11.7% and 21.8% of the starting materials to volatilization; but, did not release inorganic nitrogen [23]. An aerobic river die-away test also showed that methylpyridine biodegraded rapidly after acclimation in highly polluted natural waters maintained at 20 °C [8]. After 14 and 18 day acclimation periods, 100% of the original concn of 1 ppm of methylpyridine was removed within 2 and 4 days from the Ohio and Little Miami River waters, respectively [8]. In ground water samples collected near an oil shale facility (elevated alkylpyridine levels) no degradation occurred for 10 days after soil inoculum was added; 3% remained at 17 days and <1% at 24 days. Without soil inoculum, degradation was much slower and 19% remained at 31 days [22]. Methylpyridines are only very slowly degraded in sulfur-reducing and methanogenic ground water aquifer slurries; no degradation occurred after one month. The fastest reacting isomer (2-methyl) had 46% and <8% remaining under sulfate-reducing conditions at 3 and 8 months, respectively [17].

Abiotic Degradation: The rate constant for the vapor-phase reaction of methylpyridine with photochemically produced hydroxyl radicals has been estimated to be 1.43×10^{-12} cm^3/molecule-sec at 25 °C, which corresponds to an atmospheric half-life of about 11 days at an atmospheric concn of $5 \times 10^{+5}$ hydroxyl radicals per cm^3 [2].

Bioconcentration: Because methylpyridine is miscible in water, bioconcentration in aquatic systems is not expected to be an important fate process. Based upon the log Kows, a bioconcentration factor (log BCF) of 5 for methylpyridine has been calculated using a recommended regression-derived equation [19]. This BCF value indicates that methylpyridine should not bioconcentrate in aquatic organisms.

Soil Adsorption/Mobility: Because methylpyridine is miscible in water, soil adsorption is not expected to be an important fate process. Based on the log Kow, a Koc of 107 to 110 has been calculated for methylpyridine using a recommended regression-derived equation [19]. This range of Koc values indicates methylpyridine will be highly mobile in soil [28], and it should not partition from the water column to organic matter contained in sediments and suspended solids. However, based upon the pKa's, methylpyridine and its conjugate acids will exist among environmental media in varying proportions that are pH dependent.

Volatilization from Water/Soil: Based on the pKa's, methylpyridine and its conjugate acids will exist among environmental media in varying proportions that are pH dependent. Ions are not expected to volatilize from water. Because methylpyridine is miscible in water and based upon the Henry's Law constants, volatilization of methylpyridine from natural bodies of water and moist soils is also not expected to be an important fate process [19]. Yet, based upon the vapor pressures, methylpyridine should evaporate from dry surfaces, especially when present in high concns such as in spill situations.

Water Concentrations: DRINKING WATER: Methylpyridine was listed as a contaminant found in drinking water for a survey of US cities including

Pomona, Escondido, Lake Tahoe and Orange Co, CA, and Dallas, Washington, DC, Cincinnati, Philadelphia, Miami, New Orleans, Ottumwa, IA, and Seattle [18]. GROUND WATER: 3- and 4-Methylpyridine were detected in 3 of 3 ground water samples near a coal gasification site near Hoe Creek in northeastern WY at combined concns of 0.69, 51 and 34 ppb [27]. Leachate from a landfill in Norman, OK, was reported to contaminate nearby ground water supplies with unspecified isomers of methylpyridine [5,6].

Effluent Concentrations: The 2-, 3- and 4-isomers of methylpyridine were detected in 6 of 18 wastewater effluents from energy related processes [20]. Retort water from a shale oil processing facility in DeBeque, CO, contained 2- and 3-methylpyridine isomers at average concn of 36 ppb [20]. The process water at a coal gasification facility in Gillette, WY contained the 2- and 3-methylpyridine at an average concn of 49 ppb [20]. An underground coal gasification site contained 2-methylpyridine at an average concn of 120 ppb [20]. Wastewater from coal gasification at the Grand Fork's Energy Technology Center, ND, was reported to contain 3- and 4-methylpyridine at an estimated combined concn of 2.53 mg/L [9]. Wastewater effluent from a shale oil facility in Queensland, Australia, was shown to contain 3- and 4-methylpyridine in a combined concn of 4 mg/L [4]. Reactor tar from a coal gasification plant contained unidentified isomers of methylpyridine at a concn ranging from 5.5 to 12.4 mg/g [26]. The dissolved air flotation effluent of a Class B oil refinery contained unidentified isomers of methylpyridine at a concn less than 1 ng/g [25]. Unspecified isomers of methylpyridine were detected in 5 and 3 of 63 industrial wastewater effluents at concns ranging from 10 to 100 ug/L and greater than 100 ug/L, respectively [21]. Unspecified isomers of methylpyridine were detected in the ground water leachate of a landfill in Norman, OK [5,6]. Methylpyridine was listed as a component of grate ash from municipal waste incinerators [16].

Sediment/Soil Concentrations:

Atmospheric Concentrations: URBAN: Methylpyridine was not detected in the air of downtown Boulder, CO, in November 1982 [14]. RURAL: Methylpyridine was not detected in air from an undeveloped location in CO in November 1982 [14]. SOURCE DOMINATED: In November 1982, 3- and 4-methylpyridine were detected in the air outside an oil shale wastewater facility of Occidental Oil Shale Inc. at Logan Wash, CO, at a combined average concn of 8 ug/m^3 [14].

Food Survey Values: 2-, 3- and 4-Methylpyridine were identified as a volatile component of boiled beef [11] and fried bacon [15].

Plant Concentrations:

Fish/Seafood Concentrations:

Animal Concentrations:

Milk Concentrations:

Other Environmental Concentrations: Methylpyridine was detected in cigarette smoke [12].

Probable Routes of Human Exposure: The most probable route of human exposure to 3-methylpyridine is by inhalation, dermal contact and ingestion. Cigarette smokers are likely to inhale methylpyridine [12], and atmospheric workplace exposures have been documented [14]. Drinking water supplies [18] and certain foods [11,15] have been shown to contain methylpyridine.

Average Daily Intake:

Occupational Exposure: The most probable human exposure to methylpyridine would be occupational exposure, which may occur through dermal contact or inhalation at places where it is produced or used. A 1982 study showed 3- and 4-methylpyridine were emitted to the air from wastewaters at a shale oil facility exposing inside workers to a combined average concn of 35 ug/m^3 [14]. Nonoccupational exposures are likely to occur among cigarette smokers [12] and populations with contaminated drinking water supplies [18]; or via the ingestion of certain food [11,15].

Body Burdens:

Methylpyridine

REFERENCES

1. Andon RJL et al; J Chem Soc, p. 3188-96 (1954)
2. Atkinson R; Intern J Chem Kin 19: 799-828 (1987)
3. Chao J et al; J Phys Chem Ref Data 12: 1033-63 (1983)
4. Dobson KR et al; Water Res 19: 849-56 (1985)
5. Dunlap WJ et al; Organic Pollutants Contributed to Groundwater by a Landfill USEPA-600/9-76-004 p. 96-110 (1976)
6. Dunlap WJ et al; Identif Anal Org Pollut 1: 453-77 (1975)
7. Eisenreich SJ et al; Environ Sci Technol 15: 30-8 (1981)
8. Ettinger LA et al; Indus Eng Chem 46: 791-3 (1954)
9. Giabbai, MF et al; Intern J Environ Anal Chem 20: 113-29 (1985)
10. Goe GL; Kirk-Othmer Encycl Chem Tech 3rd NY: Wiley Interscience 19: 454-83 (1982)
11. Golovnya RV, et al; Chem Senses Flavour 4: 97-105 (1979)
12. Guerin MR, Buchanan MV; Environmental Exposure to N-Aryl Compounds, Carcinogenic and Mutagenic Responses to Aromatic Amines and Nitroarenes p. 37-45 (1988)
13. Hansch C, Leo AJ; Medchem Project Issue No.26 Pomona College, Claremont CA (1985)
14. Hawthorne SB, Sievers RE; Environ Sci Technol 18: 483-90 (1984)
15. Ho CT et al; J Agric Food Chem 31: 336-42 (1983)
16. Junk GA, Ford CS; Chemosphere 9: 187-230 (1980)
17. Kuhn EP, Suflita JM; Environ Toxicol Chem 8:1149-58 (1989)
18. Lucas SV; GC/MS Anal of Org in Drinking Water Concentrates and Advanced Treatment Concentrates Vol 1 USEPA-600/1-84-020A (NTIS PB85-128239) p. 397 (1984)
19. Lyman WJ et al; Handbook of Chemical Property Estimation Methods NY: McGraw-Hill (1982)
20. Pelizzari E et al; ASTM Spec Tech Publ STP 686: 256-74 (1979)
21. Perry DL et al; Iden of Org Compounds in Ind Effluent Discharges USEPA-600/4-79-016 (NTIS PB-294794) p. 230 (1979)
22. Rodgers JE et al; Water Air Soil Pollut 24: 443-54 (1985)
23. Sims GK, Sommers LE; Environ Toxicol Chem 5: 503-9 (1985)
24. Sims GK, Sommers LE; Appl Environ Microbiol 51: 963-8 (1986)
25. Snider EH, Manning FS; Environ Int 7: 237-58 (1982)
26. Stetter JR et al; Environ Sci Technol 19: 924-8 (1985)
27. Stuermer DH et al; Environ Toxicol Chem 16: 582-7 (1982)
28. Swann RL et al; Res Rev 85: 16-28 (1983)

N-Methylpyrrolidone

SUBSTANCE IDENTIFICATION

Synonyms:

Structure:

CAS Registry Number: 872-50-4

Molecular Formula: C_5H_9NO

SMILES Notation: O=C1CCCN1C

CHEMICAL AND PHYSICAL PROPERTIES

Boiling Point: 202 °C

Melting Point: -23 °C

Molecular Weight: 99.13

Dissociation Constants:

Log Octanol/Water Partition Coefficient: -0.727 [6]

Water Solubility: Miscible [17]

Vapor Pressure: 0.334 mm Hg at 25 °C [17]

Henry's Law Constant: 3.16×10^{-8} atm-m^3/mol at 25 °C [15]

ENVIRONMENTAL FATE/EXPOSURE POTENTIAL

Summary: N-Methylpyrrolidone is used as a solvent in a wide variety of commercial applications. It may be released to the environment as a fugitive emission during its production and use. N-Methylpyrrolidone may be a naturally occurring compound. If released to soil, N-methylpyrrolidone has the potential to biodegrade under aerobic conditions. It is expected to display very high mobility in soil. N-Methylpyrrolidone may slowly volatilize from dry soil to the atmosphere, but it is not expected to significantly volatilize from moist soil. If released to water, screening studies indicate that N-methylpyrrolidone will biodegrade under aerobic conditions after a short lag period. A calculated bioconcentration factor of 0.16 indicates that N-methylpyrrolidone is not expected to significantly bioconcentrate in fish and aquatic organisms nor is it expected to significantly adsorb to sediment or suspended organic matter. An estimated Henry's Law constant of 1.56×10^{-8} atm-m^3/mol at 25 °C indicates that N-methylpyrrolidone is not expected to significantly volatilize from water to the atmosphere. The estimated half-life for volatilization of N-methylpyrrolidone from a model river is 2335 days. If released to the atmosphere, N-methylpyrrolidone is expected to undergo a vapor-phase reaction with photochemically produced hydroxyl radicals with an estimated half-life of 5.2 hr. It may undergo atmospheric removal by wet deposition processes. Occupational exposure to N-methylpyrrolidone may occur by inhalation or dermal contact during its production and use. The general population may be exposed to N-methylpyrrolidone by the ingestion of foods in which it is contained or by the ingestion of contaminated drinking water.

Natural Sources: N-Methylpyrrolidone may be a naturally occurring compound as it has been identified as a volatile component of roasted nuts [9].

Artificial Sources: N-Methylpyrrolidone is used as a solvent extractant for the removal of olefins and butadiene from crack gas mixtures or of aromatics from petroleum [18]. It is also used as a solvent for resins, acetylene, pigment dispersant, petroleum processing, spinning agents for polyvinyl chloride, and as a chemical intermediate [20] and has been suggested as a possible substitute for methylene chloride for use in paint stripper formulations [23]. N-Methylpyrrolidone may be released to the environment as a fugitive emission during its production, formulation or use. It may also be released to the environment in the effluent of industrial processes [2], shale retort [4] and textile finishing [7].

Terrestrial Fate: If released to soil, N-methylpyrrolidone has the potential to biodegrade under aerobic conditions [21]. An estimated soil adsorption coefficient of 9.6 [12] obtained from the calculated log octanol/water partition coefficient indicates that N-methylpyrrolidone will display very high mobility in soil [22]. Limited laboratory data also indicate that N-methylpyrrolidone will be highly mobile in soil [21]. The vapor pressure of N-methylpyrrolidone and its estimated Henry's Law constant indicate that N-methylpyrrolidone may slowly volatilize from dry soil, but it is not expected to significantly volatilize from moist soil.

Aquatic Fate: If released to water, screening studies indicate that N-methylpyrrolidone will biodegrade under aerobic conditions with a short lag period [3,14,19,24]. A calculated bioconcentration factor of 0.16 [12] obtained from the estimated log octanol/water partition coefficient indicates that N-methylpyrrolidinone is not expected to significantly bioconcentrate in fish and aquatic organisms. The estimated Henry's Law constant indicates that N-methylpyrrolidone is not expected to significantly volatilize from water to the atmosphere. The estimated half-life for volatilization of N-methylpyrrolidone from a model river 1 m deep, flowing at 1 m/sec, with a wind speed of 3 m/sec, is 2,335 days [12]. A calculated soil adsorption coefficient of 9.6 [12] from the octanol/water partition coefficient indicates that N-methylpyrrolidone will not significantly adsorb to sediment or suspended organic matter.

Atmospheric Fate: If released to the atmosphere, N-methylpyrrolidone is expected to undergo a rapid vapor-phase reaction with photochemically produced hydroxyl radicals; the half-life for this process can be estimated at 5.2 hr [1]. N-Methylpyrrolidone is not expected to undergo significant atmospheric removal by the vapor-phase reaction with ozone [5,10]. N-Methylpyrrolidone's miscibility in water suggests that it will undergo atmospheric removal by wet deposition processes.

Biodegradation: In a static die-away system using a sewage sludge seed and a semi-continuous activated sludge system, 1-methylpyrrolidone displayed 95% removal after 2 weeks and an average 7 day biodegradability of 95%, respectively [3]. N-Methylpyrrolidone at an initial concentration of 210 ppm underwent >98% removal after 24 hr when inoculated with a sewage sludge seed [13]. N-Methylpyrrolidone at an initial concentration of 150 mg/L underwent 94% removal as measured by COD after a 1 day lag period in a screening study using an activated sludge seed [14]. In a model flow-through biological treatment apparatus with an 18 hr retention time, N-methylpyrrolidone at an initial concn of 300 mg/L underwent >98% removal using an activated sewage sludge [19]. In a static screening test, N-methylpyrrolidone underwent >90% removal after a 3-5 day acclimation period using a sewage sludge seed [24]. The half-life of N-methylpyrrolidone in clay, loam and sand soils was 4.0, 8.7, and 11.5 days, respectively [21].

Abiotic Degradation: N-Methylpyrrolidone was listed as a member of reactivity class II in a five tiered system used to classify reactivity towards ozone, indicating low reactivity in this reaction [5]. In a similar classification scheme developed to aid in controlling the release of solvents in Los Angeles county, it was listed in reactivity class 2 in a three tiered system indicating intermediate reactivity [10]. A calculated rate constant for the vapor-phase reaction of N-methylpyrrolidone with photochemically produced hydroxyl radicals of $7.4 \times 10^{-11} \, cm^3$/molecule-sec [1] translates to a half-life of 5.2 hr using an average atmospheric hydroxyl radical concentration of $5 \times 10^{+5}$ molecules/cm^3 [1].

Bioconcentration: N-Methylpyrrolidone is described as being miscible with water suggesting limited potential for bioconcentration [12]. Based on the estimated log octanol/water partition coefficient, a bioconcentration factor of 0.16 can be calculated using an appropriate regression equation [12], which indicates that N-methylpyrrolidone is not expected to significantly bioconcentrate in fish and aquatic organisms.

Soil Adsorption/Mobility: N-Methylpyrrolidone had Rf values of 0.74, 0.65, 0.67, and 1.0 in silt, loam, clay and sand, respectively, in laboratory soil thin layer chromatography (TLC) experiments [21], which is consistent with significant mobility in soil. N-Methylpyrrolidone is described as being miscible with water indicating that it is not expected to significantly adsorb to soil [12]. Based on the estimated log octanol/water partition coefficient, a soil adsorption coefficient of 9.6 can be calculated using an appropriate regression equation [12] indicating very high mobility in soil [22].

Volatilization from Water/Soil: Based on the vapor pressure, N-methylpyrrolidone may slowly volatilize from dry soil to the atmosphere. The estimated Henry's Law constant indicates that N-methylpyrrolidone is not expected to significantly volatilize from water or moist soil to the atmosphere [12]. The estimated half-life for volatilization from a model river 1 m deep, flowing at 1 m/sec, with a wind speed of 3 m/sec, is 2335 days [12].

Water Concentrations: DRINKING WATER: N-Methylpyrrolidone has been qualitatively detected in US drinking water supplies [11].

Effluent Concentrations: N-Methylpyrrolidone was detected in 1 of 46 US industrial effluent samples [2]. N-Methylpyrrolidone was found in shale retort water at a concentration of 3 mg/L [4]. N-Methylpyrrolidone was detected in the raw effluent from a textile finishing plant, NC, concentration not provided [7].

Sediment/Soil Concentrations:

Atmospheric Concentrations:

Food Survey Values: N-Methylpyrrolidone has been detected as a volatile flavor component of roasted filberts [9].

Plant Concentrations:

Fish/Seafood Concentrations:

Animal Concentrations:

Milk Concentrations:

Other Environmental Concentrations:

Probable Routes of Human Exposure: Occupational exposure to N-methylpyrrolidone may occur by inhalation or dermal contact during its production or use. The general population may be exposed to N-methylpyrrolidone by the ingestion of foods in which it is contained or by the ingestion of contaminated drinking water.

Average Daily Intake:

N-Methylpyrrolidone

Occupational Exposure: NIOSH (NOES 1981-83) has statistically estimated that 71,373 workers are exposed to N-methylpyrrolidone in the US [16].

Body Burdens:

REFERENCES

1. Atkinson R; Int J Chem Kinet 19: 799-828 (1987)
2. Bursey JT, Pellizzari ED; Analysis of Industrial Wastewater for Organic Pollutants in Consent Decree Survey Res Triangle Park, NC (1982)
3. Chow ST, Ng TL; Water Res 17: 117-8 (1983)
4. Dobson KR et al; Water Res 19: 849-56 (1985)
5. Farley FF; Int Conf Photochem Oxid Pollut Control Proc 2: 713-26 (1977)
6. GEMS; Graphic Exposure Modeling System CLOGP USEPA (1987)
7. Gordon WA, Gordon, M; Trans Ky Acad Sci 42: 149-57 (1981)
8. Hine J, Mookerjee PK; J Org Chem 40: 292-8 (1975)
9. Kinlin TE et al; J Agr Food Chem 20: 1021-8 (1972)
10. Levy A; The Photochemical Smog Reactivity of Organic Solvents Adv Chem Ser Amer Chem Soc 124: 70-94 (1973)
11. Lucas SV; GC/MS Analysis of Organics in Drinking Water Concentrates and Advanced Waste Treatment Concentrates: Vol 3 USEPA-600/1-84-020, NTIS PB85-128247, Columbus, OH (1984)
12. Lyman WJ et al; Handbook of Chemical Property Estimation Methods NY: McGraw-Hill (1982)
13. Matsui S et al; Prog Water Technol 7: 645-59 (1975)
14. Matsui S et al; Wat Sci Tech 20: 201-20 (1988)
15. Meylan WM, Howard, PH; Environ Toxicol Chem 10:1283-93 (1991)
16. NIOSH; National Occupational Exposure Survey (NOES) (1989)
17. Riddick JA et al; Organic Solvents: 4th ed Wiley Interscience NY (1986)
18. Riemenschneidenr W; pp. 235-48 in Ullmann's Encyclopedia of Industrial Technology, 5th ed. Gerhartz W et al Eds. VCH Publ A5 (1985)
19. Rowe EH, Tullos LFJR; Hydrocarbon Process 59: 63-5 (1980)
20. Sax NI, Lewis RJSR; Hawley's Condensed Chemical Dictionary. Van Nostrand Reinhold Co. NY. 11th ed pp. 781-2 (1987)
21. Shaver TN; Arch Environ Contam Toxicol 13: 335-40 (1984)
22. Swann RL et al; Res Rev 85: 17-28 (1983)
23. USEPA; Testing Consent Agreement for N-Methylpyrrolidone, FR58:61814 Nov 23, (1993)
24. Zahn R, Wellens H; Z Wasser Abwasser Forsch 13: 1-7 (1980)

Oleyl Alcohol

SUBSTANCE IDENTIFICATION

Synonyms:

Structure:

CAS Registry Number: 143-28-2

Molecular Formula: $C_{18}H_{36}O$

SMILES Notation: OCCCCCCCCC=CCCCCCCCC

CHEMICAL AND PHYSICAL PROPERTIES

Boiling Point: 333 °C

Melting Point: 13-19 °C

Molecular Weight: 268.48

Dissociation Constants:

Log Octanol/Water Partition Coefficient: 7.50 [11]

Water Solubility: 0.024 mg/L at 25 °C (estimated from Kow) [8]

Vapor Pressure: 9.3 x 10^{-5} mm Hg at 25 °C (extrapolated from vapor pressures of 8 mm Hg at 195 °C [1], 13 mm Hg at 207 °C [15], and 760 mm Hg at 333-335 °C [6])

Henry's Law Constant: 4.6 x 10^{-4} atm-m^3/mol at 25 °C (estimated) [10]

ENVIRONMENTAL FATE/EXPOSURE POTENTIAL

Summary: Oleyl alcohol is a natural product in fish oils. It is used in a number of commercial applications, which may result in release to the environment in waste streams resulting from its production and use. If released to the atmosphere, oleyl alcohol may exist in both the vapor and particulate phases. Vapor-phase oleyl alcohol in the ambient atmosphere is expected to rapidly degrade by reaction with photochemically produced hydroxyl radicals (estimated half-life of 4.9 hr) and ozone radicals in the troposphere (estimated half-life of 2.1 hr). Particulate-phase oleyl alcohol may be removed from air via dry deposition. If released to soil, oleyl alcohol is expected to be immobile in soil and adsorption to soil will be important. One microbial study which used pure cultures suggests that biodegradation may be an important fate process of oleyl alcohol in soil and water, but no rate data are available. If released to water, adsorption from the water column to sediment and suspended materials and biodegradation are expected to be the predominant aquatic fate processes of oleyl alcohol. Hydrolysis and bioconcentration in aquatic organisms are not expected to be important aquatic fate mechanisms. A volatilization half-life of 7.9 hr has been estimated for a model river (1 m deep) indicating moderate volatilization from shallow rivers. The general public may be exposed to oleyl alcohol by dermal contact during the use of cosmetics in which it is contained as a cosmetic emollient and through fish consumption. Workers may be exposed to oleyl alcohol by inhalation of vapors or through eye and skin contact.

Natural Sources: Oleyl alcohol occurs in fish and marine mammal oils [13]. Oleyl alcohol is naturally contained in fish oils [15].

Artificial Sources: Oleyl alcohol's use as a chemical intermediate, automotive lubricant, defoamer, cosolvent and plasticizer for printing ink, and as a cosmetic emollient [7] indicates that release to the environment may be possible during waste disposal from its production and use.

Terrestrial Fate: An estimated Koc of $1.29 \times 10^{+4}$ [8] suggests that oleyl alcohol may adsorb strongly to soil. Based on one microbial study, oleyl alcohol was found to be utilized as the sole carbon source by bacteria, yeast,

356

and fungi [16]. Although this study provides little insight into the rate of biodegradation in soil, it suggests that biodegradation in soil may be important.

Aquatic Fate: Using the Henry's Law constant, a volatilization half-life of 7.9 hr has been estimated for a model river (1 m deep) indicating moderate volatilization from shallow rivers [8]. Alcohols are generally resistant to hydrolysis [8]. An estimated Koc of 1.29 x 10^{+4} [9] suggests that oleyl alcohol may partition from the water column to sediments and suspended materials. Based on one microbial study, oleyl alcohol was found to be utilized as the sole carbon source by bacteria, yeast, and fungi [16]. Although this study provides little insight into the rate of biodegradation in water, it suggests that biodegradation in water may be important.

Atmospheric Fate: Based upon the estimated vapor pressure, oleyl alcohol will exist in both the vapor and particulate phases in the ambient atmosphere [4]. Vapor phase oleyl alcohol is degraded in the ambient atmosphere by reaction with photochemically formed hydroxyl radicals; the half-life for this reaction in air can be estimated to be about 4.9 hr [2]. Vapor phase oleyl alcohol is also readily degraded by reaction with ozone in the troposphere; the half-life for this reaction in air can be estimated to be about 2.1 hr [2]. Particulate-phase oleyl alcohol may be removed via dry deposition.

Biodegradation: Oleyl alcohol (10 g) was found to be utilized as the sole carbon source by bacteria (*Pseudomonas*) in 10 days at 30 °C and pH 6.8-7.0. In the same study, 10 g oleyl alcohol was utilized as the sole carbon source by 3 yeasts (*Candida*, *Pichia*, and an unknown) in 10 days at 30 °C and pH 6.8-7.0. It was also utilized by 3 fungi (*Aspergillus*, *Penicillium*, and an unknown) in 20 days at 20-25 °C and pH 5.5-5.6 [16].

Abiotic Degradation: The rate constant for the vapor-phase reaction of oleyl alcohol with photochemically produced hydroxyl radicals can be estimated to be 7.85 x 10^{-11} cm^3/molecule-sec at 25 °C, which corresponds to an atmospheric half-life of about 4.9 hr at an atmospheric concn of 5 x 10^{+5} hydroxyl radicals per cm^3 [2]. The rate constant for the vapor-phase reaction of oleyl alcohol with ozone in the troposphere can be estimated to be 1.35 x 10^{-16} cm^3/molecule-sec at 25 °C, which corresponds to a half-life of about 2.1 hr at an atmospheric concn of 7 x 10^{+11} molecules per cm^3 [2,3].

Alcohols are generally resistent to environmental hydrolysis [8]; therefore, hydrolysis is not expected to be an important aquatic fate process of oleyl alcohol.

Bioconcentration: Oleyl alcohol is naturally contained in fish oils and therefore, concentrations in fish will be quite high compared to ambient water concentrations [15].

Soil Adsorption/Mobility: Based on an estimated Koc of $1.29 \times 10^{+4}$ [9], oleyl alcohol is expected to be immobile in soil and sediment [14].

Volatilization from Water/Soil: The Henry's Law constant suggests that volatilization from water may not be rapid but possibly important [8]. Based on this value, the volatilization half-life of oleyl alcohol from a model river 1 m deep flowing 1 m/sec with a wind velocity of 3 m/sec has been estimated to be approximately 7.9 hr [8].

Water Concentrations:

Effluent Concentrations:

Sediment/Soil Concentrations:

Atmospheric Concentrations:

Food Survey Values:

Plant Concentrations:

Fish/Seafood Concentrations: Oleyl alcohol is found in fish oils [15]. Oleyl alcohol was qualitatively detected in adult lake trout (*Salvelinus namaycush*) collected in 1977 from Lake Michigan in Charlevoix, MI and it was qualitatively detected in walleye (*Stizostedion v. vitreum*) from Lake St. Clair Anchor Bay, MI, and Lake Erie Dunkirk, PA [5].

Animal Concentrations:

Milk Concentrations:

Other Environmental Concentrations:

Probable Routes of Human Exposure: The general public may be exposed to oleyl alcohol by dermal contact during the use of cosmetics in which it is contained as a cosmetic emollient [7] and through fish consumption [15]. Workers may be exposed by inhalation of vapors or through eye and skin contact.

Average Daily Intake:

Occupational Exposure: NIOSH (NOES 1981-1983) has statistically estimated that 77,106 workers are potentially exposed to oleyl alcohol in the US [12].

Body Burdens:

REFERENCES

1. Aldrich; Catalog Handbook of Fine Chemicals Milwaukee, WI: Aldrich Chem Co (1990)
2. Atkinson R; Int J Chem Kinet 19: 799-828 (1987)
3. Atkinson R; Statewide Air Pollut Res Cent; Univ California, Riverside, CA p. 193 (1985)
4. Eisenreich SJ et al; Environ Sci Technol 15: 30-8 (1981)
5. Hesselberg RJ, Seelye JG; p. 49 in ADMEN Rep 82-1 Ann Arbor, MI: US Fish Wildlife Soc (1982)
6. Jurgen F et al; A1: 279-303 in Ullmann's Encycl of Indust Chem NY: VCH Publishers (1985)
7. Kuney JH; Chemcyclopedia 92 Washington DC: Amer Chem Soc p. 133 (1991)
8. Lyman WJ et al; Handbook of Chemical Property Estimation Methods. Environment Behavior of Organic Compounds. McGraw-Hill NY (1982)
9. Meylan WM et al; Environ Sci Technol 26: 1560-1567 (1992)
10. Meylan WM, Howard PH; Environ Toxicol Chem 10:1283-1293 (1991)
11. Meylan W, Howard PH; J Pharm Sci 84: 83-92 (1995)
12. NIOSH National Occupational Exposure Survey (NOES) (1983)
13. Sax NI, Lewis RJSR; Hawley's Condensed Chemical Dictionary 11th ed. NY, NY: Van Nostrand Reinhold Co p. 193 (1987)
14. Swann RL et al; Res Rev 85: 17-28 (1983)
15. Windholz M et al; The Merck Index 10th ed. Rahway, NJ: Merck & Co Inc (1983)
16. Yanagi M, Onishi G; J Soc Cosmet Chem 22: 851-65 (1971)

Pentafluoroethane

SUBSTANCE IDENTIFICATION

Synonyms:

Structure:

CAS Registry Number: 354-33-6

Molecular Formula: C_2HF_5

SMILES Notation: FC(F)C(F)(F)F

CHEMICAL AND PHYSICAL PROPERTIES

Boiling Point: -48.5 °C

Melting Point: -103 °C

Molecular Weight: 120.20

Dissociation Constants:

Log Octanol/Water Partition Coefficient: 2.30 (calculated) [12]

Water Solubility: 39 mg/L at 25 °C (estimated from log Kow) [7]

Vapor Pressure: 7065 mm Hg at 25 °C (estimated from Henry's Law constant and water solubility)

Henry's Law Constant: 3.05 atm-m^3/mol at 25 °C [9]

ENVIRONMENTAL FATE/EXPOSURE POTENTIAL

Summary: Pentafluoroethane is an anthropogenic compound that holds promise as an alternative to chlorofluorocarbons (CFCs). It may be released to the environment as a fugitive emission during its production or use. If released to soil, pentafluoroethane will rapidly volatilize from both moist and dry soil to the atmosphere. It will display moderate to high mobility in soil. If released to water, pentafluoroethane will rapidly volatilize to the atmosphere. The estimated half-life for volatilization from a model river is 3.2 hr. Pentafluoroethane will not bioconcentrate in fish and aquatic organisms nor will it adsorb to sediment or suspended organic matter. If released to the atmosphere, pentafluoroethane will undergo a slow gas-phase reaction with photochemically produced hydroxyl radicals with an estimated half-life of 6417 days. The atmospheric lifetime of pentafluoroethane has been estimated to range from 19 to 43 years. Pentafluoroethane may undergo atmospheric removal by wet deposition processes; however, any removed is expected to rapidly revolatilize to the atmosphere. Occupational exposure to pentafluoroethane may occur by inhalation or dermal contact during its production or use.

Natural Sources: Pentafluoroethane is of anthropogenic origin, and it is not known to be produced by natural sources.

Artificial Sources: Pentafluoroethane is an anthropogenic compound that may be used as a replacement for chlorofluorocarbons (CFCs) [8]; if so, it may be released to the environment as a fugitive emission during its production and use.

Terrestrial Fate: If released to soil, the estimated vapor pressure for pentafluoroethane indicates that it will rapidly volatilize from dry soil to the atmosphere. Estimated soil adsorption coefficients ranging from 142-580 [7,9] indicate that it will display moderate to high mobility in soil [11]. An estimated Henry's Law constant of 3.05 atm-m^3/mol at 25 °C [9] indicates that it will also rapidly volatilize from moist soil to the atmosphere.

Aquatic Fate: If released to water, an estimated Henry's Law constant for pentafluoroethane indicates that it will rapidly volatilize to the atmosphere. The estimated half-life for volatilization from a model river 1 m deep flowing at 1 m/sec with a wind speed of 3 m/sec is 3.2 hr [7]. Estimated bioconcentration factors ranging from 7 to 78 [7,12] indicate that pentafluoroethane will not bioconcentrate in fish and aquatic organisms. Estimated soil adsorption coefficients ranging from 142 to 580 [7,10,12] indicate that it will not adsorb to sediment or suspended organic matter.

Pentafluoroethane

Atmospheric Fate: If released to the atmosphere, pentafluoroethane will undergo a slow vapor-phase reaction with photochemically produced hydroxyl radicals. The recommended rate constant for this process of 2.5 x 10^{-15} cm³/molecule-sec [2] translates to an atmospheric half-life of 6417 days using an average atmospheric hydroxyl radical concn of 5 x 10^{+5} molecules/cm³ [2]. The atmospheric lifetime of pentafluoroethane, calculated using both 1 and 2 dimensional models, ranges from 19 to 43 yrs [6]. The estimated water solubility of pentafluoroethane indicates that it may undergo atmospheric removal by wet deposition processes; however, any removed is expected to rapidly revolatilize to the atmosphere based upon the Henry's Law constant.

Biodegradation:

Abiotic Degradation: Experimental rate constants for the vapor-phase reaction of pentafluoroethane with photochemically produced hydroxyl radicals of 4.96 x 10^{-15} cm³/molecule-sec at 294 K [1], 2.9 x 10^{-15} cm³/molecule-sec at 303 °C [3] and 5.01 x 10^{-15} cm³/molecule-sec at ambient temperature [4,5] have been reported. The recommended value of 2.5 x 10^{-15} cm³/molecule-sec [2] translates to an atmospheric half-life of 6417 days using an average atmospheric hydroxyl radical concn of 5 x 10^{+5} molecules/cm³ [2]. The atmospheric lifetime of pentafluoroethane, calculated using both 1 and 2 dimensional models, ranges from 19 to 43 yrs [6].

Bioconcentration: Estimated bioconcentration factors ranging from 7 to 78 can be calculated for pentafluoroethane based on its estimated log octanol/water partition coefficient and estimated water solubility using appropriate regression equations [7]. These values indicate that pentafluoroethane will not bioconcentrate in fish and aquatic organisms.

Soil Adsorption/Mobility: Estimated soil adsorption coefficients ranging from 142 to 580 can be calculated for pentafluoroethane based on its estimated log octanol/water partition coefficient and estimated water solubility using appropriate regression equations [7]. These values indicate that pentafluoroethane will display moderate to high mobility in soil [11].

Volatilization from Water/Soil: The estimated Henry's Law constant indicates that pentafluoroethane will rapidly volatilize from water and moist soil to the atmosphere. The estimated half-life for volatilization from a model river 1 m deep flowing at 1 m/sec with a wind speed of 3 m/sec is 3.2 hr [7]. The estimated vapor pressure of pentafluoroethane indicates that it will rapidly volatilize from dry soil to the atmosphere.

Water Concentrations:

Effluent Concentrations:

Sediment/Soil Concentrations:

Atmospheric Concentrations:

Food Survey Values:

Plant Concentrations:

Fish/Seafood Concentrations:

Animal Concentrations:

Milk Concentrations:

Other Environmental Concentrations:

Probable Routes of Human Exposure: Occupational exposure to pentafluoroethane may occur by inhalation or dermal contact during its production or use.

Average Daily Intake:

Occupational Exposure:

Body Burdens:

REFERENCES

1. Atkinson R et al; Adv Photochem 11: 375-488 (1979)
2. Atkinson R; J Chem Phys Ref Data Monograph 1 (1989)
3. Brown AC et al; Atmos Environ 24A: 2499-511 (1990)
4. Cohen N, Benson SW; J Phys Chem 91: 171-5 (1987)
5. Cohen N, Benson SW; J Phys Chem 91: 162-70 (1987)
6. Fisher DA et al; Nature 344: 508-12 (1990)
7. Lyman WJ et al; Handbook of Chemical Property Estimation Methods NY: McGraw-Hill (1982)
8. McClinden MO, Didion DA; Int J Thermophys 10: 563-76 (1989)
9. Meylan WM, Howard PH; Environ Toxicol Chem 10: 1283-93 (1991)
10. Meylan WM, Howard PH; Environ Sci Technol 26:1560-1567 (1992)
11. Swann RL et al; Res Rev 85: 17-28 (1983)
12. USEPA; PCGEMS & CLOGP (1988)

n-Pentyl Alcohol

SUBSTANCE IDENTIFICATION

Synonyms:

Structure:

CAS Registry Number: 71-41-0

Molecular Formula: $C_5H_{12}O$

SMILES Notation: OCCCCC

CHEMICAL AND PHYSICAL PROPERTIES

Boiling Point: 137.5 °C

Melting Point: -79 °C

Molecular Weight: 88.15

Dissociation Constants:

Log Octanol/Water Partition Coefficient: 1.42 [16]

Water Solubility: 22,000 mg/L [54]

Vapor Pressure: 2.8 mm Hg at 20 °C [38]

Henry's Law Constant: 1.254×10^{-5} atm-m^3/mol [33]

ENVIRONMENTAL FATE/EXPOSURE POTENTIAL

Summary: n-Pentyl alcohol occurs naturally in animal wastes and in essential oils of vegetation. It also occurs naturally in volatile components of many foods. Some of the man-made sources of n-pentyl alcohol are building construction material, paint and solvents, rendering plants, sewage treatment plants, turbine exhaust and wood pulping. Both photolysis and hydrolysis of n-pentyl alcohol should not be important in the environment. The loss of n-pentyl alcohol from soil and aquatic media may occur by aerobic and anaerobic biodegradation, but the rates for these processes are not known. The estimated log Koc value indicates it may be very mobile in soil. From the Henry's Law constant for this compound, the volatilization half-life from a model river has been estimated to be 2.8 days. Neither sorption to sediment and suspended solid in water nor bioconcentration in aquatic organisms are expected to be important in water. The half-life for the reaction of n-pentyl alcohol with photochemically generated hydroxyl radicals in the atmosphere has been estimated to be 1.5 days. The high water solubility of n-pentyl alcohol suggests partial removal from the atmosphere should occur by wet deposition. The general population is exposed to n-pentyl alcohol through food and the environment. Additional exposure of occupational groups to n-pentyl alcohol occurs from the workplace. n-Pentyl alcohol has been detected in air, human milk and adipose tissue.

Natural Sources: n-Pentyl alcohol occurs naturally in animal wastes and in essential oils of vegetation [15]. It also occurs naturally as a volatile component of many foods including blue cheese [9], cassava [13], nectarine [43], chickpea seed [37] and kiwi fruit flowers [45].

Artificial Sources: n-Pentyl alcohol is released to the environment from building construction materials [15], paint and solvents [15], rendering plants [49], sewage treatment plants [15], turbine exhaust [15], and wood pulping [7,15].

Terrestrial Fate: The lack of a hydrolyzable functional group and a chromophoric group capable of absorbing sunlight [28] in n-pentyl alcohol suggests that neither hydrolysis nor photolysis should be important in soil. n-Pentyl alcohol should biodegrade in soil under both aerobic and anaerobic conditions [30], but the rate of biodegradation in soil could not be estimated due to lack of data. Based on the log Kow and water solubility, log Koc values of 1.95 and 2.15 have been calculated by two recommended regression equations [28]. The average Koc value of 115 indicates n-pentyl alcohol will be highly to very highly mobile in soil [42].

n-Pentyl Alcohol

Aquatic Fate: The hydrolysis of n-pentyl alcohol in water may be unimportant because the compound lacks any hydrolyzable functional groups [28]. The direct photolysis may not be environmentally important due to lack of a suitable chromophoric group in the compound capable of absorbing sunlight [28]. The biodegradation tests of n-pentyl alcohol with sewage and activated sludge [18,46,48,56] indicate that aerobic biodegradation will occur in natural water. Biodegradation tests under anaerobic conditions [20,41] suggest that biodegradation of the compound should occur in anaerobic water and sediment. Based on the Henry's Law constant, the volatilization half-life of n-pentyl alcohol from a model river 1 m deep, flowing at 1 m/sec with a wind speed of 3 m/sec, has been estimated to be 2.8 days [28]. Estimated log Koc values of 1.95 and 2.15 from two recommended regression equations [28] using the log Kow and water solubility values indicate sorption to sediments and suspended solids in water should not be important. Based on a BCF estimated from Kow and a regression equation [28], bioconcentration of n-pentyl alcohol should not be an important process in aquatic organisms.

Atmospheric Fate: The direct photolysis of n-pentyl alcohol in the atmosphere should not be important since the compound does not contain any chromophoric groups that are capable of absorbing tropospheric sunlight [28]. Based on a rate constant [51] and the concn of hydroxyl radicals in the atmosphere [4], the half-life of n-pentyl alcohol due to reaction with atmospheric hydroxyl radicals can be estimated to be 1.5 days. Due to its reasonably high water solubility, removal of n-pentyl alcohol by wet deposition may also occur.

Biodegradation: In 5-day BOD tests with sewage as microbial inoculum, the oxygen consumption of n-pentyl alcohol ranged from 59 to 86.9% of the theoretical value [5,11,18,48,50]. In a Warburg test with activated sludge as inoculum, the oxygen consumption at 1 day of incubation was 28% of the theoretical value [14]. At concentrations above 300 mg/L, n-pentyl alcohol may have an inhibitory effect on the oxidative respiratory rate in the presence of activated sludge [46]. The first-order rate constants (at a constant microorganism concn) for biodegradation of n-pentyl alcohol in nonadapted activated sludge was 0.0285 per hr [56], corresponding to an aerobic biodegradation half-life of 1 day. After a lag period of 7 days, 95-100% of n-pentyl alcohol biodegraded in 49 days under anaerobic conditions with digester sludge as microbial inoculum [41]. The removal of n-pentyl alcohol at retention times of 4-5 days was 100% in anaerobic lagoons containing digester sludge or activated sludge as microbial inoculum [20]. n-Pentyl alcohol biodegraded in two soils under both aerobic and anaerobic conditions [30].

Abiotic Degradation: n-Pentyl alcohol does not contain any functional groups that are amenable to hydrolysis [28]. Similarly, n-pentyl alcohol does not contain any chromophoric groups with characteristic absorption maxima at wavelength greater than 290 nm [28] and the longest wavelength absorption band in the alcohols occur at wavelength below 200 nm [6]. Therefore, direct photolysis of n-pentyl alcohol in the environment should not be important. The rate constant for the reaction of n-pentyl alcohol with photochemically generated hydroxyl radicals in the atmosphere has been determined to be 1.08×10^{-11} cm^3/molecule-sec at 23 °C [51]. This corresponds to a half-life of 1.5 day at a daily average concn of hydroxyl radicals of $5 \times 10^{+5}$ per cm^3 [4]. The rate constant for the reaction of n-pentyl alcohol with hydroxyl radicals in aqueous solution has a maximum value of $4.8 \times 10^{+9}$ 1/M-sec at a pH of 5 [3,12]. Based on this rate constant and a value of 10-17 M for the concentration of hydroxyl radical in eutrophic water [31], the half-life of n-pentyl alcohol due to this reaction can be estimated to be 56 days. Therefore, this reaction should not be important in water.

Bioconcentration: Based on the log Kow and an estimation method [28], the bioconcentration factor for n-pentyl alcohol has been estimated to be 7.1. Therefore, bioconcentration of n-pentyl alcohol in aquatic organisms should not be important.

Soil Adsorption/Mobility: The sorption of n-pentyl alcohol in soil follows an S-type isotherm and the exponential 1/n in Freundlich equation is not equal to one [52]. In other words, besides the organic content of soil, the sorption characteristic of the n-pentyl alcohol is likely to depend on other factors such as pH and electrolyte concentration. Based on the log Kow value and water solubility, log Koc values of 1.95 and 2.15 have been calculated by two recommended regression equations [28]. The average Koc value of 115 indicates n-pentyl alcohol should be highly to very highly mobile in soil [42].

Volatilization from Water/Soil: Based on the Henry's Law constant, the volatilization half-life of n-pentyl alcohol from a model river 1 m deep, flowing at 1 m/sec with a wind speed of 3 m/sec, has been estimated to be 2.8 days [28]. The overall mass transfer coefficient during volatilization of n-pentyl alcohol from water has been determined to be 3.80×10^{-6} m/sec at a wind speed of 5.96 m/sec [29], which corresponds to a volatilization half-life of 51 hr from an average depth of 1 m [28]. Using data on water solubility, vapor pressure, log Kow and Koc of 115, the volatilization half-life from a model pond by EXAMS II simulation [47], which also considers the effect of adsorption on the volatility of a compound from water, has been estimated to be 40 days.

Water Concentrations: DRINKING WATER: n-Pentyl alcohol was identified in drinking water of unspecified location(s) in the US [23]. It was also detected in drinking water from Cincinnati, OH, in 1976/1978/1980 surveys [27] and in drinking water from Seattle, WA in a 1976 survey [27]. An unspecified isomer of pentyl alcohol was detected at a concn of 0.001 mg/L in the drinking water of Washington, DC [40]. SURFACE WATER: n-Pentyl alcohol was detected in sea water near Kitakyushu area of Japan [2]. GROUND WATER: Unspecified isomer(s) of pentyl alcohol were detected at a concn of 11.66 mg/L in ground water from a sanitary landfill near Wilmington, DE [10]. Unspecified isomer(s) of pentyl alcohol were also detected at a concn of 0.065 mg/L in 1 of 2 municipal solid waste landfill leachate samples from Minnesota [39].

Effluent Concentrations: n-Pentyl alcohol was detected in the effluents from advanced waste water treatment facilities from Orange county, CA, and Washington, DC [27]. n-Pentyl alcohol is likely to be found in effluents from the following industries: plastics, synthetic resins and some elastomers; synthetic rubber; cleaning, polishing and sanitation preparation; industrial organic chemical; and petroleum refining [1].

Sediment/Soil Concentrations:

Atmospheric Concentrations: n-Pentyl alcohol was detected in air over the southern Black Forest in southwestern Germany [21]. It was also detected in indoor air of schools in Stockholm, Sweden [34].

Food Survey Values: n-Pentyl alcohol has been detected in volatile components of nectarine [43], fried bacon [19] and roasted filberts [22],

Chickpea (*Cocer arietimum*) [37], kiwi fruit flower [45], Japanese apples [53], Cassava products [13] and Dalieb (*Borassus aethiopum*), an edible fruit of deciduous palm grown in Sudan [17]. Unspecified isomer(s) of pentyl alcohol were detected in volatile components of fried chicken [44]. n-Pentyl alcohol was detected at a mean concn of 0.008 mg/kg in dry beans, at 0.019 mg/kg in dry split peas and at 0.095 mg/kg in dry lentils [26].

Plant Concentrations:

Fish/Seafood Concentrations:

Animal Concentrations:

Milk Concentrations: n-Pentyl alcohol was found in 4 of 12 mothers' milk samples collected from NJ, PA and LA [36]

Other Environmental Concentrations: Unspecified isomer(s) of pentyl alcohol were detected in the concn range 0.184-0.768 mg/kg (manure) in the volatile components of poultry manure [55].

Probable Routes of Human Exposure:

Average Daily Intake:

Occupational Exposure: NIOSH (NOES 1981-1983) has statistically estimated that 23,189 workers are potentially exposed to n-pentyl alcohol in the US [32].

Body Burdens: n-Pentyl alcohol was detected in the expired air in 4 of 8 male subjects (smoking and nonsmoking) at a concn range 0.007-0.60 ug/hr [8]. It was also detected in the expired air of nonsmoking healthy subjects [25] and the mean concn was 0.505 ng/L in 54 subjects with a percent occurrence of 22% [24]. n-Pentyl alcohol was found in 4 of 12 mothers' milk samples collected from NJ, PA and LA [36]. It was also detected in 29 of 46 adipose tissues collected during the 1982 National Human Adipose Tissue Survey [35].

REFERENCES

1. Abrams EF et al; Identification of Organic Compounds in Effluents from Industrial Sources, USEPA-560/3-75-002, Off Toxic Sub, NTIS PB-241641, Springfield, VA (1975)
2. Akiyama T et al; J UOEH 2: 285-300 (1980)
3. Anbar M, Neta P; Int J Appl Radiation and Isotopes 18: 493-523 (1967)
4. Atkinson R; Chem Rev 85: 69-201 (1985)
5. Babeu L, Vaishnav DD; J Ind Microbiol 2: 107-15 (1987)
6. Calvert JG, Pitts JN, Jr; Photochemistry NY: John Wiley & Sons, Inc p. 441 (1966)
7. Carlberg GE et al; Sci Total Environ 48: 157-67 (1986)
8. Conkle JP et al; Arch Environ Health 30: 290-95 (1975)
9. Day EA, Anderson DF; J Agric Food Chem 13: 2-4 (1965)
10. Dewalle FB, Chian ESK; J Am Water Works Assoc 73: 206-11 (1981)
11. Dias FF, Alexander M; Appl Microbiol 22: 1114-18 (1971)
12. Dorfman LM, Adams GE; Reactivity of the Hydroxyl Radical in Aqueous Solution NSRD-NBS-46, (NTIS COM-73-50623) Natl Bureau Stand, Washington DC p. 24 (1973)
13. Dougan J et al; J Sci Food Agric 34: 874-84 (1983)
14. Gerhold RM, Malaney GW; J Water Poll Control Fed 38: 562-79 (1966)
15. Graedel TE et al; Atmospheric Chemical Compounds. Orlando, FL: Academic Press, Inc p. 236 (1986)
16. Hansch C, Leo AJ; Medchem Project Issue No.26 Pomona College, Claremont CA (1985)
17. Harper DB et al; J Sci Food Agric 37: 685-88 (1986)
18. Heukelekian H, Rand MC; J Water Pollut Control Assoc 29: 1040-53 (1955)
19. Ho C-T et al; J Agric Food Chem 31: 336-42 (1983)
20. Hovious JC et al; Anaerobic Treatment of Synthetic Organic Wastes, EPA 12020 DIS 01/72, Off Res Monit Washington DC (1972)
21. Juttner F; Chemosphere 15: 985-92 (1986)
22. Kinlin TE et al; J Agric Food Chem 20: 1021-28 (1972)
23. Kopfler FC et al; Adv Environ Sci Technol 8: 419-33 (1977)
24. Krotoszynski BK et al; J Anal Toxicol 3: 225-34 (1979)
25. Krotoszynski B et al; J Chromatogr Sci 15: 239-44 (1977)
26. Lovegren NV et al; J Agric Food Chem 27: 851-53 (1979)
27. Lucas SA; GC/MS Analysis of Organics in Drinking Water Concentrates and Advanced Waste Treatment Concentrates Vol 1. USEPA-600/1-84-020a, Health Effects Research Lab. Office of Res. Dev. Research Triangle Park, NC p. 215 (1984)
28. Lyman WJ et al; Handbook of Chemical Property Estimation Methods NY: McGraw-Hill (1982)
29. Mackay D, Yeun ATK; Environ Sci Technol 17: 211-17 (1983)
30. Morris MS; Diss Abstr Int B 42: 2766 (1989)
31. Neely WB, Blau GE; Environmental Exposure from Chemicals Vol I. CRC Press, Boca Raton FL p. 207 (1985)

32. NIOSH; National Occupational Exposure Survey (NOES), NIOSH, Cincinnati, OH (1989)
33. Nirmalakhandan NN, Speece RE; Environ Sci Technol 22: 1349-57 (1988)
34. Noma E et al; Atmos Environ 22: 451-60 (1988)
35. Onstot JD et al; Characterization of HRGC/MS Unidentified Peaks from the Broad Scan Analysis of the FY82 NHATS Composites. Vol I, MRI Project No. 8823-AO1, Off Pest Toxic Sub, USEPA, Washington DC (1987)
36. Pellizzari ED et al; Bull Environ Contam Toxicol 28: 322-28 (1982)
37. Rembold H et al; J Agric Food Chem 37: 659-62 (1989)
38. Riddick JA et al; Organic Solvents Vol 2 4th ed. NY: John Wiley & Sons, p. 205-7 (1986)
39. Sabel GV, Clark TP; Waste Manag Res 2: 119-30 (1984)
40. Scheiman MA et al; Biomed Mass Spectrom 4: 209-11 (1974)
41. Shelton DR, Tiedje JM; Development of Test for Determining Anaerobic Biodegradation Potential USEPA-560/5-81-013, NTIS PB84-166495, Springfield, VA (1981)
42. Swann RL et al; Res Rev 85: 17-28 (1983)
43. Takeoka GR et al; J Agric Food Chem 36: 553-60 (1988)
44. Tang J et al; J Agric Food Chem 31: 1287-92 (1983)
45. Tatsuka K et al; J Agric Food Chem 38: 2176-80 (1990)
46. Therien N et al; Water Res 18: 905-10 (1984)
47. USEPA; EXAMS II computer simulation, USEPA, Athens, GA (1987)
48. Vaishnav DD et al; Chemosphere 16: 695-703 (1987)
49. Van Langenhove HR et al; Environ Sci Technol 16: 883-6 (1982)
50. Wagner R; Vom Wasser 47: 241-65 (1976)
51. Wallington TJ, Kurylo MJ; Int J Chem Kinet 19: 1015-23 (1987)
52. Weber JB, Miller CT; pp 305-33 in Reactions and Movement of Organic Chemicals in Soils, SSSA Special Publication No. 22 Madison, WI (1989)
53. Yajima I et al; Agric Biol Chem 48: 849-55 (1984)
54. Yalkowsky SH; Arizona Database of Aqueous Solubility, College of Pharmacy, Univ of Arizona, Tucson, AZ (1989)
55. Yasuhara A; J Chromatogr 387: 371-78 (1987)
56. Yonezawa Y, Urushikuni Y; Chemosphere 8: 139-42 (1979)

Propane

SUBSTANCE IDENTIFICATION

Synonyms:

Structure:

CAS Registry Number: 74-98-6

Molecular Formula: C_3H_8

SMILES Notation: C(C)C

CHEMICAL AND PHYSICAL PROPERTIES

Boiling Point: -42.1 °C

Melting Point: -189.7 °C

Molecular Weight: 44.09

Dissociation Constants:

Log Octanol/Water Partition Coefficient: 2.36 [22]

Water Solubility: 62.4 ppm in water at 25 °C [34]

Vapor Pressure: 7,162 mm Hg at 25 °C [17]

Henry's Law Constant: 7.07×10^{-1} atm-m^3/mol at 25 °C (calculated from the vapor pressure and water solubility)

ENVIRONMENTAL FATE/EXPOSURE POTENTIAL

Summary: Propane is a highly volatile constituent in the paraffin fraction of crude oil and natural gas. Propane gas is released to the environment via the manufacture, use and disposal of many products associated with the petroleum and natural gas industries. Extensive data show release of propane into ambient air from waste incinerators and the combustion of gasoline, natural gas and polyethylene. Photolysis, hydrolysis and bioconcentration of propane are not expected to be important environmental fate processes. Biodegradation of propane may occur in soil and water; however, volatilization is expected to be the dominant fate process. To a lesser extent, adsorption may also occur. A Koc range of 450 to 460 indicates a medium mobility class in soil for propane. In aquatic systems, propane may partition from the water column to organic matter contained in sediments and suspended materials. A Henry's Law constant of 7.07 x 10^{-1} atm-m^3/mol at 25 °C suggests extremely rapid volatilization of propane from environmental waters. The volatilization half-lives from a model river and a model pond, the latter considers the effect of adsorption, have been estimated to be 1.9 hr and 2.3 days, respectively. Propane is expected to exist entirely in the vapor phase in ambient air. Reactions with photochemically produced hydroxyl radicals in the atmosphere have been shown to occur (average half-life of 13 days). Data also suggest that nighttime reactions with radical species and nitrogen oxides may contribute to the atmospheric transformation of propane. The most probable route of human exposure to propane is by inhalation. Extensive monitoring data indicate propane is a widely occurring atmospheric pollutant.

Natural Sources: Propane is a constituent in the paraffin fraction of crude oil and natural gas [59].

Artificial Sources: Propane is released to the environment via the manufacture, use and disposal of many products associated with the petroleum [3,46] and natural gas industries [3]. The combustion of gasoline [38,40,53,61], natural gas [38] and polyethylene [23] releases propane to the atmosphere. Waste incinerators may also release propane into the air [13].

Terrestrial Fate: Photolysis or hydrolysis [33] of propane is not expected to be important in soils. The biodegradation of propane may occur in soils; however, primarily volatilization is expected to be the dominant fate process. To a lesser extent, adsorption may occur. A calculated Koc range of 450 to 460 [32] indicates a medium mobility class for propane in soils [56]. Based upon an estimated Henry's Law constant of 7.07 x 10^{-1}

atm-m^3/mol at 25 °C [33], propane is expected to rapidly volatilize from most surface soils [32].

Aquatic Fate: Photolysis or hydrolysis [32] of propane in aquatic systems is not expected to be important. The bioconcentration factor (log BCF) for propane has been estimated to range from 1.56 to 1.78 [32] suggesting bioconcentration is not an important factor in aquatic systems. Biodegradation of propane may occur in aquatic environments; however, volatilization is expected to be the dominant fate process. To a lesser extent, adsorption may occur. An estimated range for Koc from 450 to 460 [32] indicates propane may partition from the water column to organic matter [56] contained in sediments and suspended materials. An estimated Henry's Law constant of 7.07 x 10^{-1} atm-m^3/mol at 25 °C suggests extremely rapid volatilization of propane from environmental waters [32]. Based on this Henry's Law constant, the volatilization half-life from a model river has been estimated to be 1.9 hr [32]. The volatilization half-life from a model pond, which considers the effect of adsorption, can be estimated to be about 2.3 days [58].

Atmospheric Fate: Based on a vapor pressure of 7162 mm Hg at 25 °C [17], propane is expected to exist entirely in the vapor phase in ambient air [19]. Propane gas does not absorb UV light in the environmentally significant range, >290 nm [54], and should not undergo direct photolysis in the atmosphere. Vapor-phase reactions with photochemically produced hydroxyl radicals in the atmosphere have been shown to occur. Rate constants for propane were measured [8,15,16] to be about 1.22 x 10^{-12} cm^3/molecule-sec at 25 °C[4], respectively, which correspond to atmospheric half-lives of about 13 days at an atmospheric concn of 5 x 10^{+5} hydroxyl radicals per cm^3. Experimental data showed that 3.6% of the propane fraction in a dark chamber reacted with nitrogen oxide to form the corresponding alkyl nitrate [5,7], suggesting nighttime reactions with radical species and nitrogen oxides may contribute to the atmospheric transformation of propane.

Biodegradation: Within 24 hr, propane was oxidized to its corresponding methyl ketone, acetone [24,42], and the corresponding alcohols, 1-propanol and 2-propanol, by cell suspensions of over 20 methyltrophic organisms isolated from lake water and soil samples. After 192 hr, the trace concn of propane contained in gasoline remained unchanged for both a sterile control and a mixed culture sample collected from ground water contaminated with gasoline [26]. The average propane utilization by microflora of 5 soils was 23 and 32% for single and mixed alkanes, respectively [12]. The respective gas exchange and degradation rate constants were 0.67×10^{-5} cm^2/ sec and 0.033 day^{-1} for propane contained in a model estuarine ecosystem at 10 °C and a salinity of 30 ppt, the corresponding biodegradation half-life ranged from 33 to 99 days [11]. At 20 °C and a salinity of 30 ppt, the respective gas exchange and degradation rate constants were 0.92×10^{-5} cm^2/sec and 0.120 day^{-1}; the corresponding biodegradation half-life for n-propane ranged from 7 to 9 days [11].

Abiotic Degradation: Alkanes are generally resistant to hydrolysis [32]. Based on data for isooctane and n-hexane, propane is not expected to absorb UV light in the environmentally significant range, >290 nm [54]. Therefore, propane probably will not undergo hydrolysis or direct photolysis in the environment. An air sample's propane concn of 140 ppbC was not reduced within 6 hr of irradiation by natural sunlight in downtown Los Angeles, CA [28]. The rate constants for the vapor-phase reaction of propane with photochemically produced hydroxyl radicals was measured to be 2.0×10^{-12} [15], 1.49×10^{-12} [16], 1.22×10^{-12} [4] and 1.20×10^{-12} [8] cm^3/molecule-sec at 27 [15,16], 25 [4] and 22 [8] °C, respectively, which correspond to atmospheric half-lives of about 8.0 [15], 10.8 [16], 13.2 [4] and 13.4 [8] days at an atmospheric concn of $5 \times 10^{+5}$ hydroxyl radicals per cm^3. The photooxidation of propane by ozone in air is not expected to be environmentally important [6]. Experimental data showed that 7.7% of the propane fraction in a dark chamber reacted with nitrogen oxides to form the corresponding alkyl nitrate [5,7], suggesting nighttime reactions with radical species and nitrogen oxides may contribute to the atmospheric transformation of propane.

Bioconcentration: Based upon the water solubility and the log Kow, the bioconcentration factor for propane has been calculated, using recommended regression-derived equations, to be 1.56 and 1.78, respectively [32]. These bioconcentration factor values do not indicate that bioconcentration in aquatic organisms is important.

Soil Adsorption/Mobility: Based on the water solubility and the log Kow, the Koc of propane has been calculated, using various regression-derived equations, to range between 450 and 460 [32]. These Koc values indicate a medium soil mobility class for propane [56].

Volatilization from Water/Soil: The Henry's Law constant indicates extremely rapid volatilization from environmental waters [32]. The volatilization half-life from a model river (1 m deep flowing 1 m/sec with a wind speed of 3 m/sec) has been estimated to be 1.9 hr [32]. The volatilization half-life from a model pond, which considers the effect of adsorption, has been estimated to be 2.3 days [58].

Water Concentrations: SEAWATER: All 8 near-surface seawater samples from the intertropical Indian Ocean contained propane at concns ranging from 2.53 to 14.66 nl of gas/L [10].

Effluent Concentrations: Flue gases from a waste incinerator at Babylon, Long Island, NY were found to emit propane at concns generally less than 0.5 ppm [13]. Propane is a product of gasoline [38,40,53,61], natural gas [38] and polyethylene [23] combustion. The average exhaust from 67 gasoline fueled vehicles was found to contain propane at a concns of 0.1% by weight [40]. The average concn of propane for the exhaust of 46 automobiles was 2.6, 1.6 and 2.2 weight % of total hydrocarbon according to the federal test procedure, hot soak test and the New York City cycle, respectively [53]. Propane from car exhaust ranged in concn from 0.02 to 0.05 ppmV with an average for 8 samples of 0.03 ppmV [38]. A Texaco refinery located in Tulsa, OK, was attributed with emissions to the surrounding atmosphere where the propane concn was measured to be 95.5 and 189.8 ppbC for two min before and after 1:33 PM [3]. The propane content of the air downwind of a Mobil natural gas facility in Rio Blanco, CO, was 465.3 ppbC [3]. Underwater hydrocarbon vent discharges from offshore oil production platforms were found to contain propane concn in the vapor phase at 2000 umol/L of gas [46].

Sediment/Soil Concentrations: Propane was detected in 10 of 10 sediment samples from Walvis Bay of the Namibian shelf of SW Africa at concns of 15.0, 8.8, 8.5, 9.9, 12.3, 16.5, 6.4, 7.8, 7.3, and 4.0 ng/g [60]. Sediments from the Bering Sea contained propane gas at concn ranging from 4 to 150 nL/L [29].

Propane

Atmospheric Concentrations: URBAN: The average propane concn for 2 samples per 4 sites in Tulsa, OK, was 43.3 ppbC with a range of 4.6 to 189.8 ppbC [3]. The propane concn for 6 sites in Rio Blanco, CO averaged 81.6 ppbC with a range from 3.2 to 465.3 [30]. Propane was detected in 21 of 21 air samples from Houston, TX, ranging in concn from 13.0 to 592.4 ppm with an average of 108.1 ppm [30]. The arithmetic and geometric means were 12.2 and 10.1 ppbC, respectively, for the atmospheric propane content at urban locations in New England [14]. The average propane concn in the air at the 6th floor of the Cooper Union Building in New York City, NY, was 9, 9 and 4 ppbC for 19, 12 and 10 samples taken at 6:00-9:00 AM, 9:00-11:00 AM and 1:00-3:00 PM, respectively, in July 1978 [2]. The average propane concn in the air at the 82nd floor of the Empire State Building in New York City, NY, was 4, 6 and 4 ppbC for 18, 21 and 17 samples taken at 6:00-9:00 AM, 9:00-11:00 AM and 1:00-3:00 PM, respectively, in July 1978 [2]. At street level at the Empire State and World Trade Buildings in Manhattan, NY, the average propane concns of 4 samples was 10 ppbC in July 1978 [2]. In 1975 the average propane concn of 14 air samples taken between 05:30-08:30 and 12:30-15:30 EST at the World Trade Center in New York City, NY, were 24 and 12 ppbC, respectively [2]. In 1975 the average propane concns of 11 and 8 air samples taken between 5:30-8:30 AM and 12:30-3:30 PM at the Interstate Sanitation Commission in New York City, NY, were 39 and 25 ppbC, respectively [2]. Propane was detected at an average concn of 14.8 ug/m^3 for 5 samples collected at the 82nd floor of the World Trade Center in New York City between 5:00 AM - 5:30 PM August 23, 1977 [1]. URBAN: The ground level atmospheric concn of propane at 13:25 was 23 ppb and 166 ppb at 08:00 for Huntington Park, CA [47]. At 1500 ft the propane concn was 13 ppb at 07:43 and at 08:07 at a height of 2200 ft the propane concn was 9 ppb [47]. The propane concn ranged from 11 to 99 ppbv at a downtown Los Angeles location for the Fall of 1981 [20]. The propane concn at 1100 ft just east of Antioch, CA, was 7.0 ug/m^3, at 1000 ft near Pittsburg, CA, was 7.5 ug/m^3, at 1100 ft over Carquinez Strait, CA, was 5.0 ug/m^3 and at 1000 ft over San Pablo Bay, CA, was 1.5 ug/ m^3 [50]. According to the National Ambient Volatile Organic Compounds (VOCs) Database, the median urban atmospheric concn of propane is 5.733 ppbV for 541 samples [52]. URBAN: Propane was detected in the atmospheres of Pretoria, Johannesburg and Durban, South Africa [31]. Propane was identified in the ambient air of Sydney, Australia [37], ranging in concn from 0.3 to 44.9 ppbV with an average concn of 7.2 ppbV [39]. Propane was detected at an average concn of 197.0 ppbC in the atmosphere over the British Columbia Research Council Laboratory at the University of British Columbia [55]. The average propane concns in the air over Tokyo, Japan, in 1980 and 1981 were 2.6 and 5.4 ppb for 66 and 192 samples, respectively [57]. At Deuselbach, Hunsruck in Germany, the atmospheric propane concn was 0.72 ppb for October 23, 1983 [45]. The minimum, maximum and average propane concns in the ambient air of Bombay, India, were 1.6, 131.4 and 22.7 ppb, respectively [36]. SUBURBAN: According to the National Ambient Volatile Organic Compounds (VOCs) Database, the median suburban atmospheric concn of propane is 10.480 ppbV for 225 samples [52]. The propane concn was 3.0, 2.0 and 2.0 ug/m^3 at 10, 15 and 40 mi downwind of Janesville, WI, 8-14-78 [51]. RURAL: At a rural site near Duren, Germany, the atmospheric propane concn was 2.8 ppb for March 1984 [45]. The respective median, minimum and maximum atmospheric concns of propane for 5 rural locations in NC ranged from 3.4 to 8.6, 2.0 to 6.4, and 4.4 to 14.4 ppb [48]. The atmospheric concns of propane for Jones State Forest, TX, ranged from 12.8 to 34.9 ppb with an average of 21.8 ppb for 10 samples [49]. According to the National Ambient Volatile Organic Compounds (VOCs) Database, the median rural atmospheric concn of propane is 1.076 ppbV for 36 samples [52]. The arithmetic and geometric means were 10.3 and 6.6 ppbC, respectively, for the atmospheric propane content at rural locations in New England [14]. REMOTE: According to the National Ambient Volatile Organic Compounds (VOCs) Database, the median remote atmospheric concn of propane is 0.439 ppbV for 10 samples [52]. For 9 samples collected over a 30 hour period, the average propane concn in the Smokey Mountains, NC was 8.1 ppbC with a range from 5.0 to 13.3 ppbC [3]. On August 27, 1976, the average propane concn for air over Lake Michigan at altitudes of 2000, 2500 and 3000 ft was 9.7 ppbV [35]. On August 28, 1976, the average propane concn for air over Lake Michigan at altitudes of 1000 and 1500 ft was 1.3 ppbV [35]. The air over the Norwegian Arctic had an average propane concn for 5 samples from Bear Island, 2 from Hopen and 2 from Spitsbergen of 87.1 ppt/volume in July 1982 and 2156 ppt/volume in the spring of 1983 [25]. All 27 air samples from the intertropical Indian Ocean contained propane at concns ranging from 0.10 to 0.76 ppbV [10]. On April 20 to May 10, 1980, during the french research flight of stratoz II, the mixing ration of propane at altitudes between 800 and 200 mb was measured to range from less than 0.05 and 1.0 ppb for 110 samples of air collected between the latitudes of 60 deg S [18]. In March 1985, the average ambient air concn of propane in the Guyana tropical forest was 0.095 and 0.070 ppbV at ground level and 120 cm in height, respectively [9]. SOURCE DOMINATED: According to the National Ambient Volatile Organic Compounds (VOCs) Database, the median source dominated and indoor atmospheric concn of propane is 6.483 ppbV for 54 samples, respectively [52]. A Texaco refinery located in Tulsa, OK, was attributed with emissions to the surrounding atmosphere where the propane concn was measured to be 95.5 and 189.8 ppbC for 2 min before and after 1:33 PM [3]. The propane content of the air downwind of a Mobil natural gas facility in Rio Blanco, CO, was 465.3 ppbC [3]. The arithmetic and geometric means were 20.4 and 14.9 ppbC, respectively, for the atmospheric propane content at polluted rural locations in New England [14].

Food Survey Values:

Plant Concentrations:

Fish/Seafood Concentrations:

Animal Concentrations:

Milk Concentrations:

Other Environmental Concentrations:

Probable Routes of Human Exposure: The most probable route of human exposure to propane is by inhalation. Atmospheric workplace exposures have been documented [21,27,44]. Propane is a highly volatile compound and monitoring data indicate that it is a widely occurring atmospheric pollutant.

Average Daily Intake: According to the National Ambient Volatile Organic Compounds (VOCs) Database, the median urban atmospheric concn of propane is 5.733 ppbV for 541 samples [52]. Based upon this figure and the value for average daily inhalation by a human adult of 20 m^3 of air, the average daily intake of propane via air is 115 mg.

Propane

Occupational Exposure: NIOSH (NOES 1981-1983) has estimated that 235,193 workers are potentially exposed to propane in the US [41]. A 1984 study showed propane was emitted from gasoline, exposing outside operators at the refineries to an average air concn of 0.379 mg/ m³; propane was detected in 35 of 56 samples [44]. Transport drivers were exposed to propane at an atmospheric concn of 0.445 mg/m³ and propane was detected in 48 of 49 samples [44]. Gas station attendants were exposed to propane at an atmospheric concn of 1.116 mg/m³ and propane was detected in 48 of 49 samples [44]. Attendants at a high volume service station in eastern PA were exposed to levels of propane ranging from 0.1 to 0.2 ppm for 17 of 18 air samples [27]. Workers at Shell Oil Co gasoline bulk handling facilities were exposed to vapors that contained propane at concn of 0.8% by weight, 21.2% by volume of total hydrocarbons [21]. Exposures to total hydrocarbons at one of the facilities exceeded 240 ppm for 5% of the sampling time [21].

Body Burdens: Propane was detected in 1 of 12 samples of mother's breast milk from the cities of Bayonne, NJ, Jersey City, NJ, Bridgeville PA, and Baton Rouge, LA [43].

REFERENCES

1. Altwicker ER, Whitby RA; Sampling, Sample Prep and Measurement of Specific Non-methane Hydrocarbons 72 Ann Meet Air Pollut Contr Asssoc (1979)
2. Altwicker ER et al; J Geophys Res 85: 7475-87 (1980)
3. Arnts RR, Meeks SA; Atmos Environ 15: 1643-51 (1981)
4. Atkinson R et al; Internat J Chem Kin 14: 781-8 (1982)
5. Atkinson R et al; Preprints Div Environ Chem 23: 173-6 (1983)
6. Atkinson R, Carter WPL; Chem Rev 84: 437-70 (1984)
7. Atkinson R et al; J Phys Chem 86: 4563-9 (1982)
8. Baulch DL et al; J Phys Chem Ref Data 13: 1259-1380 (1984)
9. Bonsang B et al; Geophys Res Let 14: 1250-3 (1987)
10. Bonsang B et al; J Atmos Chem 6: 3-20 (1988)
11. Bopp RF et al; Org Geochem 3: 9-14 (1981)
12. Brisbane RG, Ladd JN; J Gen Appl Microbiol 14: 447-50 (1968)
13. Carotti AA, Kaiser ER; J Air Pollut Contr Assoc 22: 224-53 (1972)
14. Colbeck I, Harrison RM; Atmos Environ 19: 1899-904 (1985)
15. Cox RA et al; Environ Sci Technol 14: 57-61 (1980)
16. Darnall KR et al; J Phys Chem 82: 1581-4 (1978)
17. Daubert TE, Danner RP; Data Compilation, Tables of Properties of Pure Cmpds, Design Inst for Phys Prop Data, Am Inst for Phys Prop Data, NY NY(1985)

18. Ehhalt DH et al; J Atmos Chem 3: 29-52 (1985)
19. Eisenreich SJ et al; Environ Sci Technol 15: 30-8 (1981)
20. Grosjean D, Fung K; J Air Pollut Control Assoc 34: 537-43 (1984)
21. Halder CA et al; Am Ind Hyg Assoc J 47: 164-72 (1986)
22. Hansch C, Leo AJ; Medchem Project Issue No. 26 Pomona College, Claremont CA (1985)
23. Hodgkin JH et al; J Macromol Sci Chem 17: 35-43 (1982)
24. Hou CT et al; Appl Environ Microbiol 46: 178-84 (1983)
25. Hov O et al; Geophys Res Lett 11: 425-8 (1984)
26. Jamison VW et al; pp. 187-96 in Proc Int Biodeg Symp 3rd Sharpley JM and Kapalan AM (eds) Essex Eng (1976)
27. Kearney CA, Dunham DB; Am Ind Hyg Assoc J 47: 535-9 (1986)
28. Kopczynski SL et al; Environ Sci Technol 6: 342-7 (1972)
29. Kvenvolden KA, Redden GD; Geochimica et Cosmochimica Acta 44: 1145-50 (1980)
30. Lonneman WA et al; Hydrocarbons in Houston Air USEPA-600/3-79/018 p. 44 (1979)
31. Louw CW et al; Atmos Environ 11: 703-17 (1977)
32. Lyman WJ et al; Handbook of Chemical Property Estimation Methods NY: McGraw-Hill (1982)
33. Mackay D, Shiu WY; J Phys Chem Ref Data 19: 1175-99 (1981)
34. McAuliffe C; J Phys Chem 70: 1267-75 (1966)
35. Miller MM, Alkezweeny AJ; Ann NY Acad Sci 338: 219-32 (1980)
36. Mohan Rao AM, Panditt GG; Atmos Environ 2: 395-401 (1988)
37. Mulcahy MFR et al; Paper IV p 17 in Occurrence Contr Photochem Pollut, Proc Symp Workshop Sess (1976)
38. Neligan RE; Arch Environ Health 5: 581-91 (1962)
39. Nelson PF, Quigley SM; Environ Sci Technol 16: 650-5 (1982)
40. Nelson PF, Quigley SM; Atmos Environ 18: 79-87 (1984)
41. NIOSH; National Occupational Exposure Survey (NOES) (1989)
42. Patel RN et al; Appl Environ Microbiol 39: 720-6 (1980)
43. Pellizzari ED et al; Bull Environ Contam Toxicol 28: 322-8 (1982)
44. Rappaport SM et al; Appl Ind Hyg 2: 148-54 (1987)
45. Rudolph J, Khedim A; Int J Environ Anal Chem 290: 265-82 (1985)
46. Sauer TC Jr; Org Geochem 7: 1-16 (1981)
47. Scott Research Labs Inc; Atmospheric Reaction Studies in the Los Angeles Basin, NTIS PB-194-058 p 86 (1969)
48. Seila RL et al; Atmospheric Volatile Hydrocarbon Composition at Five Remote Sites in NW NC, USEPA-600/D-84-092 (1984)
49. Seila RL; Non-urban Hydrocarbons Concn in Ambient Air No of Houston TX USEPA-500/3-79-010 p38 (1979)
50. Sexton K, Westberg H; Environ Sci Tech 14: 329-32 (1980)
51. Sexton K; Environ Sci Technol 17: 402-7 (1983)
52. Shah JJ, Heyerdahl EK; National Ambient VOC Database Update USEPA 600/3-88/010 (1988)

Propane

53. Sigsby JE et al; Environ Sci Technol 21: 466-75 (1987)
54. Silverstein RM, Bassler GC; Spectrometric Id of Org Cmpd, J Wiley and Sons Inc pp. 148-69 (1963)
55. Stump FD, Dropkin DL; Anal Chem 57: 2629-34 (1985)
56. Swann RL et al; Res Rev 85: 16-28 (1983)
57. Uno I et al; Atmos Environ 19: 1283-93 (1985)
58. USEPA; EXAMS II Computer Simulation (1987)
59. USEPA; Drinking Water Criteria Document for Gasoline ECAO-CIN-D006, 8006-61-9 (1986)
60. Whelan JK et al; Geochim Cosmochim Acta 44: 1767-85 (1980)
61. Zweidinger RB et al; Environ Sci Tech 22: 956-62 (1988)

Propionic Acid

Synonyms:

Structure:

CAS Registry Number: 79-09-4

Molecular Formula: $C_3H_6O_2$

SMILES Notation: O=C(O)CC

CHEMICAL AND PHYSICAL PROPERTIES

Boiling Point: 141 °C at 760 mm Hg

Melting Point: -20.8 °C

Molecular Weight: 74.09

Dissociation Constants: pK_a = 4.87 [41]

Log Octanol/Water Partition Coefficient: 0.33 [24]

Water Solubility: miscible [41]

Vapor Pressure: 3.53 mm Hg at 25 °C [10]

Henry's Law Constant: 4.15×10^{-7} atm-m^3/mol at 25 °C [27]

ENVIRONMENTAL FATE/EXPOSURE POTENTIAL

Summary: Propionic acid may be released to the environment via effluents at sites where it is produced or used as a chemical intermediate, or as a grain and feed preservative. It is also released to the environment with the manufacture and use of coal-derived and shale oil liquid fuels and during the disposal of coal liquefication and gasification and wood preserving chemical waste byproducts. Textile mills, sewage treatment facilities, municipal and industrial landfills, hazardous waste sites, and gasoline and diesel fueled engines can release propionic acid to the environment. With a pKa of 4.87, propionic acid and its conjugate base will exist in environmental media in varying proportions that are pH dependent. Ions generally do not volatilize or adsorb to particulate matter as strongly as do their neutral counterparts. An estimated Henry's Law constant of 4.15×10^{-7} atm-m ^{3}/mol at 25 °C indicates that volatilization of propionic acid from environmental waters and moist soil should be extremely slow. Yet, propionic acid should evaporate from dry surfaces, especially when present in high concentrations such as in spill situations. The hydrolysis, photolysis and bioconcentration of propionic acid are not expected to be important fate processes. A low Koc indicates propionic acid should not partition from the water column to organic matter contained in sediments and suspended solids, and it should be highly mobile in soil. Monitoring data has shown it can leach to ground water. Biodegradation is likely to be the most important removal mechanism of propionic acid from aerobic soil and water. In the atmosphere, propionic acid is expected to exist almost entirely in the vapor phase and reactions with photochemically produced hydroxyl radicals should be important (estimated half-life of 13 days). Propionic acid is miscible in water and monitoring data has shown that physical removal from air by wet deposition (rainfall, dissolution in clouds) is an important removal mechanism. The most probable human exposure would be occupational exposure, which may occur through dermal contact or inhalation at workplaces where propionic acid, coal-derived or shale oil fuels are produced or used. The most common nonoccupational exposure is likely to result from either inhalation of urban air or from the ingestion of certain foods.

Natural Sources:

Artificial Sources: Demand for propionic acid was 130 million pounds in 1988 and is expected to increase to 145 million pounds by 1992 [8], with current use as follows: grain and feed preservatives (25%), cellulose plastics (20%), calcium and sodium propionates (18%), herbicide manufacture (18%), exports (15%) and miscellaneous uses, including butyl and pentyl propionates (4%). Butyl and pentyl propionates may be half a growing market in the paint and coatings industries in the future [8]. Consequently, propionic acid may be released to the environment via effluents where it is produced or used. Propionic acid is also released to the environment via effluents from the manufacture and use of coal-derived and shale oil liquid fuels, the disposal of coal liquefication and gasification waste by products, and wood preserving chemical waste byproducts [14,20,21,43]. Propionic acid may also be released to the aquatic environment in wastewater discharges from textile mills and sewage treatment facilities [5,39]. Municipal and industrial landfills and hazardous waste sites via leachates can release propionic acid to ground water supplies [1,4,31]. Propionic acid can be emitted to air as a component of exhaust from gasoline and diesel fueled engines [1,30].

Terrestrial Fate: With a pKa of 4.87 [41], propionic acid and its conjugate base will exist in soils in varying proportions that are pH dependent. Ions generally do not volatilize and undergo adsorption to the extent of their neutral counterparts. An estimated Henry's Law constant of 4.15×10^{-7} atm-m^3/mol at 25 °C [27] indicates that volatilization of propionic acid from moist soil should be extremely slow [32]. Yet, propionic acid should evaporate from dry surfaces, especially when present in high concentrations such as in spill situations. An estimated Koc of 36 [32] indicates propionic acid should be highly mobile in soil [44]; and monitoring data has shown it can leach to ground water [1,4,21,31,43]. Biodegradation is likely to be the most important removal mechanism of propionic acid from aerobic soil.

Aquatic Fate: With a pKa of 4.87 [41], propionic acid and its conjugate base will exist among environmental waters in varying proportions that are pH dependent. Under alkaline conditions, propionic acid is expected to exist as its conjugate base, the propionate ion [32]. Ions are not expected to volatilize from water or adsorb to particulate matter as strongly as their neutral counterparts. The miscibility of propionic acid in water suggests that volatilization, adsorption and bioconcentration are not important fate processes. This is supported by an estimated Henry's Law constant of 4.15×10^{-7} atm-m ^{3}/mol at 25 °C [27] which indicates that volatilization of propionic acid from environmental waters should be extremely slow [32]. An estimated Koc of 36 [32] indicates propionic acid should not partition from the water column to organic matter [44] contained in sediments and suspended solids; and an estimated bioconcentration factor (log BCF) of 0.02 indicates propionic acid should not bioconcentrate among aquatic organisms [32]. The hydrolysis [32], direct photolysis [6] and reaction of propionic acid with photochemically generated hydroxyl radicals in water [2] are not expected to be important fate processes. Aerobic biodegradation is likely to be the most important removal mechanism of propionic acid from aquatic systems [13].

Atmospheric Fate: Based on a vapor pressure of 3.53 mm Hg at 25 °C [10], propionic acid is expected to exist almost entirely in the vapor phase in ambient air [16]. In the atmosphere, vapor phase reactions with photochemically produced hydroxyl radicals may be important. The rate constant for propionic acid was measured to be 1.22×10^{-12} cm^3/molecule-sec at 25 °C [11], which corresponds to an atmospheric half-life of about 13 days at an atmospheric concentration of $5 \times 10^{+5}$ hydroxyl radicals per cm^3 [3]. Direct photolysis of propionic acid in air is not expected to occur. Propionic acid is miscible in water [41] and extensive monitoring data [7,28,35,48] have shown that physical removal from air by wet deposition (rainfall, dissolution in clouds) is an important removal mechanism.

Biodegradation: River die-away tests and grab sample data pertaining to the biodegradation of propionic acid in natural waters and soil were not located in the available literature. Yet, a number of aerobic biological screening studies, which utilized settled waste water, sewage, or activated sludge for inocula, have demonstrated that propionic acid is readily biodegradable [12,13,14,17,18,19,26,29,33,34,36,45,46,47]. For example, 5 day theoretical BODs of 23-55% [26], 37% [15], 40% [33] and 71% [13] have been reported. These studies indicate propionic acid should degrade rapidly under most environmental conditions. No data are available to suggest that any degradation process in water and soil, other than biodegradation, is important.

Abiotic Degradation: Carboxylic acids are generally resistant to aqueous environmental hydrolysis [32]. Therefore, hydrolysis of propionic acid is not expected to be important in the environment. The rate constant for the reaction of propionate ion with hydroxyl radicals in water at pH 9 is $4.7 \times 10^{+8}$ 1/mol-sec [2]. Assuming the hydroxyl radical concentration in full intensity sunlit natural water is 1.0×10^{-17} mole/L [37], the half-life for the photochemical reaction of propionic acid with hydroxyl radicals in water under conditions of continuous full intensity sunlight would be about 4.65 years. Therefore, this photooxidation reaction in natural waters is not expected to be important. The rate constant for the reaction of propionic acid with photochemically produced hydroxyl radicals in air has been experimentally determined to be 1.22 x 10^{-12} cm^3/molecule-sec at 25 °C [11]. Based upon an average yearly atmospheric hydroxyl radical concentration of 5.0 x 10^{+5} molecules/cm^3 in a typical atmosphere [3], the corresponding half-life for propionic acid would be about 13 days. Since low molecular weight organic acids have absorption bands at wavelengths well below the environmentally important range (>290 nm), the direct photolysis of propionic acid in air is not expected to be important [6].

Bioconcentration: Because propionic acid is miscible in water [41], bioconcentration in aquatic systems is not expected to be an important fate process. Based upon a log Kow of 0.33 [24], a bioconcentration factor (log BCF) of 0.02 for propionic acid has been calculated using a recommended regression-derived equation [32]. This BCF value also indicates propionic acid should not bioconcentrate among aquatic organisms.

Soil Adsorption/Mobility: Because propionic acid is miscible in water [41], soil adsorption is not expected to be an important fate process. Based on a log Kow of 0.33 [24], a Koc of 36 for propionic acid has been calculated using a recommended regression-derived equation [32]. This Koc value indicates propionic acid will be very highly mobile in soil [44], and it should not partition from the water column to organic matter contained in sediments and suspended solids. However, with a pKa of 4.87 [41], propionic acid and its conjugate base, the propionate ion, will exist among environmental media in varying proportions that are pH dependent. Monitoring data have shown that propionic acid can leach to ground water.

Volatilization from Water/Soil: With a pKa of 4.87 [41], propionic acid and its conjugate acid will exist among environmental media in varying proportions that are pH dependent. Ions are not expected to volatilize from water. Because propionic acid is miscible in water [41], and based upon an estimated Henry's Law constant of 4.15 x 10^{-7} atm-m^3/mol at 25 °C [27], volatilization of propionic acid from natural bodies of water and moist soils is also not expected to be an important fate process [32]. Yet, based upon a vapor pressure of 3.53 mm Hg at 25 °C [10], propionic acid can evaporate from dry surfaces, especially when present in high concentrations such as in spill situations.

Water Concentrations: SURFACE WATER: Ohio and Little Miami Rivers and Tanners Creek water contained propionic acid at concentrations ranging from 0.1 to 0.8 ug/L [39]. Water samples from Lake Kizaki, Japan, on October 28, 1977, contained propionic acid at concentrations ranging from undetected levels to 90 ugC/L [23]. GROUND WATER: Propionic acid was detected in ground water samples near a coal gasification site near Hoe Creek in northeastern WY [43]. Propionic acid has been detected in ground waters contaminated with leachates from municipal and industrial landfills, and hazardous waste sites [1,4,31]. Leachates at sites from the Netherlands, United Kingdom, Canada, France and Spain contained propionic acid at concentrations of 3.4, 1.36, 1.01, 5.25 and 0.91 g/L [31]. Wood preserving chemicals at Pensacola, FL, are responsible for propionic acid concentrations of 23.60 and 0.02 mg/L at ground water depths of 6 and 18 m, respectively [21]. RAIN/SNOW: Between March 15 and May 28, 1984, the propionate ion was detected in precipitation collected at Round and Geneva Lakes, Wisconsin at concentrations up to 2.7 umol/L [7]. Rain water at Brookhaven National Laboratory at Upton, NY, contained the propionate ion at trace concentrations [28]. Rain water at Hannover, Germany, contained propionic acid [48]. Between June 1976 and May 1977, propionic acid was detected in precipitation collected at Hubbard Brook, NH, at concentrations ranging from 0.3 umol/75 cm ppt to 0.8 umol/94 cm ppt [35]. In July 1975, propionic acid was detected in precipitation collected at Ithaca, NY, at a mean concentration of 1.0 uequiv/L [35]. Between 1969 and 1971, propionic acid was detected in precipitation collected at Voronezh, USSR, at concentrations ranging from undetected levels to 73.7 mequiv/L [35].

Effluent Concentrations: Propionic acid may be released to the aquatic environment in wastewater discharges from industry and sewage treatment facilities. In response to the June 1976 consent decree, the EPA surveyed the wastewaters of 46 industrial categories for 129 priority pollutants. Propionic acid was detected in 2 of 21 industrial categories of wastewater effluents [5]. Extract from the wastewater of a textile mill contained propionic acid at an average concentration of 38,144 mg/L [5]. Extract from the wastewater of a publicly owned sewage treatment works contained propionic acid at an average concentration of 5911 mg/L [5]. Primary effluents from 3 sewage treatment facilities contained propionic acid at concentrations from 16 to 3800 ug/L [39]. Secondary effluents from 4 sewage treatment facilities contained propionic acid at concentrations from

1.2 to 68 ug/L [39]. Coal gasification facilities can release propionic acid to ground water [43]. Propionic acid was detected in the wastewater effluent of a coal gasification facility located at the Grand Fork's Energy Technology Center, ND, at an estimated concentration of 64 mg/L [20]. In addition, wastewater effluent from a shale oil facility in Queensland, Australia, was shown to contain propionic acid at a concentration of 130 mg/L [14]. The disposal of waste byproducts from the production of wood preserving chemicals at Pensacola, FL, was responsible for the release of propionic acid to ground water [21]. Municipal and industrial landfills and hazardous waste sites via leachates can release propionic acid to ground water supplies [1,4,31]. Propionic acid was emitted in the exhaust from gasoline and diesel fueled engines in Los Angeles, CA, from July to September, 1984 at concentrations ranging from 1.22 to 19 ppb [30].

Sediment/Soil Concentrations: The propionate ion was detected in the sediments from 2 of 3 sampling stations of Loch Eil, Scotland [38]. At one station, the average concentrations at depths of 0 to 3, 3 to 6 and 6 to 12 cm were 30.6 ug/g, 0.5 ug/g and trace quantities, respectively [38]. At the other, the average concentrations at depths of 0 to 3, 3 to 6, 6 to 9 and 9 to 12 cm were 59.0, 15.9, 1.1 and 0.5 ug/g, respectively [38].

Atmospheric Concentrations: The average and maximum propionic acid concentrations for the ambient air over the Netherlands in 1980 was reported to be 0.15 and 2.0 ppb [22]. Propionic acid was detected in the ambient air of Los Angeles, CA, from July to September, 1984 at concentrations ranging from 0.019 to 0.305 ppb [30]. On September 24-25, 1984, propionic acid was detected in the ambient air of Los Angeles, CA, at an average concentration of 0.139 ppb for 3 samples with a high and low concentration of 0.154 and 0.126 ppb [30].

Food Survey Values: Propionic acid has been qualitatively detected as a volatile component of baked potatoes [9] and cooked meats [42]. Dalieb fruit (*Borassus aethiopum* L.) contained propionic acid at an average concentration of 84 mg/kg [25]. Propionic acid occurs in dairy products in trace amounts. It occurs naturally in swiss cheese at levels that may be as high as 1%. Propionic acid was isolated from boiled beef in a slurry, and in dry cured ham; quantities were not identified.

Plant Concentrations:

Fish/Seafood Concentrations: Two samples of Mussels (*Mytilus edulis*) from the Oarai Coast, Japan, contained propionic acid at a concentration of 2.73 and 0.50 ug/g [49].

Animal Concentrations:

Milk Concentrations:

Other Environmental Concentrations:

Probable Routes of Human Exposure: The most probable routes of human exposure to propionic acid are by inhalation, dermal contact and ingestion. Urban atmospheres containing propionic acid have been documented [22,30]. Certain foods [9,25,42] and mussels [49] have been shown to contain propionic acid [25].

Average Daily Intake:

Occupational Exposure: The most probable human exposure to propionic acid would be occupational exposure, which may occur through dermal contact or inhalation at places where it is produced or used. NIOSH (NOES 1981-1983) has statistically estimated that 23,167 workers are potentially exposed to propionic acid in the US [40]. The most common nonoccupational exposure is likely to result from either inhalation of urban air or the ingestion of certain foods.

Body Burdens:

REFERENCES

1. Albaiges J et al; Wat Res 20: 1153-9 (1986)
2. Anbar M, Neta P; Int J Appl Radiation Isotopes 18: 493-523 (1967)
3. Atkinson R; Chem Rev 85: 69-290 (1985)
4. Burrows WD, Rowe RS; J Water Pollut Control Fed 47: 92-3 (1975)
5. Bursey JT, Pellizzari ED; Analysis of Industrial Wastewater for Organic Pollutants in Consent Decree Survey. Contract No 68-03-2867. Athens, GA: USEPA Environ Res Lab (1982)
6. Calvert JG, Pitts JN Jr; Photochemistry John Wiley & Sons Inc. NY pp. 427-30 (1966)
7. Chapman et al; Atmos Environ 20: 1717-27 (1986)

Propionic Acid

8. CMR 1988; Chemical Marketing Reporter Chemical Profile: Propionic acid Aug 1 (1988)
9. Coleman EC et al; J Agric Food Chem 29: 42-8 (1981)
10. Daubert TE, Danner RP; Data Compilation, Tables of Properties of Pure Cmpds, Design Inst for Phys Prop Data, Am Inst for Phys Prop Data, NY (1985)
11. Daugaut et al; Int J Chem Kinet 20:331-8 (1988)
12. Dawson PSS, Jenkins SH; Sew Ind Wastes 22: 490-507 (1950)
13. Dias FF, Alexander M; Appl Microbiol 22: 1114-8 (1971)
14. Dobson KR et al; Water Res 19: 849-56 (1985)
15. Dore M et al; Trib Cebedeau 28: 3-11 (1975)
16. Eisenreich SJ et al; Environ Sci Technol 15: 30-8 (1981)
17. Gaffney PE, Heukelekian H; Sew Indust Wastes 30: 503 (1958)
18. Gaffney PE, Heukelekian H; Sew Indust Wastes 30: 673-9 (1958)
19. Gaffney PE, Heukelekian H; J Water Pollut Control Fed 33: 1169-83 (1961)
20. Giabbai, MF et al; Inter J Environ Anal Chem 20: 113-29 (1985)
21. Goerlitz DF et al; Environ Sci Technol 19: 995-61 (1985)
22. Guichert R, Schulting FL; Sci Total Environ 43: 193-219 (1985)
23. Hama T, Handa N; Rikusiugaka Zasshi 42: 8-19 (1981)
24. Hansch C, Leo AJ; Medchem Project Issue No 26. Claremont CA: Pomona College (1985)
25. Harper DB et al: J Sci Food Agric 37: 685-8 (1986)
26. Heukelekian H, Rand MC; J Water Pollut Control Assoc 29: 1040-53 (1955)
27. Hine J, Mookerjee PK; J Org Chem 40: 292-8 (1975)
28. Hoffman WA Jr, Tanner RL; Detection of Organic Acids in Atmospheric Precipitation BNL-51922 NTIS DE86 005294 Brookhaven National Lab Environ Chem Div Dept Appl Sci pp. 21 (1986)
29. Ishikawa S et al; Water Res 13: 681-5 (1979)
30. Kawamura K et al; Environ Sci Technol 19: 1082-6 (1985)
31. Lema JM et al; Water Air Soil Pollut 40: 223-50 (1988)
32. Lyman WJ et al; Handbook of Chemical Property Estimation Methods NY: McGraw-Hill p. 4-9, 5-4, 6-3, 7-4, 15-15 to 15-29 (1982)
33. Malaney GW, Gerhold RM; J Water Pollut Control Fed 41: R18- R33 (1969)
34. Malaney GW, Gerhold RM; Proc 17th Indust Waste Conf Purdue Univ Ext Ser 112: 249-57 (1962)
35. Mazurek MA, Simoneitt BRT; Organic Components in Bulk and Wet-only Precipitation CRC Critical Review Environ Control 16: 140 (1986)
36. McKinney RE et al; Sew Indust Wastes 28: 547-57 (1956)
37. Mill T; Science 207: 886-7 (1980)
38. Miller D et al; Mar Biol 50: 375-83 (1979)
39. Murtaugh JJ, Bunch RL; J Water Pollut Contr Fed 37: 410-5 (1965)
40. NIOSH; National Occupational Exposure Survey (NOES) (1989)
41. Riddick JA et al; Organic Solvents John Wiley & Sons Inc. NY (1984)
42. Shibamoto T et al; J Agric Food Chem 29: 57-63 (1981)
43. Stuermer DH et al; Environ Toxicol Chem 16: 582-7 (1982)
44. Swann RL et al; Res Rev 85: 16-28 (1983)

45. Takemoto S et al; Suishitsu Odaku Kenkyu 4: 80-90 (1981)
46. Thom NS, Agg AR; Proc R Soc Lond B 189: 347-57 (1975)
47. Urano K, Kato Z; J Hazardous Materials 13: 147-59 (1986); Yonezawa Y et al; Kogai Shigen Kenkyusho Iho 12: 85-91 (1982)
48. Winkeler HD et al; Vom Wasser 70: 107-17 (1988)
49. Yasuhara A, Morita M; Chemosphere 16: 2559-65 (1987)

1-Propoxy-2-propanol

SUBSTANCE IDENTIFICATION

Synonyms:

Structure:

CAS Registry Number: 569-01-3

Molecular Formula: $C_6H_{14}O_2$

SMILES Notation: O(CC(O)C)CCC

CHEMICAL AND PHYSICAL PROPERTIES

Boiling Point: 148-149 °C at 730 mm Hg

Melting Point:

Molecular Weight: 118.2

Dissociation Constants:

Log Octanol/Water Partition Coefficient: 0.621 (estimated) [11]

Water Solubility: Miscibility with water [2]

Vapor Pressure: 1.7 mm Hg at 20 °C [2]

Henry's Law Constant: 3.46×10^{-8} atm-m^3/mol [7]

ENVIRONMENTAL FATE/EXPOSURE POTENTIAL

Summary: 1-Propoxy-2-propanol may be released to the environment in effluents and emissions from its manufacturing plants, in spills during transport of bulk quantities, and from the land disposal of unused solvent formulations that contain the compound. If released to soil, it will be expected to exhibit very high mobility, based upon the reported infinite solubility of the compound in water and an estimated Koc of 3.0 calculated from an estimated log Kow. It is not known whether biodegradation will be an important environmental pathway in soil. It should not be subject to volatilization from moist near-surface soil based upon a Henry's Law constant of 3.46 x 10^{-8} atm-m^3/mol, which has been estimated using a structure-based estimation method. However, it may volatilize from dry near-surface soil and other dry surfaces based upon its vapor pressure of 1.7 mm Hg at 20 °C. If released to water, it will not be expected to adsorb to sediment or suspended particulate matter or to bioconcentrate in aquatic organisms based upon its estimated Koc and BCF, respectively, calculated from the estimated log Kow. It is not known whether biodegradation will be an important environmental pathway in water. It will not hydrolyze or directly photolyze in environmental water. It should not be subject to volatilization from surface waters based upon the estimated Henry's Law constant. If 1-propoxy-2-propanol is released to the atmosphere, it can be expected to exist mainly in the vapor-phase in the ambient atmosphere based on its vapor pressure. The estimated half-life for vapor-phase reaction with photochemically produced hydroxyl radicals is 15 hr based upon an estimated rate constant for this process. It will not be susceptible to direct photolysis in the atmosphere. Based upon its miscibility with water, the compound may be susceptible to removal from the atmosphere by washout. Occupational exposure may occur through inhalation of contaminated air and dermal contact with solutions containing this solvent.

Natural Sources:

Artificial Sources: 1-Propoxy-2-propanol may be released to the environment in effluents and emissions from its manufacturing plants, in spills during transport of bulk quantities, and from the land disposal of unused solvent formulations that contain this compound [2].

Terrestrial Fate: If 1-propoxy-2-propanol is released to soil, it will be expected to exhibit very high mobility [10], based upon the reported infinite solubility of the compound in water and an estimated Koc of 3.0 [5] from the estimated log Kow. It is not known whether biodegradation will be an important environmmental pathway in soil. It should not be subject to volatilization from moist near-surface soil based upon the Henry's Law constant. However, it may volatilize from dry near-surface soil and other dry surfaces based upon its vapor pressure.

Aquatic Fate: If 1-propoxy-2-propanol is released to water, it will not be expected to adsorb to sediment or suspended particulate matter or to bioconcentrate in aquatic organisms based upon its estimated Koc and BCF, respectively, calculated [5] from the measured log Kow, and its miscibility with water [2]. It is not known whether biodegradation will be an important environmental pathway. It will not hydrolyze [5] or directly photolyze [9] in environmental water. It should not be subject to volatilization from surface waters based upon the estimated Henry's Law constant.

Atmospheric Fate: If 1-propoxy-2-propanol is released to the atmosphere, it can be expected to exist mainly in the vapor phase based upon its vapor pressure. The estimated rate constant for vapor-phase reaction with photochemically produced hydroxyl radicals is 25.39×10^{-12} cm^3/molecule-sec at 25 °C [1,6], which corresponds to an atmospheric half-life of 15 hr at an atmospheric concentration of $5 \times 10^{+5}$ hydroxyl radicals per cm^3 [1]. 1-Propoxy-2-propanol will not be susceptible to direct photolysis in the atmosphere because alcohols and ethers do not absorb light at wavelengths >290 nm [9]. Based upon its miscibility with water, the compound may be susceptible to removal from the atmosphere by washout.

Biodegradation:

Abiotic Degradation: The rate constant for the vapor-phase reaction of 1-propoxy-2-propanol with photochemically produced hydroxyl radicals has been estimated to be 25.39×10^{-12} cm^3/molecule-sec at 25 °C [1,6], which corresponds to an atmospheric half-life of 15 hr at an atmospheric concentration of $5 \times 10^{+5}$ hydroxyl radicals per cm^3 [1]. Hydrolysis of 1-propoxy-2-propanol will not be important in the environment since ethers and alcohols are not susceptible to hydrolysis under environmental conditions [5]. Since alcohols and ethers do not absorb light at wavelengths >290 nm, it will not be susceptible to direct photolysis in sunlight [9].

Bioconcentration: Based upon the estimated log Kow, a BCF of 1.7 has been calculated using a recommended regression equation [5]. Based upon this estimated BCF, 1-propoxy-2-propanol will not be expected to bioconcentrate in aquatic organisms.

Soil Adsorption/Mobility: Based upon the estimated log Kow, a Koc of 3.0 has been estimated using a recommended regression equation [5]. Based upon this estimated Koc, 1-propoxy-2-propanol will be expected to exhibit very high mobility in soil [10].

Volatilization from Water/Soil: Volatilization of 1-propoxy-2-propanol from environmental waters will not be an important process [5] based upon the Henry's Law constant.

Water Concentrations:

Effluent Concentrations: 1-Propoxy-2-propanol was tentatively and qualitatively detected in 1 of 16 samples of concentrate derived from large volume (>400 gallons) samples of advanced treatment concentrate from 1 of 6 cities (Dallas, TX, November 1974) [4].

Sediment/Soil Concentrations:

Atmospheric Concentrations:

Food Survey Values:

Plant Concentrations:

Fish/Seafood Concentrations:

Animal Concentrations:

Milk Concentrations:

Other Environmental Concentrations:

Probable Routes of Human Exposure: Occupational exposure to 1-propoxy-2-propanol may occur through inhalation of contaminated air and dermal contact with solutions containing this solvent.

1-Propoxy-2-propanol

Average Daily Intake:

Occupational Exposure: NIOSH (NOHS 1972-1974) has statistically estimated that 2,434 workers are potentially exposed to 1-propoxy-2-propanol in the US [8].

Body Burdens:

REFERENCES

1. Atkinson R; Environ Toxicol Chem 7: 435-42 (1988)
2. Brown ES et al; Kirk-Othmer Encycl Chem Tech 3rd ed. NY: Wiley 11: 953 (1980)
3. Eisenreich SJ et al; Environ Sci Technol 15: 30-8 (1981)
4. Lucas SV; GC/MS Anal of Org in Drinking Water Concentrates and Advanced Treatment Concentrates Vol 1 USEPA-600/1-84-020a (NTIS PB85-128221) p. 152 (1984)
5. Lyman WJ et al; Handbook of Chemical Property Estimation Methods NY: McGraw-Hill (1982)
6. Meylan WM, Howard PH; Chemosphere 26:2293-2299 (1993)
7. Meylan W, Howard PH; Environ Toxicol Chem 10: 1283-93 (1991)
8. NIOSH; The National Occupational Hazard Survey (NOHS) (1974)
9. Silverstein RM et al; Spectrometric Id of Org Cmpd, J Wiley & Sons Inc 3rd ed. p. 239 (1974)
10. Swann RL et al; Res Rev 85: 17-28 (1983)
11. USEPA; CLOGP3 PCGEMS Graphical Exposure Modeling System (1986)

n-Propylbenzene

SUBSTANCE IDENTIFICATION

Synonyms:

Structure:

CAS Registry Number: 103-65-1

Molecular Formula: C_9H_{12}

SMILES Notation: c(cccc1)(c1)CCC

CHEMICAL AND PHYSICAL PROPERTIES

Boiling Point: 159.2 °C

Melting Point: -99.2 °C

Molecular Weight: 120.19

Dissociation Constants:

Log Octanol/Water Partition Coefficient: 3.57 [17]

Water Solubility: 52.2 mg/L at 25 °C [40]

Vapor Pressure: 3.42 mm Hg at 25 °C [9]

Henry's Law Constant: 0.0105 atm-m³/mol at 25 °C [34]

ENVIRONMENTAL FATE/EXPOSURE POTENTIAL

Summary: n-Propylbenzene occurs naturally in petroleum and bituminous coal. It is released to the atmosphere in emissions from combustion sources such as incinerators, gasoline engines, and diesel engines. Solvent evaporation, landfill leaching, and general use of asphalt also release it to the environment. If released to the atmosphere, n-propylbenzene will degrade in the vapor phase by reaction with photochemically produced hydroxyl radicals (estimated half-life of 2.7 days). If released to soil or water, n-propylbenzene will probably biodegrade. The results of various biodegradation screening studies suggest that n-propylbenzene can biodegrade in the environment. Biodegradation is the only identifiable degradation process in soil. Photosensitized photolysis may contribute to degradation in water. Volatilization may be the dominant removal mechanism in many environmental surface waters. Occupational exposure to n-propylbenzene occurs through dermal contact and inhalation of vapor. The general population is continually exposed to n-propylbenzene through inhalation since it occurs ubiquitously in the atmosphere.

Natural Sources: n-Propylbenzene occurs as a natural constituent in petroleum [1] and bituminous coal [35].

Artificial Sources: Incineration of organic, petroleum or coal wastes and combustion of fuels will release n-propylbenzene to the atmosphere [1]. General uses of asphalt and naphtha and use as a solvent will release n-propylbenzene to the environment [1]. n-Propylbenzene is emitted in the exhaust from gasoline and diesel engines [16]. Outboard motors and motor boats have been identified as sources of n-propylbenzene emissions to water [18,28]. n-Propylbenzene can be released to the environment in leachates and vapor emissions from landfills [13,52].

Terrestrial Fate: The results of various biodegradation screening studies indicate that n-propylbenzene can biodegrade in the environment. Biodegradation is the only identifiable degradation process in soil. Measured and estimated Koc values ranging from 495 to 725 suggest that n-propylbenzene has low to medium soil mobility [1-3]; therefore, some leaching in soil is possible. The detection of n-propylbenzene in landfill leachates demonstrates that leaching can occur.

Aquatic Fate: The results of various biodegradation screening studies suggest that n-propylbenzene can biodegrade in the environment. In addition to biodegradation, photosensitized photolysis may contribute to the environmental degradation of n-propylbenzene in natural water [11]. Volatilization will be an important transport process. Volatilization half-lives (which exclude adsorption) of 3.3 and 39 hr can be estimated for a model river and model environmental pond, respectively [25,42]; if maximum predictable adsorption is included in the pond simulation, the volatilization half-life increases to 10 days [42]. Volatilization half-lives of 1.3-19 days have been predicted for the Narragansett Bay near RI where volatilization is expected to be the major removal process [49]. Aquatic hydrolysis is not an important fate process.

Atmospheric Fate: Based upon the vapor pressure, n-propylbenzene is expected to exist almost entirely in the vapor phase in the ambient atmosphere [12]. The dominant degradation process in the atmosphere is the vapor-phase reaction with photochemically produced hydroxyl radicals, which has an estimated half-life of 2.7 days [3]. The detection of n-propylbenzene in rainwater samples [20] suggests that physical removal from the atmosphere by wet deposition is possible.

Biodegradation: A batch system die-away test using artificial seawater, a 10-day incubation period, and an inoculum of coastal water from the North Sea found n-propylbenzene to undergo fast biooxidation (actual rates not reported) [44]. Theoretical BODs of 21.8-43.7% were measured using 3 different activated sludges in Warburg respirometers and 7.5 days of incubation [27]. n-Propylbenzene was readily oxidized (8-day theoretical BOD of 34.4%) in Warburg respirometer studies using an activated sludge that had been acclimated to aniline [26]. n-Propylbenzene was readily oxidized (1-day and 8-day theoretical BODs of 8.4 and 27.8%, respectively) in Warburg respirometer studies using an activated sludge that had been acclimated to benzene [26]. A 5-day theoretical BOD of 25.5% was observed in a mixed microbial culture degradation study [4]. A 5-day theoretical BOD of 2.3-2.5% (standard dilution technique) and a 6-hr theoretical BOD of 0.8% (Warburg respirometer) were measured for n-propylbenzene; however, the initial concns of n-propylbenzene may have been sufficiently high to be toxic to the microbial populations [5].

Abiotic Degradation: The rate constant for the vapor-phase reaction of n-propylbenzene with photochemically produced hydroxyl radicals has been experimentally determined to be 6.0×10^{-12} cm^3/molecule-sec at 25 °C which corresponds to an atmospheric half-life of about 2.7 days at an atmospheric concn of $5 \times 10^{+5}$ hydroxyl radicals per cm^3 [3]. Alkylated benzenes are generally resistant to aqueous environmental hydrolysis [25];

therefore, n-propylbenzene is not expected to chemically hydrolyze in environmental waters. n-Propylbenzene did not directly photolyze in experiments using pure aqueous seawater solutions and simulated sunlight [11]; however, addition of an anthraquinone photosensitizer resulted in n-propylbenzene degradation and a formation of 1-phenyl-1-propanone, 1-phenyl-1-propanol, and benzaldehyde [11].

Bioconcentration: Based upon the water solubility, the BCF for n-propylbenzene can be estimated to be 66 from a regression-derived equation [25]. Based upon the measured log Kow, the Koc for n-propylbenzene can be estimated to be 304 from a regression-derived equation [25].

Soil Adsorption/Mobility: A Koc of 725 was measured for n-propylbenzene using a surface sediment collected from the Tamar estuary [47]. A similar Koc of 676 was measured in a humic acid column via HPLC [39]. Adsorption percentages ranging from 0.16 to 5.58% were measured in soil column studies using three different soil types and a sludge sample [19]. Based upon the water solubility, the Koc for n-propylbenzene can be estimated to be 495 from a regression-derived equation [25]. The measured and estimated Koc values suggest that n-propylbenzene has medium to low soil mobility [38].

Volatilization from Water/Soil: The value of Henry's Law constant indicates that volatilization from environmental waters can be rapid [25]. Using the Henry's Law constant, the volatilization half-life from a model river (1 m deep flowing 1 m/sec with a wind velocity of 3 m/sec) can be estimated to be about 3.3 hr [25]. Volatilization half-life from a model environmental pond can be estimated to be about 39 hr [42]. However, both of these half-life estimates neglect the potentially important effects of adsorption to sediment and suspended materials; when maximum adsorption effects are included in the model pond simulation, the volatilization half-life increases to nearly 10 days [42]. Results of mesocosm studies simulating the Narragansett Bay indicate that volatilization is the major removal process from seawater [49]; volatilization half-lives of 1.3-19 days were estimated for summer, spring and winter seasons [49]. n-Propylbenzene's vapor pressure suggests that evaporation from dry surfaces will occur.

Water Concentrations: DRINKING WATER: Results of the USEPA

Groundwater Supply Survey (finished water supplies that use ground water sources) found that n-propylbenzene was detected in only 1 of 945 sources (concn of 0.98 ug/L in the one source) that were surveyed from throughout the US [50]. Drinking water samples collected from Miami, FL and Cincinnati, OH, during 1974 and 1975 had respective n-propylbenzene concns of 0.05 and 0.01 ug/L [43]. A drinking water sample from Cincinnati, OH, in February 1980 had a n-propylbenzene concn of 33 ng/L [8]. SURFACE WATER: The n-propylbenzene concn in Lake Constance varied from 4 to 132 ng/L; concns were observed to increase with increasing boat traffic [18]. GROUND WATER: A well water sample collected near a landfill in Delaware had an n-propylbenzene concn of 0.5 ug/L [10]. SEAWATER: n-Propylbenzene was detected (concn not reported) in seawater samples collected from the Narragansett Bay near RI during summer and winter monitoring between 1979 and 1981 [48]. The concn of n-propylbenzene ranged from 0.2-2.9 ng/L (avg 1.1) at a coastal site in MA during a 15 month monitoring period between 1977 and 1978 [15]. RAIN/SNOW: Trace levels (concn not reported) of n-propylbenzene were detected in rainwater collected in Los Angeles, CA, on March 26, 1982 [20].

Effluent Concentrations: n-Propylbenzene has been qualitatively detected in various wastewaters from the following industries: petroleum refining, textile mills, auto and other laundries, plastics manufacturing, and publicly owned treatment works [6]. Based upon dynamometer tests, the avg n-propylbenzene emission rate from gasoline-powered engines is 1.2 mg/km [16]. Leachate from 4 hazardous waste landfills in Germany contained n-propylbenzene levels of 10-700 ug/L [13]. An n-propylbenzene concn of 69 mg/m^3 was detected in gas emissions from a landfill in Great Britain [52]. An aqueous effluent from a US petroleum refinery had an n-propylbenzene concn of 13 ng/g [37].

Sediment/Soil Concentrations: n-Propylbenzene was not detected (detection limit 5-23 ppb) in sediments collected from the Duwamish River Delta (Puget Sound, WA) [45].

Atmospheric Concentrations: URBAN/SUBURBAN: An evaluated database of US air monitoring data for the years 1970-1987 contains the following data for n-propylbenzene (concn is reported as daily median conc): suburban sites - 213 samples, 0.123 ppb; urban sites - 520 samples,

0.167 ppb [36]. Air samples collected in Tulsa, OK, in July 1978 contained n-propylbenzene levels of 0-0.9 ppb [2]; samples from Rio Blanco County, CO, (July 1978) had levels of 0.5-1.6 ppb [2]. Ambient air samples collected in Long Beach, Burbank, Azuza, Los Angeles and Inglewood, CA (date not reported) contained n-propylbenzene levels of 0.001-0.011 ppm [29]. The avg concn in the ambient air of Sydney, Australia, between September 1979 and June 1980 (140 samples) was 0.4 ppb [30]. The n-propylbenzene concn in Los Angeles, CA, air ranged from 1-3 ppb during September-November 1981 monitoring [14]. URBAN/SUBURBAN: The avg concns of n-propylbenzene in the air of Vienna, Austria, during October 1986 to February 1987 monitoring were 0.6 ppb (background), 2.2 ppb (suburbs), and 4.8 ppb (streets) [22]. An avg n-propylbenzene concn of 2 ppb was monitored in Los Angeles, CA, air during the fall of 1966 [2]. Levels ranging from 0.2-61.1 ppb were detected in the ambient air of Houston, TX, during 1973 and 1974 monitoring [24]. RURAL/REMOTE: An evaluated database of US air monitoring data for the years 1970-1987 contains the following data for n-propylbenzene (concn is reported as daily median concn): remote sites - 2 samples, 0.056 ppb; rural sites - 2 samples, 0.056 ppb [36]. Samples from the Smokey Mountains (September 1978) had levels of 0-0.6 ppb [2].

Food Survey Values: n-Propylbenzene has been qualitatively identified as a volatile constituent of chickpea flour [33] and roasted filbert nuts [21].

Plant Concentrations:

Fish/Seafood Concentrations:

Animal Concentrations:

Milk Concentrations:

Other Environmental Concentrations: n-Propylbenzene has been qualitatively detected in latex paint [41]. It is present in gasoline at 0.61 wt % [46].

Probable Routes of Human Exposure: Occupational exposure to n-propylbenzene occurs through dermal contact and inhalation of vapor [32];

absorption takes place by inhalation and in small quantities through intact skin [32].

Average Daily Intake: AIR: Assuming an avg n-propylbenzene concn of 0.605-0.822 ug/m^3 in urban-suburban outdoor air [36] and an inhalation rate of 20 m^3/day, the average daily intake is 12.1-16.44 ug/day.

Occupational Exposure: Mean air levels of 0.1-0.2 ppm n-propylbenzene were detected inside two US factories involved in spray painting and glueing [51]. Air samples collected inside a tire retreading factory in Italy contained n-propylbenzene levels of 0-15 ug/m^3 [51]; a shoe sole factory (vulcanization area) had a level of 30-300 ug/m^3 [51].

Body Burdens: In the body, following oxidation of its side chain, n-propylbenzene is converted into benzoic acid, conjugated with glycine and excreted in the urine as hippuric acid [32]. n-Propylbenzene was qualitatively detected in 8 of 46 samples of human adipose tissue analyzed during the EPA National Human Adipose Tissue Survey in fiscal year 1982 [31].

REFERENCES

1. Abrams EF et al; Identification of Organic Compounds in Effluents from Industrial Sources USEPA-560/3-75-002 p. 134 (1975)
2. Arnts RR, Meeks SA; Atmos Environ 15: 1643-51 (1981)
3. Atkinson R; J Chem Phys Ref Data Monograph No. 1 p. 229 (1989)
4. Babeu L, Vaishnav DD; J Indus Microbiol 2: 107-15 (1987)
5. Bogan RH, Sawyer CN; Sewage Indust Wastes 27: 917-28 (1955)
6. Bursey JT, Pellizzari ED; Analysis of Industrial Wastewater for Organic Pollutants in Consent Decree Survey. Contract No. 68-03-2867. Athens, GA: USEPA Environ Res Lab p. 79, 90 (1982)
7. Cocheo V et al; Amer Ind Hyg Assoc J 44: 521-7 (1983)
8. Coleman EW et al; Arch Environ Contam Toxicol 13: 171-8 (1984)
9. Daubert TE, Danner RP; Physical and Thermodynamic Properties of Pure Chemicals: Data Compilation, NY: Hemisphere Pub Corp (1989)
10. DeWalle FB, Chian ESK; J Amer Water Works Assoc 73: 206-11 (1981)
11. Ehrhardt M, Petrick G; Marine Chem 15: 47-58 (1984)
12. Eisenreich SJ et al; Environ Sci Technol 15: 30-8 (1981)
13. Foerst C et al; Intern J Environ Chem 37: 287-93 (1989)
14. Grosjean D, Fung K; J Air Pollut Control Assoc 34: 537-43 (1984)
15. Gschwend PM et al; Environ Sci Technol 16: 31-8 (1982)

16. Hampton CV et al; Environ Sci Technol 17: 699-708 (1983)
17. Hansch C, Leo AJ; Medchem Project Issue No 19. Claremont CA: Pomona College (1981)
18. Juttner F; Z Wasser-Abwasser-Forsch 21: 36-9 (1988)
19. Kanatharana P, Grob RL; J Environ Sci Health A18: 59-77 (1983)
20. Kawamura K, Kaplan IR; Environ Sci Technol 17: 497-501 (1983)
21. Kinlin TE et al; J Agric Food Chem 20: 1021 (1972)
22. Lanzerstorfer C, Puxbaum H; Water, Air, and Soil Pollut 51: 345-55 (1990)
23. Lonneman WA et al; Environ Sci Technol 2: 1017-20 (1968)
24. Lonneman WA et al; Hydrocarbons in Houston Air. USEPA-600/3-79-018 Research Triangle Park, NC: USEPA (1979)
25. Lyman WJ et al; Handbook of Chemical Property Estimation Methods Washington, DC: Amer Chem Soc (1990)
26. Malaney GW; J Water Pollut Control Fed 32: 1300-11 (1960)
27. Marion CV, Malaney GW; pp. 297-308 in Proc 18th Ind Waste Conf, Eng Bull Purdue Univ, Eng Ext Ser (1964)
28. Montz WE Jr et al; Arch Environ Contam Toxicol 11: 561-5 (1982)
29. Neligan RE et al; The Gas Chromatographic Determination of Aromatic Hydrocarbons in the Atmosphere ACS Natl Mtg p. 118-21 (1965)
30. Nelson PF, Quigley SM; Environ Sci Technol 16: 650-5 (1982)
31. Onstat JD et al; Characterization of HRGC/MS Unidentified Peaks from the Broad Scan Analysis of the FY82 NHATS Composites. Vol I. EPA Contract No. 68-02-4252. Washington, DC: USEPA (1987)
32. Parmeggiani L; Encyl Occup Health & Safety 3rd ed Geneva, Switzerland: International Labour Office p. 1074-5 (1983)
33. Rembold H et al; J Agric Food Chem 37: 659-62 (1989)
34. Sanemasa I et al; Bull Chem Soc Jap 55: 1054-62 (1982)
35. Schobert HH; The Chemistry of Hydrocarbon Fuels London: Butterworths & Co Ltd p. 50 (1990)
36. Shah JJ, Heyerdahl EK; National Ambient Volatile Organic Compounds (VOCs) Database Update USEPA/600/3-88-010(a) Research Triangle Park, NC: USEPA p. 53 (1988)
37. Snider EH, Manning FS; Environ Intern 7: 237-58 (1982)
38. Swann RL et al; Res Rev 85: 23 (1983)
39. Szabo G et al; Chemosphere 21: 729-39 (1990)
40. Tewari YB et al; J Chem Eng Data 27: 451-4 (1982)
41. Tichenor BA, Mason MA; JAPCA 38: 264-8 (1988)
42. USEPA; EXAMS II Computer Simulation (1987)
43. USEPA; Preliminary Assessment of Suspected Carcinogens in Drinking Water. Interim Report to Congress. Washington, DC (1975)
44. Van Der Linden AC; Dev Biodegrad Hydrocarbons 1: 165-200 (1978)
45. Varanasi U et al; Environ Sci Technol 19: 836-41 (1985)
46. Verschueren. Handbook Environmental Data Organic Chemicals p.1026 (1983)
47. Vowles PD, Mantoura RFC; Chemosphere 16: 109-16 (1987)
48. Wakeham SG et al; Can J Fish Aquat Sci 40: 304-21 (1983)

n-Propylbenzene

49. Wakeham SG et al; Environ Sci Technol 17: 611-7 (1983)
50. Westrick JJ et al; J Amer Water Works Assoc 76: 52-9 (1984)
51. Whitehead LW et al; Amer Ind Hyg Assoc J 45: 767-772 (1984)
52. Young P, Parker A; ASTM Spec Tech Publ 851(Hazard Ind Waste Manage Test): 24-41 (1984)

2-Pyridinecarbonitrile

SUBSTANCE IDENTIFICATION

Synonyms:

Structure:

CAS Registry Number: 100-70-9

Molecular Formula: $C_6H_4N_2$

SMILES Notation: C(#N)c(nccc1)c1

CHEMICAL AND PHYSICAL PROPERTIES

Boiling Point: 227 °C

Melting Point: 29 °C

Molecular Weight: 104.11

Dissociation Constants:

Log Octanol/Water Partition Coefficient: 0.50 [5]

Water Solubility: 99,424 mg/L at 25 °C (estimated from Kow) [6]

Vapor Pressure: 0.5 mm Hg at 25 °C (estimated from Henry's Law constant and water solubility)

Henry's Law Constant: 6.81×10^{-8} atm-m^3/mol at 25 °C (estimate) [7]

ENVIRONMENTAL FATE/EXPOSURE POTENTIAL

Summary: 2-Pyridinecarbonitrile is released to the environment via effluents at sites where it is produced or used as a chemical intermediate.

2-Pyridinecarbonitrile

Information pertaining to the biodegradation of 2-pyridinecarbonitrile in soil and water was not located in the available literature. With pKa's between about 5.0 and 6.0 for most alkyl and allyl pyridines, 2-pyridinecarbonitrile and its conjugate acid may exist among environmental media in varying proportions that are pH dependent. Ions generally do not volatilize. The Henry's Law constant indicates that volatilization of 2-pyridinecarbonitrile from environmental waters and moist soil should not be an important fate process. Yet, 3-pyridinecarbonitrile may evaporate from dry surfaces, especially when present in high concns such as in spill situations. In aquatic systems, 2-pyridinecarbonitrile is not expected to bioconcentrate. A low Koc indicates 2-pyridinecarbonitrile should not partition from the water column to organic matter contained in sediments and suspended solids and should be highly mobile in soil. In the atmosphere, 2-pyridinecarbonitrile is expected to primarily exist in the vapor phase. Reactions with photochemically produced hydroxyl radicals should be slow (estimated half-life of 188 days). In addition, 2-pyridinecarbonitrile has the potential to be physically removed from air by wet deposition. The most probable human exposure would be occupational exposure, which may occur through dermal contact or inhalation at workplaces where 2-pyridinecarbonitrile is produced or used as a chemical intermediate.

Natural Sources:

Artificial Sources: 2-Pyridinecarbonitrile is an important chemical intermediate [4]. Consequently, 2-pyridinecarbonitrile may be released to the environment via effluents at sites where it is produced or used.

Terrestrial Fate: Information pertaining to the biodegradation of 2-pyridinecarbonitrile in soil was not located in the available literature. With pKa's between about 5.0 and 6.0 for most alkyl and pyridincs [1,4], 2-pyridinecarbonitrile and its conjugate acid may exist among environmental media in varying proportions that are pH dependent. Ions generally do not volatilize. Based upon the Henry's Law constant, volatilization of 2-pyridinecarbonitrile from moist soils is not expected to be an important fate process [6]. 2-Pyridinecarbonitrile may evaporate from dry surfaces, especially when present in high concns such as in spill situations. An estimated Koc of 45 [6] indicates 2-pyridinecarbonitrile should be highly mobile in soil [11].

Aquatic Fate: Information pertaining to the biodegradation of 2-pyridinecarbonitrile in aquatic systems was not located in the available literature. With pKa's between about 5.0 and 6.0 for most alkyl and allyl pyridines [1,4], 2-pyridinecarbonitrile and its conjugate acid may exist among environmental media in varying proportions that are pH dependent. The ratio of 2-pyridinecarbonitrile to its conjugate acid should increase with increasing pH [6]. Ions are not expected to volatilize from water. Based upon the Henry's Law constant, volatilization of 2-pyridinecarbonitrile from natural bodies of water is not expected to be an important fate process [6]. An estimated Koc of 45 [6] indicates 2-pyridinecarbonitrile should not partition from the water column to organic matter [11] contained in sediments and suspended solids; an estimated bioconcentration factor (log BCF) of 0.15 indicates 2-pyridinecarbonitrile should not bioconcentrate among aquatic organisms [6].

Atmospheric Fate: Based on the vapor pressure, 2-pyridinecarbonitrile is expected to primarily exist in the vapor phase in ambient air [3]. In the atmosphere, vapor-phase reactions with photochemically produced hydroxyl radicals are expected to be slow. The rate constant for 2-pyridinecarbonitrile was estimated to be 8.54×10^{-14} cm^3/molecule-sec at 25 °C, which corresponds to an atmospheric half-life of about 188 days at an atmospheric concn of $5 \times 10^{+5}$ hydroxyl radicals per cm^3 [2]. The estimated water solubility indicates that physical removal from air by rainfall and dissolution in clouds may occur.

Biodegradation:

Abiotic Degradation: The rate constant for the vapor-phase reaction of 2-pyridinecarbonitrile with photochemically produced hydroxyl radicals has been estimated to be 8.54×10^{-14} cm^3/molecule-sec at 25 °C, which corresponds to an atmospheric half-life of about 188 days at an atmospheric concn of $5 \times 10^{+5}$ hydroxyl radicals per cm^3 [2].

Bioconcentration: Based upon the log Kow, a bioconcentration factor (log BCF) of 0.15 for 2-pyridinecarbonitrile has been calculated using a recommended regression-derived equation [6]. This BCF value indicates 2-pyridinecarbonitrile should not bioconcentrate among aquatic organisms.

410

Soil Adsorption/Mobility: Based on the log Kow, a Koc of 45 for 2-pyridinecarbonitrile has been calculated using a recommended regression-derived equation [6]. This Koc value indicates 2-pyridinecarbonitrile will be highly mobile in soil [11], and it should not partition from the water column to organic matter contained in sediments and suspended solids. With pKa's between about 5.0 and 6.0 for most alkyl and allyl pyridines [1,4], 2-pyridinecarbonitrile and its conjugate acid may exist among environmental media in varying proportions that are pH dependent.

Volatilization from Water/Soil: With pKa's between about 5.0 and 6.0 for most alkyl and allyl pyridine [1,4], 2-pyridinecarbonitrile and its conjugate acid may exist among environmental media in varying proportions that are pH dependent. Ions are not expected to volatilize from water. Based upon the Henry's Law constant, volatilization of 2-pyridinecarbonitrile from natural bodies of water and moist soils is not expected to be an important fate process [6]. Yet, the vapor pressure indicates 2-pyridinecarbonitrile may evaporate from dry surfaces, especially when present in high concns such as in spill situations.

Water Concentrations:

Effluent Concentrations: 2-Pyridinecarbonitrile was detected in 1 of 21 industrial wastewater effluents at a concn between 10 and 100 ug/L [10].

Sediment/Soil Concentrations:

Atmospheric Concentrations: 2-Pyridinecarbonitrile was qualitatively listed as a contaminant of the ambient air at Houston, TX [9].

Food Survey Values:

Plant Concentrations:

Fish/Seafood Concentrations:

Animal Concentrations:

Milk Concentrations:

Other Environmental Concentrations:

Probable Routes of Human Exposure: The most probable route of human exposure to 2-pyridinecarbonitrile is by inhalation and dermal contact.

Average Daily Intake:

Occupational Exposure: The most probable human exposure to 2-pyridinecarbonitrile would be occupational exposure, which may occur through dermal contact or inhalation at places where it is produced or used. NIOSH (NOES 1981-1983) has estimated that 311 workers are potentially exposed to 2-pyridinecarbonitrile in the US [8].

Body Burdens:

REFERENCES

1. Albert A et al; J Chem Soc 1948: 2240-9 (1948)
2. Atkinson R; Intern J Chem Kin 19: 799-828 (1987)
3. Eisenreich SJ et al; Environ Sci Technol 15: 30-8 (1981)
4. Goe GL; Kirk-Othmer Encycl Chem Tech 3rd NY: Wiley Interscience 19: 454-83 (1982)
5. Hansch C, Leo AJ; Medchem Project Issue No 26. Claremont CA: Pomona College (1985)
6. Lyman WJ et al; Handbook of Chemical Property Estimation Methods NY: McGraw-Hill (1982)
7. Meylan W, Howard PH; Environ Toxicol Chem 10: 1283-93 (1991)
8. NIOSH; National Occupational Exposure Survey (NOES) (1989)
9. Pelizzari ED; Development of Analytical Techniques for Measuring Ambient Atmospheric Carcinogenic Vapors. USEPA 600/2-75-076 (1975)
10. Perry DL et al; Ident of Org Compounds in Ind Effluent discharges USEPA-600/4-79-016 (NTIS PB-294794) p. 230 (1979)
11. Swann RL et al; Res Rev 85: 16-28 (1983)

3-Pyridinecarbonitrile

SUBSTANCE IDENTIFICATION

Synonyms:

Structure:

CAS Registry Number: 100-54-9

Molecular Formula: $C_6H_4N_2$

SMILES Notation: C(#N)c(cccn1)c1

CHEMICAL AND PHYSICAL PROPERTIES

Boiling Point: 245 °C

Melting Point: 50-52 °C

Molecular Weight: 104.11

Dissociation Constants:

Log Octanol/Water Partition Coefficient: 0.36 [5]

Water Solubility: 78,976 mg/L at 25 °C (estimated from Kow) [6]

Vapor Pressure: 0.4 mm Hg at 25 °C (estimated from Henry's Law constant and water solubility)

Henry's Law Constant: 6.81 x 10^{-8} atm-m^3/mol at 25 °C (estimate) [7]

ENVIRONMENTAL FATE/EXPOSURE POTENTIAL

Summary: 3-Pyridinecarbonitrile may be released to the environment via effluents at sites where it is produced or used as a chemical intermediate.

413

Information pertaining to the biodegradation of 3-pyridinecarbonitrile in soil and water was not located in the available literature. With pKa's between about 5.0 and 6.0 for most alkyl and allyl pyridines, 3-pyridinecarbonitrile and its conjugate acid may exist among environmental media in varying proportions that are pH dependent. Ions generally do not volatilize. The Henry's Law constant indicates that volatilization of 3-pyridinecarbonitrile from environmental waters and moist soil should not be an important fate process. Yet, 3-pyridinecarbonitrile may evaporate from dry surfaces, especially when present in high concns such as in spill situations. In aquatic systems, 3-pyridinecarbonitrile is not expected to bioconcentrate. A low Koc indicates 3-pyridinecarbonitrile should not partition from the water column to organic matter contained in sediments and suspended solids and should be highly mobile in soil. In the atmosphere, 3-pyridinecarbonitrile is expected to primarily exist in the vapor phase. Reactions with photochemically produced hydroxyl radicals should be slow (estimated half-life of 188 days). In addition, 3-pyridinecarbonitrile has the potential to be physically removed from air by wet deposition. The most probable human exposure would be occupational exposure, which may occur through dermal contact or inhalation at workplaces where 3-pyridinecarbonitrile is produced or used as a chemical intermediate.

Natural Sources:

Artificial Sources: 3-Pyridinecarbonitrile is commercially the most important chemical intermediate of the cyanopyridines [4]. Consequently, 3-pyridinecarbonitrile may be released to the environment via effluents at sites where it is produced or used.

Terrestrial Fate: Information pertaining to the biodegradation of 3-pyridinecarbonitrile in soil was not located in the available literature. With pKa's between about 5.0 and 6.0 for most alkyl and alkyl pyridines [1,4], 3-pyridinecarbonitrile and its conjugate acid may exist among environmental media in varying proportions that are pH dependent. Ions generally do not volatilize. Based upon the Henry's Law constant, volatilization of 3-pyridinecarbonitrile from moist soils is not expected to be an important fate process [6]. 3-Pyridinecarbonitrile may evaporate from dry surfaces, especially when present in high concns such as in spill situations. An estimated Koc of 37 [6] indicates 3-pyridinecarbonitrile should be highly mobile in soil [8].

Aquatic Fate: Information pertaining to the biodegradation of 3-pyridinecarbonitrile in aquatic systems was not located in the available literature. With pKa's between about 5.0 and 6.0 for most alkyl and allyl pyridines [1,4], 3-pyridinecarbonitrile and its conjugate acid may exist among environmental media in varying proportions that are pH dependent. The ratio of 3-pyridinecarbonitrile to its conjugate acid should increase with increasing pH [6]. Ions are not expected to volatilize from water. Based upon the Henry's Law constant, volatilization of 3-pyridinecarbonitrile from natural bodies of water is not expected to be an important fate process [6]. An estimated Koc of 37 [6] indicates 3-pyridinecarbonitrile should not partition from the water column to organic matter [8] contained in sediments and suspended solids; and an estimated bioconcentration factor (log BCF) of 0.04 indicates 3-pyridinecarbonitrile should not bioconcentrate among aquatic organisms [6].

Atmospheric Fate: Based on the vapor pressure, 3-pyridinecarbonitrile is expected to primarily exist in vapor phase in ambient air [3]. In the atmosphere, vapor-phase reactions with photochemically produced hydroxyl radicals are expected to be slow. The rate constant for 3-pyridinecarbonitrile was estimated to be 8.54×10^{-14} cm^3/molecule-sec at 25 °C, which corresponds to an atmospheric half-life of about 188 days at an atmospheric concn of $5 \times 10^{+5}$ hydroxyl radicals per cm^3 [2]. The estimated water solubility indicates that physical removal from air by rainfall and dissolution in clouds, may occur.

Biodegradation:

Abiotic Degradation: The rate constant for the vapor-phase reaction of 3-pyridinecarbonitrile with photochemically produced hydroxyl radicals has been estimated to be 8.54×10^{-14} cm^3/molecule-sec at 25 °C, which corresponds to an atmospheric half-life of about 188 days at an atmospheric concn of $5 \times 10^{+5}$ hydroxyl radicals per cm^3 [2].

Bioconcentration: Based upon the log Kow, a bioconcentration factor (log BCF) of 0.04 for 3-pyridinecarbonitrile has been calculated using a recommended regression-derived equation [6]. This BCF value indicates 3-pyridinecarbonitrile should not bioconcentrate among aquatic organisms.

3-Pyridinecarbonitrile

Soil Adsorption/Mobility: Based on the log Kow, a Koc of 37 for 3-pyridinecarbonitrile has been calculated using a recommended regression-derived equation [6]. This Koc value indicates 3-pyridinecarbonitrile will be highly mobile in soil [8], and it should not partition from the water column to organic matter contained in sediments and suspended solids. With pKa's between about 5.0 and 6.0 for most alkyl and allyl pyridines [1,4], 3-pyridinecarbonitrile and its conjugate acid may exist among environmental media in varying proportions that are pH dependent.

Volatilization from Water/Soil: With pKa's between about 5.0 and 6.0 for most alkyl and allyl pyridines [1,4], 3-pyridinecarbonitrile and its conjugate acid may exist among environmental media in varying proportions that are pH dependent. Ions are not expected to volatilize from water. Based upon the Henry's Law constant, volatilization of 3-pyridinecarbonitrile from natural bodies of water and moist soils is not expected to be an important fate process [6]. The vapor pressure indicates 3-pyridinecarbonitrile may evaporate from dry surfaces, especially when present in high concns such as in spill situations.

Water Concentrations:

Effluent Concentrations:

Sediment/Soil Concentrations:

Atmospheric Concentrations:

Food Survey Values:

Plant Concentrations:

Fish/Seafood Concentrations:

Animal Concentrations:

Milk Concentrations:

Other Environmental Concentrations:

3-Pyridinecarbonitrile

Probable Routes of Human Exposure: The most probable route of human exposure to 3-pyridinecarbonitrile is by inhalation and dermal contact.

Average Daily Intake:

Occupational Exposure: The most probable human exposure to 3-pyridinecarbonitrile would be occupational exposure, which may occur through dermal contact or inhalation at places where it is produced or used.

Body Burdens:

REFERENCES

1. Albert A et al; J Chem Soc 1948: 2240-9 (1948)
2. Atkinson R; Intern J Chem Kin 19: 799-828 (1987)
3. Eisenreich SJ et al; Environ Sci Technol 15: 30-8 (1981)
4. Goe GL; Kirk-Othmer Encycl Chem Tech 3rd NY: Wiley Interscience 19: 454-83 (1982)
5. Hansch C, Leo AJ; Medchem Project Issue No 26. Claremont CA: Pomona College (1985)
6. Lyman WJ et al; Handbook of Chemical Property Estimation Methods NY: McGraw-Hill (1982)
7. Meylan W, Howard PH; Environ Toxicol Chem 10: 1283-93 (1991)
8. Swann RL et al; Res Rev 85: 16-28 (1983)

2-Pyridineethanol

SUBSTANCE IDENTIFICATION

Synonyms:

Structure:

CAS Registry Number: 103-74-2

Molecular Formula: C_7H_9NO

SMILES Notation: n(c(ccc1)CCO)c1

CHEMICAL AND PHYSICAL PROPERTIES

Boiling Point: 118-121 °C at 15 mm Hg

Melting Point:

Molecular Weight: 123.16

Dissociation Constants:

Log Octanol/Water Partition Coefficient: 0.12 [5]

Water Solubility: 77,400 mg/L at 25 °C [10]

Vapor Pressure: 7.03×10^{-5} mm Hg at 25 °C (calculated from Henry's Law constant and water solubility)

Henry's Law Constant: 1.47×10^{-10} atm-m^3/mol at 25 °C [9]

ENVIRONMENTAL FATE/EXPOSURE POTENTIAL

Summary: 2-Pyridineethanol may be released to the environment via effluents at sites where it is produced or used as a chemical intermediate.

Information pertaining to the biodegradation of 2-pyridineethanol in soil and water was not located in the available literature. With pKa's between 5.0 and 6.5 for most alkyl and allyl pyridines, 2-pyridineethanol and its conjugate acid should exist in environmental media in varying proportions that are pH dependent. Ions generally do not volatilize. The Henry's Law constant indicates that volatilization of 2-pyridineethanol from environmental waters and moist soil should not be an important fate process. In aquatic systems, 2-pyridineethanol is not expected to bioconcentrate; however, its presence has been qualitatively noted in fish. A low Koc indicates 2-pyridineethanol should not partition from the water column to organic matter contained in sediments and suspended solids. It should be highly mobile in soil and it may leach to ground water. In the atmosphere, 2-pyridineethanol is expected to exist in both the vapor and particulate phases, and reactions with photochemically produced hydroxyl radicals may be important (estimated half-life of 2.7 days). In addition, 2-pyridineethanol may be physically removed from air by wet deposition. The most probable human exposure would be occupational exposure, which may occur through dermal contact at workplaces where 2-pyridineethanol is produced or used. Nonoccupational exposures may occur during recreational activities at contaminated bodies of water or from the ingestion of contaminated fish.

Natural Sources:

Artificial Sources: 2-Pyridineethanol is an important chemical intermediate [3]. The first commercial end use of 2-pyridineethanol was in the manufacture of tire cord [3]. Consequently, 2-pyridineethanol may be released to the environment via effluents at sites where it is produced or used.

Terrestrial Fate: Information pertaining to the biodegradation of 2-pyridineethanol in soil was not located in the available literature. With pKa's between 5.0 and 6.5 for most alkyl and allyl pyridines [3], 2-pyridineethanol and its conjugate acid should exist in soils in varying proportions that are pH dependent. Ions generally do not volatilize. Based upon the Henry's Law constant, volatilization of 2-pyridineethanol from moist soils is also not expected to be an important fate process [7]. An estimated Koc of 28 [7] indicates 2-pyridineethanol should be highly mobile in soil [11], and it may leach to ground water.

Aquatic Fate: Information pertaining to the biodegradation of 2-pyridineethanol in aquatic systems was not located in the available literature. With pKa's between 5.0 and 6.5 for most alkyl and allyl pyridines [3], 2-pyridineethanol and its conjugate acid should exist among environmental waters in varying proportions that are pH dependent. The ratio of 2-pyridineethanol to its conjugate acid should increase with increasing pH [7]. Ions are not expected to volatilize from water. Based upon the Henry's Law constant, volatilization of 2-pyridineethanol from natural bodies of water is not expected to be an important fate process [7]. An estimated Koc of 28 [7] indicates 2-pyridineethanol should not partition from the water column to organic matter [11] contained in sediments and suspended solids. An estimated bioconcentration factor (log BCF) of -0.14 indicates 2-pyridineethanol should not bioconcentrate in aquatic organisms [7]; however, its presence has been qualitatively noted in fish [6].

Atmospheric Fate: Based on the vapor pressure, 2-pyridineethanol is expected to exist in both the vapor and particulate phases in ambient air [2]. In the atmosphere, vapor-phase reactions with photochemically produced hydroxyl radicals may be important. The rate constant for 2-pyridineethanol was estimated to be 6.02×10^{-12} cm^3/molecule-sec at 25 °C [8], which corresponds to an atmospheric half-life of about 2.7 days at an atmospheric concn of $5 \times 10^{+5}$ hydroxyl radicals per cm^3 [1]. The estimated water solubility indicates that physical removal from air by rainfall and dissolution in clouds may occur.

Biodegradation:

Abiotic Degradation: The rate constant for the vapor-phase reaction of 2-pyridineethanol with photochemically produced hydroxyl radicals has been estimated to be 6.02×10^{-12} cm^3/molecule-sec at 25 °C [8], which corresponds to an atmospheric half-life of about 2.7 days at an atmospheric concn of $5 \times 10^{+5}$ hydroxyl radicals per cm^3 [1].

Bioconcentration: Based upon a log Kow of 0.12 [5], a log bioconcentration factor of -0.14 for 2-pyridineethanol has been estimated using a recommended regression-derived equation [7]. This BCF value indicate 2-pyridineethanol should not bioconcentrate among aquatic organisms.

Soil Adsorption/Mobility: Based on a log Kow of 0.12 [5], a Koc of 28 for 2-pyridineethanol has been calculated using a recommended regression-derived equation [7]. This Koc value indicates 2-pyridineethanol will be highly mobile in soil [11], and it should not partition from the water column to organic matter contained in sediments and suspended solids. With pKa's between 5.0 and 6.5 for most alkyl and allyl pyridines [3], 2-pyridineethanol and its conjugate acid should exist among environmental media in varying proportions that are pH dependent.

Volatilization from Water/Soil: With pKa's between 5.0 and 6.5 for most alkyl and allyl pyridines [3], 2-pyridineethanol and its conjugate acid should exist among environmental media in varying proportions that are pH dependent. Ions are not expected to volatilize from water. Based upon the Henry's Law constant, volatilization of 2-pyridineethanol from natural bodies of water and moist soils is not expected to be an important fate process [7].

Water Concentrations: SURFACE WATER: 2-Pyridineethanol was detected in water sampled from the central basin of Lake Ontario [4].

Effluent Concentrations:

Sediment/Soil Concentrations:

Atmospheric Concentrations:

Food Survey Values:

Plant Concentrations:

Fish/Seafood Concentrations: 2-Pyridineethanol was detected in adult Lake trout (*Salvelinus namaycush*) and/or walleye (*Stizostediion v. vitreum*) collected from Lake St. Clair [6].

Animal Concentrations:

Milk Concentrations:

2-Pyridineethanol

Other Environmental Concentrations:

Probable Routes of Human Exposure: The most probable route of human exposure to 2-pyridineethanol is by dermal contact and ingestion.

Average Daily Intake:

Occupational Exposure: The most probable human exposure to 2-pyridineethanol would be occupational exposure, which may occur through dermal contact at places where it is produced or used. Nonoccupational exposures may occur during recreational activities at contaminated bodies of water [4] or from the ingestion of contaminated fish [4].

Body Burdens:

REFERENCES

1. Atkinson R; Intern J Chem Kin 19: 799-828 (1987)
2. Eisenreich SJ et al; Environ Sci Technol 15: 30-8 (1981)
3. Goe GL; Kirk-Othmer Encycl Chem Tech 3rd NY: Wiley Interscience 19: 454-83 (1982)
4. Great Lakes Water Quality Board; Inventory Chem Subst Id Great Lakes Ecos p. 195 (1983)
5. Hansch C, Leo AJ; Medchem Project Issue No 26. Claremont CA: Pomona College (1985)
6. Hesselberg RJ, Seelye JG; Id Org Cmpd Great L Fishes ADMEN Rep 82-1 US Fish Wildlife Soc Great L Fishery Lab p49 (1982)
7. Lyman WJ et al; Handbook of Chemical Property Estimation Methods NY: McGraw-Hill p. 4-9, 5-4, 6-3, 15-16 (1982)
8. Meylan WM, Howard PH; Chemosphere 26:2293-2299 (1993)
9. Meylan W, Howard PH; Environ Toxicol Chem 10: 1283-93 (1991)
10. PCCHEM; PCGEMS Graphical Exposure Modeling System USEPA (1987)
11. Swann RL et al; Res Rev 85: 16-28 (1983)

Safrole

SUBSTANCE IDENTIFICATION

Synonyms: 5-Allyl-1,3-Benzodioxole; 5-(2-Propenyl)-1,3-benzodioxole

Structure:

CAS Registry Number: 94-59-7

Molecular Formula: $C_{10}H_{10}O_2$

SMILES Notation: O(c(c(O1)cc(c2)CC=C)c2)C1

CHEMICAL AND PHYSICAL PROPERTIES

Boiling Point: 234.5 °C at 760 mm Hg

Melting Point: 11.2 °C

Molecular Weight: 162.18

Dissociation Constants:

Log Octanol/Water Partition Coefficient: 3.45 (estimated) [10]

Water Solubility: 76 mg/L at 25 °C (estimated) [7]

Vapor Pressure: 1.8 x 10^{-3} mm Hg at 25 °C (estimated from the boiling point [1,7]

Henry's Law Constant: 9 x 10^{-6} atm-m^3/mol at 25 °C (fragment constant estimation method [8])

Safrole

ENVIRONMENTAL FATE/EXPOSURE POTENTIAL

Summary: Safrole may be released to the environment during the manufacturing process or during subsequent use as a chemical intermediate in the production of piperonal and piperonyl butoxide. It may also be released as a result of its use in a variety of consumer products, including soaps, perfumes and medicines. The primary environmental fate of safrole is likely to be biodegradation. Experimental biodegradation data are not available; however, safrole is expected to biodegrade based upon its chemical structure. It is not expected to undergo hydrolysis, photolysis or to strongly adsorb to soils or sediments. The compound is not expected to volatilize significantly. The volatilization half-lives from a model river and a model lake are estimated to be 5 and 42 days, respectively. Any safrole which does enter the atmosphere will react with hydroxyl radicals with an estimated half-life of 5 hours.

Natural Sources: Safrole is a constituent of several essential oils. Sassafras oil contains up to 93%. Lesser quantities occur in essential oils from nutmeg, mace, ginger, cinnamon and black pepper, usually in the range of <1-10% of the oil. Star anise oil obtained from the Japanese tree *Illicium anisatum* Linn. also contains about 6% safrole, whereas no safrole was detected in star anise oil obtained from the Chinese tree Illicium [4].

Artificial Sources: Safrole is used in soaps, as an antiseptic [2], as a topical anesthetic [6], in perfumery and in the manufacture of heliotropin (piperonal) [2,6]. The extent to which safrole is released to the environment as a result of its production or from these uses cannot be estimated from the available information but it is expected to be significant.

Terrestrial Fate: Based on the moderate value of the estimated Koc (661 [7]), safrole is not expected to adsorb strongly to soils and sediments. Slow migration of safrole to ground water is anticipated based on the moderate estimated water solubility and moderate Koc value. Although no data are available, safrole contains no hydrolyzable functional group and, therefore, is not expected to hydrolyze appreciably [7]. Biodegradation of the olefin bond or aromatic ring may be a significant process.

Aquatic Fate: When released to the water, safrole is not expected to hydrolyze [7]. No information on aqueous biodegradation or photolysis of safrole is available. Based upon the Henry's Law constant, the volatilization half-lives from a model river and a model lake are estimated to be 5 and 42 days, respectively [7].

Atmospheric Fate: If safrole is released to the atmosphere, it can be expected to exist mainly in the vapor phase in the ambient atmosphere [3] based on the estimated vapor pressure. The overall half-lives for the vapor phase reactions of safrole with photochemically produced hydroxyl radicals and ozone have been estimated to be 5 and 23 hours at atmospheric concentrations of $5 \times 10^{+5}$ hydroxyl radicals per cm^3 and $7 \times 10^{+11}$ ozone molecules per cm^3 [9]. It may be susceptible to direct photolysis in the atmosphere based upon its absorption of light at wavelengths >290 nm for the structurally similar isosafrole [13].

Biodegradation: No data are available concerning the biodegradation of safrole in culture or in soil and water in the environment. It is expected, however, that safrole may biodegrade in the environment based upon its chemical structure.

Abiotic Degradation: After 4 days of irradiation at 253.7 nm from a 10-W low pressure mercury lamp at 20 °C, safrole in acetonitrile yielded 1-cyclopropyl-3,4-(methylenedioxy)benzene. 41% of the safrole did not react [11]. This photolytic reaction may not be relevant to the environmental fate of safrole due to the low wavelength used. No information concerning the wavelength of maximum absorbance of safrole is available. When released to the atmosphere, safrole will react with hydroxyl radicals with an estimated half-life of 5 hours [9]. Hydrolysis is not expected to be an important environmental process due to the lack of hydrolyzable functional groups [7].

Bioconcentration: The estimated octanol/water partition coefficient was used to estimate a bioconcentration factor of 250 [7]. It is expected, therefore, that safrole will not significantly bioconcentrate.

Soil Adsorption/Mobility: The estimated log octanol/water partition coefficient was used to estimate a Koc of 661 [7]. Based on this Koc value and the estimated water solubility of safrole, medium migration in soil and moderate adsorption to sediments is expected.

Volatilization from Water/Soil: The estimated value for the Henry's Law constant indicates that volatilization will occur only slowly from water, at a rate controlled by the slow diffusion of the compound through the air. Based upon the Henry's Law constant, the volatilization half-life from a model river (1 m deep, flowing 1 m/sec, wind velocity of 3 m/sec) is estimated as approximately 5 days [7]. The volatilization half-life from a model lake (1 m deep, flowing 0.05 m/sec, wind velocity of 0.5 m/sec) is estimated as approximately 42 days [7]. Volatilization from soil should also be slow due to the Henry's Law constant and a moderate tendency of safrole to adsorb to soils.

Water Concentrations:

Effluent Concentrations:

Sediment/Soil Concentrations:

Atmospheric Concentrations:

Food Survey Values:

Plant Concentrations: The commercial preparation of Xi-Xin made from the whole plant of *Asarum insigne* contains safrole [15]. The oil of *Asarum heterotropoides*, an ingredient of the chinese traditional medicine, hsi-hsin, contains safrole as a main component [16]. Safrole was identified in *Michelia hedyosperma*, a traditional medicine [17]. Leaf oil of *Nectandra falcifolia* oil contains safrole [14]. Safrole was identified in the oil of Shin-I, which is made from the dried buds of magnolia (*Salicifolia maxim*) [5].

Fish/Seafood Concentrations:

Animal Concentrations:

Milk Concentrations:

Other Environmental Concentrations:

Probable Routes of Human Exposure:

Average Daily Intake:

Occupational Exposure: NIOSH (NOES 1981-1983) has statistically estimated that 6,475 workers are potentially exposed to safrole in the US [12].

Body Burdens:

REFERENCES

1. Furia TE, Bellanca N; Fenaroli's Handbook of Flavor Ingredients 2nd ed CRC Press Cleveland p.515 (1975)
2. Hawley GG; The Condensed Chemical Dictionary 10th ed Van Nostrand Reinhold New York p.906 (1981)
3. Eisenreich SJ et al; Environ Sci Technol 15: 30-8 (1981)
4. IARC; Monograph on the Evaluation of the Carcinogenic Risk of Chemicals to Man 10: 235 (1976)
5. Kikuchi T et al; Wakanyaku Shinpojumu (Kiroku) 14: 101-4 (1981)
6. Kirk-Othmer Encycl Chem Technol; 3rd ed John Wiley & Sons New York 13: 62 (1981)
7. Lyman WJ et al; Handbook of Chemical Property Estimation Methods. Environmental behavior of organic compounds McGraw-Hill New York (1982)
8. Meylan W, Howard PH; Environ Toxicol Chem 10: 1283-93 (1991)
9. Meylan WM, Howard PH; Chemosphere 26:2293-2299 (1993)
10. Meylan W, Howard PH; J Pharm Sci 84: 83-92 (1995)
11. Mihara S, Shibamoto T; J Agric Food Chem 30: 1215-8 (1982)
12. NIOSH; National Occupational Exposure Survey (1983)
13. Sadtler; UV No. 2308 (1960)
14. Talenti ECJ et al; Essenze Deriv Agrum 51(2): 121-8 (1981)
15. Tian Z, Lou Z; Yaoxue Tongbao 16 (8): 59 (1981)
16. Tien C et al; Yao Hsueh T'ung Pao 16(2): 53 (1981)
17. Wu S et al; Chung Ts'ao Yao 12(2): 8-10 (1981)

Sulfolane

SUBSTANCE IDENTIFICATION

Synonyms:

Structure:

CAS Registry Number: 126-33-0

Molecular Formula: $C_4H_8O_2S$

SMILES Notation: O=S(=O)(CCC1)C1

CHEMICAL AND PHYSICAL PROPERTIES

Boiling Point: 285 °C

Melting Point: 27.4-27.8 °C

Molecular Weight: 120.16

Dissociation Constants:

Log Octanol/Water Partition Coefficient: -0.77 [5]

Water Solubility: Miscible [14]

Vapor Pressure: 0.0062 mm Hg at 27.6 °C [4]

Henry's Law Constant: 4.85×10^{-6} atm-m^3/mol at 25 °C [11]

ENVIRONMENTAL FATE/EXPOSURE POTENTIAL

Summary: The release of sulfolane to the environment may occur as a result of using this compound as an extractive solvent. The hydrolysis of sulfolane should not be important in the environment. Based on standard

428

BOD test data, the biotransformation of sulfolane in soil and water may be slow but biological treatment simulation tests suggest that biodegradation may be rapid. The estimated low Koc value indicates that adsorption of sulfolane to suspended solids and sediments in water should not be important and the compound should be highly mobile in soil. Due to its low vapor pressure and high water solubility, the volatilization of the compound from soil and water should not be important (estimate half-lives from a model river and lake are 8 days and 65 days, respectively). The bioconcentration of sulfolane in aquatic organisms should not be important based upon experimental data and high water solubility. In the atmosphere, the reaction of sulfolane with photochemically produced hydroxyl radicals may be an important process. The half-life of this reaction has been estimated to be about 1 day. Because of its high water solubility, atmospheric removal through wet deposition appears likely. Exposure of workers is likely to occur from using sulfolane as an extractive solvent and the most probable route of exposure is by skin absorption.

Natural Sources:

Artificial Sources: Sulfolane is usually present in wastewaters from petroleum refining and gas works [3].

Terrestrial Fate: Hydrolysis of sulfolane should not be important in soil since it is chemically stable [8] and does not have any hydrolyzable functional groups [10]. Based on 5-day BOD tests in water [2], the biodegradation of sulfolane in soil may not be important although biological treatment simulations suggest that sulfolane may biodegrade. The estimated Koc of 9 based on the log Kow and a regression equation [10] indicates that sulfolane should be highly mobile in soil [16]. The low vapor pressure and high water solubility indicate that volatilization from soil should not be important.

Aquatic Fate: Hydrolysis of sulfolane in water should not be important because the compound is chemically stable [8] and does not have any hydrolyzable functional groups [10]. The aerobic biodegradation study with sewage as microbial inoculum suggests that the compound should be stable towards biodegradation [2]; but the compound was found to biodegrade fast in an aerated lagoon with activated sludge as microbial inoculum [3]. The low vapor pressure and the high water solubility suggest that volatilization will not be important in water. An estimated Koc of 9 derived from the log Kow and a regression equation [10] indicates that adsorption to suspended solids and sediments in water should not be important [16]. Both the estimated bioconcentration factor of 0.15 and the Japanese MITI test [15] suggest that bioconcentration of sulfolane in aquatic organisms should not be important.

Atmospheric Fate: Based on an estimation method [12], the rate constant for the reaction of sulfolane with photochemically produced hydroxyl radicals has been estimated to be 16.5982×10^{-12} cm^3/molecule-sec [12]. Assuming the daily average concn of hydroxyl radicals in the atmosphere as $5 \times 10^{+5}$ per cm^3 [12], the half-life for this reaction has been estimated to be about 1 day. Since the compound is miscible in water, wet deposition should partly remove the compound from the atmosphere.

Biodegradation: In a 5-day standard BOD test with or without acclimated sewage as microbial inoculum, no biodegradation of sulfolane was observed [2,13]. In a standard Japanese MITI test with activated sludge as microbial inoculum, biodegradation equivalent to 0-29% theoretical BOD was observed in 14 days for a 100 mg/L aqueous sulfolane solution [7]. On the other hand, more than 99% of sulfolane at an initial concn of 100 mg/L biodegraded by activated sludge cultures in 1 day in a batch die-away test [3]. In a bench scale aerated lagoon reactor, more than 90% of sulfolane at an initial concn of 20-80 mg/L biodegraded at hydraulic retention times of 2-2.4 days [3]. It was concluded that sulfolane can be biodegraded by activated sludge cultures [3]. Inorganic sulfate has been identified as the biodegradation product of sulfolane [3].

Abiotic Degradation: Hydrolysis of sulfolane in the environment should not be important because it is chemically stable [8] and does not have any hydrolyzable functional groups [10]. The rate constant for the reaction of vapor phase sulfolane with hydroxyl radicals has been estimated to be 16.6×10^{-12} cm^3/molecule-sec by an estimation method [12]. Assuming the daily average concn of hydroxyl radicals in the atmosphere of $5 \times 10^{+5}$ per cm^3 [10], the half-life for this reaction has been estimated to be about 1 day.

Bioconcentration: In a test method developed by MITI in Japan, the bioconcentration of sulfolane in aquatic organisms was found to be unimportant [15]. A BCF value of 0.15 estimated from the log Kow and a regression equation [10], also suggests that bioconcentration should not be important.

Soil Adsorption/Mobility: A log Koc value of 0.96 has been estimated for sulfolane from its Kow value and a regression equation [10]. This low Koc indicates that adsorption of the compound to suspended solids and sediments in water should be very low and it should be highly mobile in soil [16].

Volatilization from Water/Soil: Based on the value of Henry's Law constant, the volatilization half-life of sulfolane from a model river 1 m deep flowing 1 m/sec with a wind velocity of 5 m/sec can be estimated to be approximately 8 days [10]. The volatilization half-life from a model lake has been estimated to be 65 days.

Water Concentrations:

Effluent Concentrations: Sulfolane was detected in effluents of advanced waste water treatment plants in Orange County, CA [9].

Sediment/Soil Concentrations:

Atmospheric Concentrations:

Food Survey Values:

Plant Concentrations:

Fish/Seafood Concentrations:

Animal Concentrations:

Milk Concentrations:

Other Environmental Concentrations:

Probable Routes of Human Exposure: Since sulfolane is used as a solvent in many extraction processes including aromatic hydrocarbons from oil refinery, fatty acids and polymerization, the most probable route of exposure is by skin absorption [6].

Average Daily Intake:

Occupational Exposure: Workers using sulfolane during extractive processes [6] are the most likely people for exposure to this compound.

Body Burdens:

REFERENCES

1. Atkinson R; Environ Toxicol Chem 7: 435-42 (1988)
2. Bridie AL et al; Water Res 13: 627-30 (1979)
3. Chou CC, Swatloski RA; Proc Ind Waste Conf 37: 559-66 (1983)
4. Daubert TE, Danner RP; Physical and Thermodynamic Properties of Pure Chemicals, Design Inst Phys Prop Data, Amer Inst Chem Eng, Vol 2. Hemisphere Publishing Corp, NY (1991)
5. Hansch C, Leo AJ; Medchem Project Issue No.26 Pomona College, Claremont CA (1985)
6. Hawley GG; The Condensed Chemical Dictionary, 10th ed. Van Nostrand Reinhold Co. NY p. 980 (1981)
7. Kawasaki M; Ecotoxic Environ Safety 4: 444-54 (1980)
8. Lindstrom M, Williams R; Kirk-Othmer Encycl Chem Tech Grayson M (Ed), 3rd ed. NY: John Wiley & Sons 21: 961-68 (1983)
9. Lucas SV; GC/MS Analysis of Organics in Drinking Water Concentrates and Advanced Waste Treatment Concentrates: Vol 1. USEPA-600/1-84-020A, NTIS PB85-128221, Springfield, VA (1884) (1) Hawley GG; The Condensed Chemical Dictionary, 10th ed. Van Nostrand Reinhold Co. NY p. 980 (1981)
10. Lyman WJ et al; Handbook of Chemical Property Estimation Methods NY: McGraw-Hill (1982)
11. Meylan W, Howard PH; Environ Toxicol Chem 10: 1283-93 (1991)
12. Meylan WM, Howard PH; Chemosphere 26:2293-2299 (1993)
13. Niemi GJ et al; Environ Toxicol Chem 6: 515-27 (1987)
14. Riddick JA et al; Organic Solvents Vol 2 4th ed. John Wiley & Sons NY p. 686-87 (1986)
15. Sasaki S; pp. 283-98 in Transformation and biological effects, Hutzinger O et al (Eds), Pergamon Press, Oxford (1978)
16. Swann RL et al; Res Rev 85: 17-28 (1983)

1,1,1,2-Tetrachloro-2,2-difluoroethane

SUBSTANCE IDENTIFICATION

Synonyms: R112a

Structure:

CAS Registry Number: 76-11-9

Molecular Formula: $C_2Cl_4F_2$

SMILES Notation: FC(F)(C(Cl)(Cl)Cl)Cl

CHEMICAL AND PHYSICAL PROPERTIES

Boiling Point: 91.5 °C at 760 mm Hg

Melting Point: 40.6 °C

Molecular Weight: 203.82

Dissociation Constants:

Log Octanol/Water Partition Coefficient: 3.41 [10]

Water Solubility: Based 100 mg/L at 25 °C [12]

Vapor Pressure: 54.9 mm Hg at 25 °C [4]

Henry's Law Constant: 0.147 atm-m³/mol at 25 °C (calculated from vapor pressure and water solubility)

ENVIRONMENTAL FATE/EXPOSURE POTENTIAL

Summary: If released to soil, 1,1,1,2-tetrachloro-2,2-difluoroethane (R112a) will rapidly volatilize from soil surfaces or leach through soil. If released to water, essentially all 1,1,1,2-tetrachloro-2,2-difluoroethane is expected to be lost by volatilization (half-life 4 hours from a model river). If released to the atmosphere (troposphere), 1,1,1,2-tetrachloro-2,2-difluoroethane will not degrade. It will gradually diffuse into the stratosphere (diffusion half-life 20 years) where it will either photolyze or react with singlet oxygen. The stratospheric lifetime is predicted to be on the order of several decades. Due to its persistence in the troposphere, long distance transport from its emission sources is expected to occur.

Natural Sources:

Artificial Sources: 1,1,1,2-Tetrachloro-2,2-Difluoroethane may be released to the environment from the disposal of refrigeration units in which this compound was used [3].

Terrestrial Fate: If released to soil, 1,1,1,2-tetrachloro-2,2-difluoroethane will rapidly volatilize from soil surfaces or leach through soil. Chemical hydrolysis and biodegradation will not be important fate processes.

Aquatic Fate: If released to water, essentially all 1,1,1,2-tetrachloro-2,2-difluoroethane is expected to be lost by volatilization (half-life 4 hours from a model river). Chemical hydrolysis, biodegradation, bioaccumulation and adsorption to sediments are not expected to be important fate processes in water.

Atmospheric Fate: Based on the vapor pressure, 1,1,1,2- tetrachloro-2,2-difluoroethane (R112a) is expected to exist almost entirely in the vapor phase in the atmosphere [7]. The moderate water solubility of 1,1,1,2-tetrachloro-2,2-difluoroethane suggests that some loss by wet deposition may occur, but any loss by this mechanism would probably be returned to the atmosphere by volatilization. R112a is not expected to degrade in the troposphere. As a result, diffusion into the stratosphere would be the ultimate removal process. The half-life for tropospheric to stratospheric diffusion of compounds is generally 20 years [5]. In the stratosphere 1,1,1,2-tetrachloro-2,2-difluoroethane may undergo direct photolysis, producing chlorine atoms which in turn participate in the catalytic removal of stratospheric ozone, or it may react with singlet oxygen. The stratospheric lifetime is predicted to be on the order of several decades. Due to its persistence in the troposphere, long distance transport from its emission sources is expected to occur.

Biodegradation:

Abiotic Degradation: Chemical hydrolysis of 1,1,1,2-tetrachloro-2,2-difluoroethane (R112a) is not expected to be an environmentally important fate process [6]. 1,1,1,2-Tetrachloro-2,2-difluoroethane is inert to chemical degradation in the troposphere [1,9,11]. Upon diffusion into the stratosphere this compound either will photolyze slowly to release chlorine atoms which in turn participate in the catalytic removal of stratospheric ozone or it will slowly react with singlet oxygen [2,9]. By analogy to other chlorofluorocarbons, 1,1,1,2-tetrachloro-2,2,-difluoroethane is predicted to have a stratospheric lifetime on the order of several decades [2].

Bioconcentration: Based on the water solubility, a bioconcentration factor (BCF) of 46 was estimated for 1,1,1,2-tetrachloro-2,2-difluoroethane (R112a) [8]. This bioconcentration factor value suggests that 1,1,1,2-tetrachloro-2,2-difluoroethane would not bioaccumulate in aquatic organisms.

Soil Adsorption/Mobility: Soil adsorption coefficients (Koc) of 347 [8] and 513 [12] for 1,1,1,2-tetrachloro-2,2-difluoroethane were estimated using a linear regression equation and the water solubility. These Koc values suggests that R112a would have medium mobility in soil and that it would be moderately adsorbed to suspended solids and sediments in water [13].

Volatilization from Water/Soil: The value of the Henry's Law constant suggests that R112a would volatilize rapidly from all bodies of water and from moist soil surfaces [8]. Based on this value the volatilization half-life of R112a from a model river 1 m deep flowing 1 m/sec with a wind velocity of 3 m/sec has been estimated to be 4 hours [8].

Water Concentrations:

Effluent Concentrations:

1,1,1,2-Tetrachloro-2,2-difluoroethane

Sediment/Soil Concentrations:

Atmospheric Concentrations:

Food Survey Values:

Plant Concentrations:

Fish/Seafood Concentrations:

Animal Concentrations:

Milk Concentrations:

Other Environmental Concentrations:

Probable Routes of Human Exposure: 1,1,1,2-tetrachloro-2,2-difluoroethane can affect the body if inhaled or if it comes in contact with the eyes or skin. It can also affect the body if swallowed.

Average Daily Intake:

Occupational Exposure:

Body Burdens:

REFERENCES

1. Atkinson R; Chem Rev 85: 69-201 (1985)
2. Chou CC et al; J Phys Chem 82: 1-7 (1978)
3. Cooper KW, Hickman KE; Kirk-Othmer Encycl Chem Tech 3rd ed NY: Wiley 20: 78-107 (1982)
4. Daubert TE, Danner RP; Physical and Thermodynamic Properties of Pure Chemicals, AIChE, Hemisphere Publ Co., NY (1989)
5. Dilling WL; Environmental Risk Analysis for Chemicals; Conway RA ed NY: Van Nostrand Reinhold Co pp. 154-97 (1982)
6. Du Pont de Nemours Co; Freon Products Information B-2; A98825 12/80 (1980)
7. Eisenreich SJ et al; Environ Sci Tech 15: 30-8 (1981)
8. Lyman WJ et al; Handbook of Chemical Property Estimation Methods NY: McGraw-Hill (1982)

1,1,1,2-Tetrachloro-2,2-difluoroethane

9.   Makide T et al; Chem Lett 4:355-8 (1979)
10. Meylan W, Howard PH; J Pharm Sci 84: 83-92 (1995)
11. Meylan WM, Howard PH; Chemosphere 26:2293-2299 (1993)
12. Roy WR, Griffin RA; Environ Geol Water Sci 7: 241-7 (1985)
13. Swann RL et al; Res Rev 85: 17-28 (1983)

1,1,1,2-Tetrafluoroethane

SUBSTANCE IDENTIFICATION

Synonyms:

Structure:

CAS Registry Number: 811-97-2

Molecular Formula: $C_2H_2F_4$

SMILES Notation: FCC(F)(F)F

CHEMICAL AND PHYSICAL PROPERTIES

Boiling Point: -26.5 at 736 mm Hg

Melting Point: -101 °C

Molecular Weight: 102.03

Dissociation Constants:

Log Octanol/Water Partition Coefficient: 1.274 (estimated) [12]

Water Solubility: 67 mg/L at 25 °C [12]

Vapor Pressure: 4730 mm Hg at 25 °C (estimated) [12]

Henry's Law Constant: 1.53 atm-m^3/mol at 25 °C [9]

ENVIRONMENTAL FATE/EXPOSURE POTENTIAL

Summary: 1,1,1,2-Tetrafluoroethane is an anthropogenic compound that holds potential as an alternative to chlorofluorocarbons (CFCs). It may be released to the environment as a fugitive emission during its production or

438

use. If released to soil, 1,1,1,2-tetrafluoroethane will rapidly volatilize from either moist or dry soil to the atmosphere. It will display moderate to high mobility in soil. If released to water, 1,1,1,2-tetrafluoroethane will rapidly volatilize to the atmosphere. The estimated half-life for volatilization from a model river is 3.0 hr. 1,1,1,2-Tetrafluoroethane will not bioconcentrate in fish and aquatic organisms nor will it adsorb to sediment and suspended organic matter. If released to the atmosphere, 1,1,1,2-tetrafluoroethane will undergo a very slow vapor-phase reaction with photochemically produced hydroxyl radicals with an estimated half-life of 187 days. The atmospheric lifetime of 1,1,1,2-tetrafluoroethane has been estimated to range from 12.5 to 24 years. 1,1,1,2-Tetrafluoroethane may also undergo atmospheric removal by wet deposition processes; however, any removed is expected to rapidly revolatilize to the atmosphere. Occupational exposure to 1,1,1,2-tetrafluoroethane may occur by inhalation or dermal contact during its production or use.

Natural Sources: 1,1,1,2-Tetrafluoroethane is of anthropogenic origin, and it is not known to be produced by natural sources.

Artificial Sources: 1,1,1,2-Tetrafluoroethane is an anthropogenic compound that may be used as a replacement for chlorofluorocarbons (CFCs) [8]; if so, it may be released to the environment as a fugitive emission during its production or use.

Terrestrial Fate: If released to soil, an estimated vapor pressure for 1,1,1,2-tetrafluoroethane of 4730 mm Hg at 25 °C [12] indicates that it will rapidly volatilize from dry soil to the atmosphere. Estimated soil adsorption coefficients ranging from 117-432 [6,10,12] indicate that it will display moderate to high mobility in soil [11]. An estimated Henry's Law constant of 1.53 atm-m^3/mol at 25 °C [9] indicates that 1,1,1,2-tetrafluoroethane will also rapidly volatilize from moist soil to the atmosphere.

Aquatic Fate: If released to water, the estimated Henry's Law constant indicates that it will rapidly volatilize to the atmosphere. The estimated half-life for volatilization from a model river 1 m deep flowing at 1 m/sec with a wind speed of 3 m/sec is 3.0 hr [6]. Estimated bioconcentration factors ranging from 5 to 58 [3,6,12] indicate that 1,1,1,2-tetrafluoroethane will not bioconcentrate in fish and aquatic organisms. Estimated soil adsorption coefficients ranging from 117-432 [6,12] indicate that it will not adsorb to sediment or suspended organic matter.

Atmospheric Fate: If released to the atmosphere, 1,1,1,2-tetrafluoroethane will undergo a slow vapor-phase reaction with photochemically produced hydroxyl radicals. The recommended rate constant for this process of 8.54 x 10^{-15} cm^3/molecule-sec [1] translates to an atmospheric half-life of 1878 days using an average atmospheric hydroxyl radical concn of 5 x 10^{+5} molecules/cm^3 [1]. The atmospheric lifetime of 1,1,1,2-tetrafluoroethane, calculated using both 1 and 2 dimensional models, ranges from 12.5 to 24 yrs [5]. The estimated water solubility of 1,1,1,2-tetrafluoroethane indicates that it may undergo atmospheric removal by wet deposition processes; however, any removed is expected to rapidly revolatilize to the atmosphere.

Biodegradation:

Abiotic Degradation: Experimental rate constants for the vapor-phase reaction of 1,1,1,2-tetrafluoroethane with photochemically produced hydroxyl radicals of 5.2 x 10^{-15} cm^3/molecule-sec at ambient temperature [7], 6.9 x 10^{-15} cm^3/molecule-sec at 301 K [2] and 8.32 x 10^{-15} cm^3/molecule-sec at ambient temperature [3,4] have been reported. The recommended value of 8.54 x 10^{-15} cm^3/molecule-sec [1] translates to an atmospheric half-life of 1878 days using an average atmospheric hydroxyl radical concn of 5 x 10^{+5} molecules/cm^3 [1]. The atmospheric lifetime of 1,1,1,2-tetrafluoroethane, calculated using both 1 and 2 dimensional models, ranges from 12.5 to 24 yrs [5].

Bioconcentration: Estimated bioconcentration factors ranging from 5 to 58 can be calculated for 1,1,1,2-tetrafluoroethane based on its estimated log octanol/water partition coefficient, 1.274 [12], and estimated water solubility, 67 mg/L at 25 °C [6], in turn estimated from its estimated Henry's Law constant [9] and estimated vapor pressure [12], using appropriate regression equations [6]. These values indicate that 1,1,1,2-tetrafluoroethane will not bioconcentrate in fish and aquatic organisms.

Soil Adsorption/Mobility: Estimated soil adsorption coefficients ranging from 117 to 432 can be calculated for 1,1,1,2-tetrafluoroethane [6] based on its estimated log octanol/water partition coefficient and estimated water solubility. These values indicate that 1,1,1,2-tetrafluoroethane will display moderate to high mobility in soil [11].

Volatilization from Water/Soil: An estimated Henry's Law constant of 1.53 atm-m^3/mol at 25 °C [9] indicates that 1,1,1,2-tetrafluoroethane will rapidly volatilize from water and moist soil to the atmosphere. The estimated half-life for volatilization from a model river 1 m deep flowing at 1 m/sec with a wind speed of 3 m/sec is 3.0 hr [6]. The estimated vapor pressure of 1,1,1,2-tetrafluoroethane indicates that it will rapidly volatilize from dry soil to the atmosphere.

Water Concentrations:

Effluent Concentrations:

Sediment/Soil Concentrations:

Atmospheric Concentrations:

Food Survey Values:

Plant Concentrations:

Fish/Seafood Concentrations:

Animal Concentrations:

Milk Concentrations:

Other Environmental Concentrations:

Probable Routes of Human Exposure: Occupational exposure to 1,1,1,2-tetrafluoroethane may occur by inhalation or dermal contact during its production or use.

Average Daily Intake:

Occupational Exposure:

1,1,1,2-Tetrafluoroethane

Body Burdens:

REFERENCES

1. Atkinson R; J Chem Phys Ref Data Monograph 1 (1989)
2. Brown AC et al; Atmos Environ 24A: 2499-511 (1990)
3. Cohen N, Benson SW; J Phys Chem 91: 171-5 (1987)
4. Cohen N, Benson SW; J Phys Chem 91: 162-70 (1987)
5. Fisher DA et al; Nature 344: 508-12 (1990)
6. Lyman WJ et al; Handbook of Chemical Property Estimation Methods NY: McGraw-Hill Chapt 4, 5 & 15 (1982)
7. Makide Y, Rowland FS; Proc Natl Acad Sci USA 78: 5933-7 (1981)
8. McClinden MO, Didion DA; Int J Thermophys 10: 563-76 (1989)
9. Meylan WM, Howard PH; Environ Toxicol Chem 10: 1283-93 (1991)
10. Meylan WM et al; Environ Sci Technol 26: 1560-1567 (1992)
11. Swann RL et al; Res Rev 85: 17-28 (1983)
12. USEPA; PCGEMS and CLOGP (1988)

442

Tetrahydrofurfuryl Alcohol

SUBSTANCE IDENTIFICATION

Synonyms:

Structure:

CAS Registry Number: 97-99-4

Molecular Formula: $C_5H_{10}O_2$

SMILES Notation: O(C(CC1)CO)C1

CHEMICAL AND PHYSICAL PROPERTIES

Boiling Point: 178 °C

Melting Point: < -80 °C

Molecular Weight: 102.15

Dissociation Constants:

Log Octanol/Water Partition Coefficient: -0.11 (estimated) [9]

Water Solubility: Miscible [14]

Vapor Pressure: 0.80 mm Hg at 25 °C [3]

Henry's Law Constant: 4.09 x 10^{-9} atm-m^3/mol [8]

ENVIRONMENTAL FATE/EXPOSURE POTENTIAL

Summary: Tetrahydrofurfuryl alcohol may be released to the atmosphere through evaporation in its use as a solvent in lacquers, dyes, resins, pesticides, industrial cleaners, and stripping formulations. If released to the atmosphere, it will degrade in the vapor phase by reaction with photochemically produced hydroxyl radicals (estimated half-life of 13 hours). If released to soil or water, tetrahydrofurfuryl alcohol is expected to degrade via biodegradation. Two biodegradation screening studies have found tetrahydrofurfuryl alcohol to be readily biodegradable. Leaching in soil is possible since tetrahydrofurfuryl alcohol is miscible in water. Occupational exposure to tetrahydrofurfuryl alcohol occurs through dermal contact and inhalation of vapor.

Natural Sources:

Artificial Sources: Tetrahydrofurfuryl alcohol use as a solvent in lacquers, dyes, resins, pesticides, industrial cleaners, and stripping formulations [7] will result in releases to the atmosphere through evaporation.

Terrestrial Fate: The dominant environmental fate process for tetrahydrofurfuryl alcohol in soil is probably biodegradation. Two biodegradation screening studies have found tetrahydrofurfuryl alcohol to be readily biodegradable [13,15]. Leaching in soil is possible since tetrahydrofurfuryl alcohol is miscible in water. Some evaporation may occur from dry surfaces.

Aquatic Fate: The dominant environmental fate process for tetrahydrofurfuryl alcohol in water is probably biodegradation. Two biodegradation screening studies have found tetrahydrofurfuryl alcohol to be readily biodegradable [13,15]. Aquatic hydrolysis, volatilization, adsorption to sediment, and bioconcentration are not expected to be environmentally important.

Atmospheric Fate: Based upon a vapor pressure of 0.80 mm Hg at 25 °C [3], tetrahydrofurfuryl alcohol is expected to exist entirely in the vapor phase in the ambient atmosphere [4]. It will degrade in the ambient atmosphere by reaction with photochemically produced hydroxyl radicals (estimated half-life of 13 hours) [2]. Physical removal from air via wet deposition is possible since tetrahydrofurfuryl alcohol is miscible in water.

Biodegradation: Tetrahydrofurfuryl alcohol was considered readily biodegradable from the results of a screening test using an adapted activated

444

sludge inoculum in which 96.1% of initial tetrahydrofurfuryl alcohol (based upon COD) was degraded in 120 hours of incubation [13]. Using a respirometric dilution method and a sewage inocula, tetrahydrofurfuryl alcohol had a theoretical BOD of 82.9% over a 5-day incubation period [15].

Abiotic Degradation: The rate constant for the vapor-phase reaction of tetrahydrofurfuryl alcohol with photochemically produced hydroxyl radicals has been estimated to be 2.886 x 10^{-11} cm^3/molecule-sec at 25 °C which corresponds to an atmospheric half-life of about 13 hr at an atmospheric concn of 5 x 10^{+5} hydroxyl radicals per cm^3 [2]. Alcohols and ethers are generally resistant to aqueous environmental hydrolysis [6]; therefore, tetrahydrofurfuryl alcohol is not expected to hydrolyze in environmental waters.

Bioconcentration: Tetrahydrofurfuryl alcohol is miscible in water [14]; this suggests that bioconcentration in aquatic organisms will not be important environmentally.

Soil Adsorption/Mobility: Tetrahydrofurfuryl alcohol is miscible in water [14]; this suggests that tetrahydrofurfuryl alcohol will be very mobile in soil and will probably leach.

Volatilization from Water/Soil: The Henry's Law constant for tetrahydrofurfuryl alcohol can be estimated to be 4.09 x 10^{-9} atm-m^3/mol using a structure estimation method [8]. This value of Henry's Law constant indicates that tetrahydrofurfuryl alcohol is essentially nonvolatile from water [6]; therefore, volatilization from water will not be important.

Water Concentrations:

Effluent Concentrations:

Sediment/Soil Concentrations:

Atmospheric Concentrations:

Food Survey Values: Tetrahydrofurfuryl alcohol has been identified as a constituent of coffee aroma [1].

Plant Concentrations:

Fish/Seafood Concentrations:

Animal Concentrations:

Milk Concentrations:

Other Environmental Concentrations: Tetrahydrofurfuryl alcohol has been identified as a constituent in flue-cured tobacco [5].

Probable Routes of Human Exposure: Occupational exposure to tetrahydrofurfuryl alcohol occurs through dermal contact and inhalation of vapor [12].

Average Daily Intake:

Occupational Exposure: NIOSH (NOES 1981-1983) has statistically estimated that 79,915 workers are potentially exposed to tetrahydrofurfuryl alcohol in the US [11].

Body Burdens:

REFERENCES

1. Aeschbacher HU et al; Food Chem Toxicol 4: 227-32 (1989)
2. Atkinson R; J Inter Chem Kinet 19: 799-828 (1987)
3. Daubert TE, Danner RP; Physical and Thermodynamic Properties of Pure Chemicals: Data Compilation, NY: Hemisphere Pub Corp (1989)
4. Eisenreich SJ et al; Environ Sci Technol 15: 30-8 (1981)
5. Lloyd RA et al; Tob Sci 20: 125-33 (1976)
6. Lyman WJ et al; Handbook of Chemical Property Estimation Methods Washington, DC: Amer Chem Soc pp. 15-15 to 15-29 (1990)
7. McKillip WJ; Ullmann's Encycl Industr Chem 5th ed. NY: VCH Publ A12: 128-9 (1989)
8. Meylan W, Howard PH; Environ Toxicol Chem 10: 1283-93 (1991)
9. Meylan W, Howard PH; J Pharm Sci 84: 83-92 (1995)

10. Meylan W, Howard PH; Environ Toxicol Chem 10: 1283-93 (1991)
11. NIOSH; National Occupational Exposure Survey (NOES) (1983)
12. Parmeggiani L; Encyl Occup Health & Safety 3rd ed Geneva, Switzerland: International Labour Office p. 931-2 (1983)
13. Pitter P; Water Res 10: 231-5 (1976)
14. Riddick JA et al; Organic Solvents: Physical Properties and Methods of Purification. Techniques of Chemicals 4th ed. NY: Wiley-Interscience p. 693 (1986)
15. Wagner R; Vom Wasser 47: 241-65 (1976)

Tetranitromethane

SUBSTANCE IDENTIFICATION

Synonyms:

Structure:

CAS Registry Number: 509-14-8

Molecular Formula: CN_4O_8

SMILES: N(=O)(=O)C(N(=O)(=O))(N(=O)(=O))N(=O)(=O)

CHEMICAL AND PHYSICAL PROPERTIES

Boiling Point: 126 °C

Melting Point: 13.8 °C

Molecular Weight: 196.04

Dissociation Constants:

Log Octanol/Water Partition Coefficient: -0.791 (estimated) [4]

Water Solubility: 8.5 x 10^{+4} mg/L (calculated from log Kow) [7]

Vapor Pressure: 8.4 mm Hg at 20 °C [2]

Henry's Law Constant: 2.55 x 10^{-5} atm-m³/mol (calculated from vapor pressure and water solubility)

ENVIRONMENTAL FATE/EXPOSURE POTENTIAL

Summary: Tetranitromethane may be released to the environment as a result of its manufacture and use as a rocket fuel, diesel fuel booster, organic reagent, and as an explosive in admixture with toluene. It may also be released to the environment as a result of the production of TNT since it is a byproduct in the production of this explosive. If tetranitromethane is released to soil, it will be expected to be very highly mobile. Based upon a measured vapor pressure of 8.4 mm Hg at 20 °C, volatilization from dry near-surface soil or other surfaces may be important processes. No data were located concerning biodegradation or hydrolysis of tetranitromethane in soil. If released to water, it will be expected to be subject to volatilization based upon an estimated Henry's Law constant. The volatilization half-life from a model river (1 m deep flowing 1 m/sec with a wind speed of 3 m/sec) has been estimated to be 2.1 days. The volatilization half-life from a model pond has been estimated to be 24 days. It will not be expected to adsorb to sediment or suspended particulate matter based upon the estimated Koc or to bioconcentrate in aquatic organisms based upon an estimated BCF of 0.15. No data were located concerning biodegradation, direct photolysis, hydrolysis or oxidation of tetranitromethane in aqueous media. If it is released to the atmosphere, it will be expected to exist almost entirely in the vapor phase based upon its vapor pressure. It should not be susceptible to photooxidation via vapor phase reaction with photochemically produced hydroxyl radicals. Exposure to tetranitromethane will be primarily occupational, via inhalation and dermal routes.

Natural Sources:

Artificial Sources: Tetranitromethane may be released to the environment as a result of its manufacture and use. It has been investigated as a rocket fuel, diesel fuel booster, organic reagent, and as an explosive in admixture with toluene [6,8]. It has been proposed to be a component of war gases [8]. It may be released as a result of the production of TNT since it is a byproduct in the production of this explosive [10].

Terrestrial Fate: If tetranitromethane is released to soil, it will be expected to be very highly mobile based upon [11] an estimated Koc of 8.8 [4,7]. Based upon a measured vapor pressure of 8.4 mm Hg at 20 °C [2], volatilization from dry near-surface soil or other surfaces may be important processes. No data were located concerning biodegradation or hydrolysis of tetranitromethane in soil.

Tetranitromethane

Aquatic Fate: If tetranitromethane is released to water, it will be expected to be subject to volatilization based upon the estimated Henry's Law constant. It will not be expected to adsorb to sediment or suspended particulate matter based upon an estimated Koc of 8.8 [4,7] or to bioconcentrate in aquatic organisms based upon an estimated BCF of 0.15 [4,7]. No data were located concerning biodegradation, direct photolysis, hydrolysis or oxidation of tetranitromethane in the environment.

Atmospheric Fate: If tetranitromethane is released to the atmosphere, it will be expected to exist almost entirely in the vapor phase [5] based upon the reported vapor pressure. It should not be susceptible to photooxidation via vapor phase reaction with photochemically produced hydroxyl radicals [1].

Biodegradation:

Abiotic Degradation: Tetranitromethane has been estimated to be inert to reaction with photochemically produced hydroxyl radicals in the atmosphere [1]. Since tetranitromethane absorbs UV light at wavelengths >290 nm [3], it may be susceptible to direct photolysis.

Bioconcentration: An estimated BCF of 0.15 can be calculated [7] from the estimated log Kow. This estimated BCF indicates that tetranitromethane will not be expected to bioconcentrate in aquatic organisms.

Soil Adsorption/Mobility: An estimated Koc of 8.8 can be calculated [7] from the estimated log Kow. This estimated Koc indicates that tetranitromethane will not be expected to adsorb to soils, sediment, and suspended particulate matter and that the compound will be expected to exhibit very high mobility in soil [11].

Volatilization from Water/Soil: Based upon the estimated Henry's Law constant, the volatilization half-life from a model river (1 m deep flowing 1 m/sec with a wind speed of 3 m/sec) has been estimated to be 2.1 days [7]. The volatilization half-life from a model pond, which considers the effect of adsorption, has been estimated to be 24 days [12]. Based upon its vapor pressure, volatilization of tetranitromethane from surfaces and near-surface dry soil may be important processes.

Water Concentrations:

Effluent Concentrations:

Sediment/Soil Concentrations:

Atmospheric Concentrations:

Food Survey Values:

Plant Concentrations:

Fish/Seafood Concentrations:

Animal Concentrations:

Milk Concentrations:

Other Environmental Concentrations:

Probable Routes of Human Exposure: Exposure to tetranitromethane will be primarily occupational, via inhalation and dermal routes of exposure.

Average Daily Intake:

Occupational Exposure: NIOSH (NOES 1981-1983) has statistically estimated that 1,445 workers are potentially exposed to tetranitromethane in the US [9].

Body Burdens:

REFERENCES

1. Atkinson R; Intern J Chem Kinetics 19: 799-828 (1987)
2. Boublik T et al Vapor Pressures of Pure Substances. Elsevier NY (1984)
3. Carpenter BH et al; Specific Air Pollutants From Munitions Processing and Their Atmospheric Behavior Vol 3 Final Report, NTIS AD-A060147 Research Triangle Park, NC: Res Triangle Park, pp. 139 (1977)
4. CLOGP3; PCGEMS Graphical Exposure Modeling System USEPA (1986)

Tetranitromethane

5. Eisenreich SJ et al; Environ Sci Technol 15: 30-8 (1981)
6. Hawley GG; Condensed Chemical Dictionary 10th ed NY: Van Nostrand Reinhold p. 1010 (1981)
7. Lyman WJ et al; Handbook of Chemical Property Estimation Methods NY: McGraw-Hill pp. 2-14, 4-9, 5-5, 15-15 to 15-29 (1982)
8. Merck; The Merck Index 10th ed Rahway, NJ: Merck & Co p. 9061 (1983)
9. NIOSH; The National Occupational Exposure Survey (NOES) (1983)
10. Ryon MG et al; Database Assessment of the Health and Environmental Effects of Munition Production Waste Products. Final Rpt ORNL-6018 NTIS DE84-016512 pp. 217 (1984)
11. Swann RL et al; Res Rev 85: 17-28 (1983)
12. USEPA; EXAMS II Computer Simulation (1987)

1,1,1-Trichloro-2,2,2-trifluoroethane

SUBSTANCE IDENTIFICATION

Synonyms:

Structure:

CAS Registry Number: 354-58-5

Molecular Formula: $C_2Cl_3F_3$

SMILES Notation: FC(F)(F)C(Cl)(Cl)Cl

CHEMICAL AND PHYSICAL PROPERTIES

Boiling Point: 45.8 °C

Melting Point: 14.2 °C

Molecular Weight: 187.38

Dissociation Constants:

Log Octanol/Water Partition Coefficient: 3.09 (estimated) [6]

Water Solubility: 21 mg/L at 25 °C (estimated from log Kow) [5]

Vapor Pressure: 360 mm Hg at 25 °C [2]

Henry's Law Constant: 3.88×10^{-2} atm-m^3/mol at 25 °C (estimated) [7]

ENVIRONMENTAL FATE/EXPOSURE POTENTIAL

Summary: 1,1,1-Trichloro-2,2,2-trifluoroethane is an anthropogenic compound that is used in various commercial applications. It may be released to the environment as a fugitive emission during its production, formulation and use. If released to soil, 1,1,1-trichloro-2,2,2-trifluoroethane is expected to display moderate adsorption to soil. It is expected to rapidly volatilize from both dry and moist soil to the atmosphere. If released to water, 1,1,1-trichloro-2,2,2-trifluoroethane is not expected to bioconcentrate in fish and aquatic organisms. It is expected to rapidly volatilize from water to the atmosphere. The estimated half-life for volatilization from a model river is 4 hr. 1,1,1-trichloro-2,2,2-trifluoroethane is not expected to adsorb to sediment and suspended organic matter. If released to the atmosphere, 1,1,1-trichloro-2,2,2-trifluoroethane is expected to persist for long periods of time. It is not expected to undergo atmospheric removal by gas-phase oxidation reactions with photochemically produced hydroxyl radicals and ozone. It is expected to undergo atmospheric removal by wet deposition; however, any removed by this process is expected to revolatilize rapidly. Occupational exposure to 1,1,1-trichloro-2,2,2-trifluoroethane may occur by inhalation or dermal contact during its production, formulation or use.

Natural Sources: 1,1,1-Trichloro-2,2,2-trifluoroethane is an anthropogenic compound and it is not known to occur in nature.

Artificial Sources: 1,1,1-Trichloro-2,2,2-trifluoroethane is used as a solvent, and degreasing and cleaning agent [9] and therefore, may be released to the environment as a fugitive emission during it production, formulation, or use. It may also be released to the environment in the effluent from landfills [13].

Terrestrial Fate: If released to soil, estimated soil adsorption coefficients of 818 and 1142 [5] obtained from its estimated water solubility and estimated log octanol/water partition coefficient, respectively, indicate that 1,1,1-trichloro-2,2,2-trifluoroethane will display moderate mobility in soil [12], and it has the potential to leach. Based on its vapor pressure and its estimated Henry's Law constant, 1,1,1-trichloro-2,2,2-trifluoroethane is expected to rapidly volatilize from dry and moist soil, respectively, to the atmosphere.

Aquatic Fate: If released to water, estimated bioconcentration factors of 110 and 131 [5] obtained from its estimated water solubility and estimated log octanol/water partition coefficient, respectively, indicate that 1,1,1-trichloro-2,2,2-trifluoroethane is not expected to bioconcentrate in fish and aquatic organisms. Based on its estimated Henry's Law constant, 1,1,1-

trichloro-2,2,2-trifluoroethane is expected to rapidly volatilize from water to the atmosphere. The estimated half-life for volatilization from a model river 1 m deep, flowing at 1 m/sec with a wind speed of 3 m/sec, is 4 hr [5]. Estimated soil adsorption coefficients of 818 and 1142 [5] indicate that 1,1,1-trichloro-2,2,2-trifluoroethane will not appreciably adsorb to sediment and suspended organic matter.

Atmospheric Fate: If released to the atmosphere, 1,1,1-trichloro-2,2,2-trifluoroethane is expected to persist for long periods of time. It is not expected to undergo atmospheric removal by hydroxyl radicals and ozone [1], nor is it expected to undergo direct photochemical degradation since it does not absorb UV light above 290 nm [10]. The estimated water solubility of 1,1,1-trichloro-2,2,2-trifluoroethane indicates that it may undergo atmospheric removal by wet deposition; however, any 1,1,1-trichloro-2,2,2-trifluoroethane removed by this process is expected to rapidly revolatilize to the atmosphere because of the high Henry's Law constant.

Biodegradation:

Abiotic Degradation:

Bioconcentration: Based on the estimated water solubility, a bioconcentration factor of 110 can be obtained using an appropriate regression equation [5]. From the estimated log octanol/water partition coefficient, a bioconcentration factor of 131 can be obtained [5]. These values indicate that 1,1,1-trichloro-2,2,2-trifluoroethane is not expected to bioconcentrate in fish and aquatic organisms.

Soil Adsorption/Mobility: Based on the water solubility, a soil adsorption coefficient of 818 can be obtained using an appropriate regression equation [5]. From the estimated log octanol/water partition coefficients, a soil adsorption coefficient of 1142 can be obtained [5]. These values indicate that 1,1,1-trichloro-2,2,2-trifluoroethane will display moderate mobility in soil [12].

Volatilization from Water/Soil: Based on the vapor pressure, 1,1,1-trichloro-2,2,2-trifluoroethane is expected to rapidly volatilize from dry soil to the atmosphere. Its estimated Henry's Law constant indicates that 1,1,1-trichloro-2,2,2-trifluoroethane is expected to rapidly volatilize from water and moist soil to the atmosphere. Based on its Henry's Law constant, the estimated half-life for volatilization of 1,1,1-trichloro-2,2,2-trifluoroethane from a model river 1 m deep flowing at 1 m/sec with a wind speed of 3 m/sec is 4 hr [5].

Water Concentrations:

Effluent Concentrations: 1,1,1-Trichloro-2,2,2-trifluoroethane was detected as a component of landfill gases from sites in the UK at a measured concn of 177 mg/m^3 in underground probes [13].

Sediment/Soil Concentrations: 1,1,1-Trichloro-2,2,2-trifluoroethane was detected in sediment samples at the site of an illegal dump site, The Valley of the Drums, KY, 1979 at a concn of 5.4 mg/L [11].

Atmospheric Concentrations: RURAL/REMOTE: 1,1,1-Trichloro-2,2,2-trifluoroethane was qualitatively detected in rural air samples, date and location not provided [4]. The concn of 1,1,1-trichloro-2,2,2-trifluoroethane in air samples collected at the South Pole, January 1985, was 11 ppb [8].

Food Survey Values:

Plant Concentrations:

Fish/Seafood Concentrations:

Animal Concentrations:

Milk Concentrations:

Other Environmental Concentrations:

Probable Routes of Human Exposure: Occupational exposure of 1,1,1-trichloro-2,2,2-trifluoroethane may occur by inhalation or dermal contact during its production, formulation, or use.

Average Daily Intake:

1,1,1-Trichloro-2,2,2-trifluoroethane

Occupational Exposure:

Body Burdens: 1,1,1-Trichloro-2,2,2-trifluoroethane was detected in the expired air of 6 out of 10 air samples taken from 8 smoking and nonsmoking male volunteers from Texas, at expiration rates of 2.6-580 ug/kg [3].

REFERENCES

1. Atkinson R; J Chem Phys Ref Data Monographs 1 (1989)
2. Boublik T et al; The Vapor Pressures of Pure Substances Amsterdam: Elsevier (1984)
3. Conkle JP et al; Arch Environ Health 30: 290-5 (1975)
4. Harsch DE et al; J Air Pollut Control Fed 29: 975-6 (1979)
5. Lyman WJ et al; Handbook of Chemical Property Estimation Methods. Environment Behavior of Organic Compounds. McGraw-Hill NY (1982)
6. Meylan W, Howard PH; J Pharm Sci 84: 83-92 (1995)
7. Meylan W, Howard PH; Environ Toxicol Chem 10: 1283-93 (1991)
8. Rasmussen RA et al; Science 211: 285-7 (1981)
9. Siegemund G et al; A11: 349-92 in Ullmann's Encycl of Indust Chem NY: VCH Publishers (1988)
10. Silverstein RM, Bassler GC; Spectrometric Identification of Organic Compounds NY,NY: Wiley pp. 148-69 (1963)
11. Stonebraker RD, Smith AJ; pp. 1-10 in Control Haz Mater Spills Proc Natl Conf Nashville, TN (1980)
12. Swann RL et al; Res Rev 85: 17-28 (1982)
13. Young P, Parker A; pp. 24-41 in ASTM Spec Tech Publ 851 (1984)

2,2,4-Trimethyl-1,3-pentanediol

SUBSTANCE IDENTIFICATION

Synonyms:

Structure:

CAS Registry Number: 144-19-4

Molecular Formula: $C_8H_{18}O_2$

SMILES Notation: OCC(C(O)C(C)C)(C)C

CHEMICAL AND PHYSICAL PROPERTIES

Boiling Point: 234 °C at 737 mm Hg

Melting Point: 51-52 °C

Molecular Weight: 146.23

Dissociation Constants:

Log Octanol/Water Partition Coefficient: 1.24 [2]

Water Solubility: 19,000 mg/L at 25 °C [5]

Vapor Pressure: 0.0171 mm Hg at 25 °C [6]

Henry's Law Constant: 7.16×10^{-7} atm-m^3/mol at 25 °C (estimation) [9]

ENVIRONMENTAL FATE/EXPOSURE POTENTIAL

Summary: 2,2,4-Trimethyl-1,3-pentanediol may be released to the environment in waste water effluents generated at sites of its production and use as a chemical intermediate. Use as a solvent in printing inks could result in evaporation to the atmosphere. If released to the atmosphere, it will degrade in the vapor phase by reaction with photochemically produced hydroxyl radicals (estimated half-life of 22 hours). If released to soil or water, an estimated Koc value of 19 suggests that 2,2,4-trimethyl-1,3-pentanediol will leach readily. 2,2,4-Trimethyl-1,3-pentanediol has been identified in one human breast milk sample; therefore, the infant population may be exposed through breast feeding. Occupational exposure to 2,2,4-trimethyl-1,3-pentanediol can occur through dermal contact and inhalation.

Natural Sources:

Artificial Sources: 2,2,4-Trimethyl-1,3-pentanediol's use as an intermediate in the manufacture of resins, polyesters, elastomers, polyols and foams [15] could result in its release to the environment through wastewater effluents. Use as a solvent in printing inks [15] could result in evaporation to the atmosphere.

Terrestrial Fate: An estimated Koc value of 19 [8] suggests that 2,2,4-trimethyl-1,3-pentanediol will leach readily in soil.

Aquatic Fate: Volatilization from water is slow; the volatilization half-lives from a model environmental river (1 m deep) and model pond have been estimated to be 62 days and 1.84 years, respectively [8,14]. Aquatic hydrolysis, bioconcentration and adsorption to sediment are not expected to be important fate processes.

Atmospheric Fate: Based upon an estimated vapor pressure of 0.0171 mm Hg at 25 °C [6], 2,2,4-trimethyl-1,3-pentanediol is expected to exist almost entirely in the vapor phase in the ambient atmosphere [4]. It is degraded readily in the ambient atmosphere by reaction with photochemically produced hydroxyl radicals with an estimated half-life of about 22 hr [1].

Biodegradation: Data specific to the biodegradation of 2,2,4-trimethyl-1,3-pentanediol in mixed cultures were not available, although 2,2,4-trimethyl-1,3-pentanediol was shown to biodegrade in one pure culture (*Pseudomonas aeruginosa*) study [3].

Abiotic Degradation: The rate constant for the vapor-phase reaction of 2,2,4-trimethyl-1,3-pentanediol with photochemically produced hydroxyl radicals has been estimated to be 1.72×10^{-11} cm^3/molecule-sec at 25 °C, which corresponds to an atmospheric half-life of about 22 hr at an atmospheric concn of $5 \times 10^{+5}$ hydroxyl radicals per cm^3 [1]. Glycols are generally resistant to aqueous environmental hydrolysis [8]; therefore, aqueous hydroylsis is not expected to be an important environmental fate process for 2,2,4-trimethyl-1,3-pentanediol.

Bioconcentration: Based upon a water solubility of 19,000 mg/L at 25 °C [5], the BCF for 2,2,4-trimethyl-1,3-pentanediol can be estimated to be 2.4 from a regression-derived equation [8]. This BCF value suggests that bioconcentration in aquatic organisms is not an important environmental fate process.

Soil Adsorption/Mobility: Based upon a water solubility of 19,000 mg/L at 25 °C [5], the Koc for 2,2,4-trimethyl-1,3-pentanediol can be estimated to be 19 from a regression-derived equation [8]. This Koc value suggests that 2,2,4-trimethyl-1,3-pentanediol has very high soil mobility [13].

Volatilization from Water/Soil: The Henry's Law constant for 2,2,4-trimethyl-1,3-pentanediol can be estimated to be 7.16×10^{-7} atm-m^3/mol at 25 °C using a structure estimation method [9]. This value of Henry's Law constant indicates very slow volatilization from water [8]. Based on this Henry's Law constant, the volatilization half-life from a model river (1 m deep flowing 1 m/sec with a wind velocity of 3 m/sec) can be estimated to be about 62 days [8]. Volatilization half-life from a model environmental pond can be estimated to be about 1.84 years [14].

Water Concentrations:

Effluent Concentrations: 2,2,4-Trimethyl-1,3-pentanediol was qualitatively detected in water samples collected from an advanced waste treatment facility in Orange County, CA, on February 3, 1976 [7].

Sediment/Soil Concentrations:

Atmospheric Concentrations: URBAN/SUBURBAN: 2,2,4-Trimethyl-1,3-pentanediol was detected in 3 indoor air samples at an average concn of 1.54 ppb (sampling site and date not reported) [12].

Food Survey Values:

Plant Concentrations:

Fish/Seafood Concentrations:

Animal Concentrations:

Milk Concentrations: 2,2,4-Trimethyl-1,3-pentanediol was qualitatively detected in 1 of 12 samples of human milk collected from volunteers in Bayonne, NJ, Jersey City, NJ, Bridgeville, PA, and Baton Rouge, LA [11].

Other Environmental Concentrations:

Probable Routes of Human Exposure: 2,2,4-Trimethyl-1,3-pentanediol has been identified in one human breast milk sample [11]; therefore, the infant population may be exposed through breast feeding. Occupational exposure to 2,2,4-trimethyl-1,3-pentanediol can occur through dermal contact and inhalation.

Average Daily Intake:

Occupational Exposure: NIOSH (NOES 1981-1983) has statistically estimated that 4,194 workers are potentially exposed to 2,2,4-trimethyl-1,3-pentanediol in the US [10].

Body Burdens: 2,2,4-Trimethyl-1,3-pentanediol was qualitatively detected in 1 of 12 samples of human milk collected from volunteers in Bayonne, NJ, Jersey City, NJ, Bridgeville, PA, and Baton Rouge, LA [11].

REFERENCES

1.   Atkinson R; J Inter Chem Kinet 19: 799-828 (1987)

2. Chem Inspect Test Inst; Biodegradation and bioaccumulation data of existing chemicals based on the CSCL Japan; Japan Chemical Industry Ecology - Toxicology and Information Center. ISBN 4-89074-101-1. (1992)
3. Daugherty LC; Lubrication Engin 36: 718-23 (1980)
4. Eisenreich SJ et al; Environ Sci Technol 15: 30-8 (1981)
5. Flick EW; Industrial Solvents Handbook. 4th ed Park Ridge, NJ: Noyes Data Corp p. 452 (1991)
6. GEMS; Graphical Exposure Modeling System. PCCHEM. USEPA (1987)
7. Lucas SV; GC/MS Analysis of Organics in Drinking Water Concentrates and Advanced Waste Treatment Concentrates: Volume 1. USEPA-600/1-84-020A (NTIS PB85-128221) p. 49, 183 (1984)
8. Lyman WJ et al; Handbook of Chemical Property Estimation Methods Washington, DC: Amer Chem Soc p. 4-9, 5-10, 7-4, 15-15 to 15-29 (1990)
9. Meylan W, Howard PH; Environ Toxicol Chem 10: 1283-93 (1991)
10. NIOSH; National Occupational Exposure Survey (NOES) (1983)
11. Pellizzari ED et al; Bull Environ Contam Toxicol 28: 322-8 (1982)
12. Shah JJ, Heyerdahl; National Ambient Volatile Organic Compounds (VOCs) Database Update. USEPA/600/3-88-010(a). Research Triangle Park, NC: USEPA p. 60 (1988)
13. Swann RL et al; Res Rev 85: 23 (1983)
14. USEPA; EXAMS II Computer Simulation (1987)
15. Von Bramer PT, Bowen GB; p. 314 in Ullmann's Encycl Industr Chem A1: 5th ed. NY: VCH Publ A1: 314 (1985)

Cumulative Index of Synonyms

Cumulative Index by CAS Registry Number

Cumulative Index by CAS Registry Number

Cumulative Index by CAS Registry Number

Cumulative Index by CAS Registry Number

Cumulative Index by CAS Registry Number

Cumulative Index by CAS Registry Number

Cumulative Index by CAS Registry Number

Cumulative Index by CAS Registry Number

Cumulative Index by CAS Registry Number

Cumulative Index by CAS Registry Number

Cumulative Index by Chemical Formula

Cumulative Index by Chemical Formula

Cumulative Index by Chemical Formula

Cumulative Index by Chemical Formula

Cumulative Index by Chemical Formula

Cumulative Index by Chemical Formula

Cumulative Index by Chemical Formula

Cumulative Index by Chemical Formula

Cumulative Index by Chemical Formula

Cumulative Index by Chemical Formula

Printed and bound by CPI Group (UK) Ltd, Croydon, CR0 4YY

17/10/2024

01775688-0020